IAES International Conference on Electrical Engineering, Computer Science and Informatics (EECSI 2016)

IOP Conference Series: Materials Science and Engineering
Volume 190

Semarang, Indonesia
23 – 25 November 2016

Editors:

Munawar A. Riyadi **Tole Sutikno**
Sri Arttini Dwi Prasetyowati **Deris Stiawan**

ISBN: 978-1-5108-3960-1
ISSN: 1757-8981

Printed from e-media with permission by:

Curran Associates, Inc.
57 Morehouse Lane
Red Hook, NY 12571

Some format issues inherent in the e-media version may also appear in this print version.

Printed by Curran Associates, Inc. (2017)

For permission requests, please contact the Institute of Physics
at the address below.

Institute of Physics
Dirac House, Temple Back
Bristol BS1 6BE UK

Phone: 44 1 17 929 7481
Fax: 44 1 17 920 0979

techtracking@iop.org

Additional copies of this publication are available from:

Curran Associates, Inc.
57 Morehouse Lane
Red Hook, NY 12571 USA
Phone: 845-758-0400
Fax: 845-758-2633
Email: curran@proceedings.com
Web: www.proceedings.com

Table of contents

Volume 190

IAES International Conference on Electrical Engineering, Computer Science and Informatics

23–25 November 2016, Semarang, Indonesia

Accepted papers received: 31 March 2016
Published online: 19 April 2017

Preface

Papers

IOP Conf. Series: Materials Science and Engineering **190** (2017) 011001 doi:10.1088/1757-899X/190/1/011001

Note from Editors

Munawar A Riyadi[1], SAD Prasetyowati[2], T Sutikno[3], D Stiawan[4]

[1]Department of Electrical Engineering, Diponegoro University, Tembalang, Semarang, 50275 – Indonesia
[2]Faculty of Industrial Technologi, Sultan Agung Islamic University (UNISSULA), Semarang, Indonesia
[3]Department of Electrical Engineering, Achmad Dahlan University,Yogyakarta, Indonesia
[4]Faculty of Computer Sciences, Sriwijaya University,Palembang, Indonesia

E-mail: munawar@elektro.undip.ac.id, arttini@unissula.ac.id, tole@ee.uad.ac.id, deris@unsri.ac.id

The 3rd International Conference of Electrical Engineering, Computer Science and Informatics (EECSI) 2016 was held in Semarang, Indonesia from 23th to 25th November, 2016. The conference was organized by *Universitas Islam Sultan Agung* as the host in collaboration with Universitas Diponegoro, Universitas Ahmad Dahlan and Universitas Sriwijaya, and with full technical support from IAES Indonesia Section. Authors and participants from 10 countries made the conference truly international in scope. Participants have delivered their talks of valuable research outputs that vary from many fields of electrical engineering (power electronics, telecommunication, electronics engineering, control system and signal processing) to the field of computer science and informatics. These wide range of topics have colorized this conference.

This volume of *IOP Conference Series: Materials Science and Engineering* contains selected articles from those presented in the conference. After presentation, the revised papers were peer reviewed by fellow reviewers to ensure the quality of published materials. Finally, Editors decided to select and publish as many as 49 papers. It is hoped that the presented papers can offer more insight towards broad audience.

On behalf of Editors, we appreciate enormous work of all staffs and reviewers in the preparation of this volume. We would like to express our sincere thanks to all authors and presenters for their valuable contributions. We hope to see you again in the next event of EECSI 2017 which will be held in Yogyakarta, Indonesia, next year.

Guest Editors

Munawar A Riyadi, PhD
Dr. Sri Arttini Dwi Prasetyowati,
Tole Sutikno, PhD
Deris Stiawan, PhD

IOP Conf. Series: Materials Science and Engineering **190** (2017) 011002 doi:10.1088/1757-899X/190/1/011002

Foreword from Chair of EECSI 2016

In the name of Allah, the Gracious Most Merciful

It is great pleasure to welcome out colleagues from all over the world to attend 3rd International Conference on Electrical Engineering, Computer Science, and Informatics (EECSI 2016) Conference in Semarang City, Central Java, Indonesia

EECSI 2016 provides a forum for researchers, academicians, professionals, and students from various engineering fields and cross-disciplinary working or interested in the field of Electrical Engineering, Computer Science, and Informatics especially: Power Engineering, Power Systems and Protection; Electric Power Transmission and Distribution; High Voltage Engineering and Insulation Technology; Renewable Energy Sources, Smart-grids Technologies & Applications; Energy: Policy, Security, Infrastructure, Growth and Economics; Power Electronics and Drives; Control, Automation, Instrumentation and Robotics; Information, Internet of Things and Internet Technologies; Electromagnetic Waves and Field; Circuits and Systems; Semiconductors and Applications; Microelectronics and Electronics Technologies; Electronics and Photonics; Wireless Telecommunications and Networking; Remote Sensing and Data Interpretation; Signal, Image, Video & Multimedia Processing; ICT for Electrical and Electronics Applications; Computer Network & Information Security; High Performance Computing and Communication; Databases, Data Mining and Software Engineering.

Organizing such an prestigious conference was incredibly challenging and would have been impossible without our outstanding committee, so I would like to extend my sincere appreciation to all committees and volunteers from Universitas Islam Sultan Agung, UniversitasDiponegoro, Universitas Ahmad Dahlan and UniversitasSriwijaya for providing me with much needed support, advice, assistance on all aspects of the conference. We do hope that this event will encourage the collaboration among us in the future.

I wish all find opportunity to get rewarding technical program, intellectual inspiration, renew friendships and forge innovation, and that everyone enjoys some of what Semarang has to offer. The organizing committee sincerely hopes that the EECSI 2016 will be a truly memorable experience for all participants.

Finally, may I wish you an enjoyable conference and a pleasant stay in Semarang.

<u>Imam Much IbnuSubroto, Ph.D</u>
Chair of EECSI 2016

IOP Conf. Series: Materials Science and Engineering **190** (2017) 011002 doi:10.1088/1757-899X/190/1/011002

ORGANIZING COMMITTEE OF EECSSI 2016 CONFERENCE

Steering Committee

Adam Skorek, IEEE MGA Awards and Recognation Chair (R7) Trois-Rivières, QC, Canada

Ary Setijadi Prihatmanto, IEEE Indonesia Chapter Chair (Computer Society)

Fitri Yuli Zulkifli, IEEE Indonesia Section (Technical Activity)

Pekik Argo Dahono, IEEE Indonesia Chapters Chair (EdSoc/EDS/PELS/SPS)

Soegijardjo Soegijoko, IEEE Indonesia Chapters Chair (CAS/EMBS)

Sri Arttini Dwi Prasetyowati, Universitas Islam Sultan Agung, Semarang, Indonesia

Muhammad Haddin, Universitas Islam Sultan Agung, Semarang, Indonesia

Zainudin Nawawi, Universitas Sriwijaya, Palembang, Indonesia

Siti Nurmaini, Universitas Sriwijaya, Palembang, Indonesia

Hermawan, Universitas Diponegoro, Semarang, Indonesia

Ida Ayu Dwi Giriantari, Universitas Udayana, Bali, Indonesia

Rahmat Budiarto, Surya University, Indonesia

Tumiran, Universitas Gadjah Mada, Yogyakarta, Indonesia

Chair

Imam Much Ibnu Subroto, Universitas Islam Sultan Agung, Semarang, Indonesia

Co-Chair

Tole Sutikno, Universitas Ahmad Dahlan, Yogyakarta, Indonesia

Finance Chair and Treasurer

Wiwiek Fatmawati, Universitas Islam Sultan Agung, Semarang, Indonesia

Lina Handayani, Institute of Advanced Engineering and Science

Publication Chair

Mochammad Facta, Universitas, Diponegoro, Semarang, Indonesia

Publicity Chair

Muhammad Qomaruddin, Universitas Islam Sultan Agung, Semarang, Indonesia

Deris Stiawan, Universitas Sriwijaya, Palembang, Indonesia

Public Relations Chairs

Arief Marwanto, Universitas Islam Sultan Agung, Semarang, Indonesia

Aina Musdlolifah, Universitas Gadjah Mada, Yogyakarta, Indonesia

IOP Conf. Series: Materials Science and Engineering **190** (2017) 011002 doi:10.1088/1757-899X/190/1/011002

Technical Program Chair

Munawar A. Riyadi, Universitas Diponegoro, Semarang, Indonesia

Mudrik Alaydrus (Senior Member of IEEE), Universitas Mercu Buana Jakarta, Indonesia

Special Session Chairs

Tole Sutikno, Universitas Ahmad Dahlan, Yogyakarta, Indonesia

Local Arrangement Committee

Muhammad Haddin, Universitas Islam Sultan Agung, Semarang, Indonesia

Sri Arttini Dwi Prasetyowati, Universitas Islam Sultan Agung, Semarang, Indonesia

Suryani Alifah, Universitas Islam Sultan Agung, Semarang, Indonesia

Eka Nuryanto Budisusila, Universitas Islam Sultan Agung, Semarang, Indonesia

Arief Marwanto, Universitas Islam Sultan Agung, Semarang, Indonesia

Muhammad Khosyi'in, Universitas Islam Sultan Agung, Semarang, Indonesia

Bustanul Arifin, Universitas Islam Sultan Agung, Semarang, Indonesia

Gunawan, Universitas Islam Sultan Agung, Semarang, Indonesia

Ida Widiyastuti, Universitas Islam Sultan Agung, Semarang, Indonesia

Agus Suprayitno, Universitas Islam Sultan Agung, Semarang, Indonesia

Agus Adhinugroho, Universitas Islam Sultan Agung, Semarang, Indonesia

Technical Program Members

Ali Kattan, Ishik University, Iraq

Adya Pramudita, Unika Atma Jaya, Indonesia

Angela Amphawan, Universiti Utara Malaysia, Malaysia

Arianna Mencattini, University of Rome "Tor Vergata", Italy

Auzani Jidin, Universiti Teknikal Malaysia Melaka, Melaka, Malaysia

Dwi H. Widyantoro, Institut Teknologi Bandung, Indonesia

Farzin Piltan, Sanatkadehe Sabze Pasargad Company, Iran

Faycal Djeffal, University of Batna, Batna, Algeria

Florentinus Budi Setiawan, Soegijapranata

Kristin Y. Pettersen, Norwegian University of Science and Technology, Norway

M. Sukrisno Mardiyanto, Institut Teknologi Bandung, Indonesia

Marcin Kowalczyk, Warsaw University of Technology, Warszawa, Poland

Media Anugerah Ayu, Universitas Siswa Bangsa Internasional, Indonesia

Mokhtar Beldjehem, University of Ottawa, Canada

Muhammad Abu Bakar Sidik, Universiti Teknologi Malaysia

Nidhal Bouaynaya, University of Arkansas at Little Rock, United States

Shahrin Md. Ayob, Universiti Teknologi

International Advisory Committee

IOP Conf. Series: Materials Science and Engineering **190** (2017) 011003 doi:10.1088/1757-899X/190/1/011003

Peer review statement

All papers published in this volume of *IOP Conference Series: Materials Science and Engineering* have been peer reviewed through processes administered by the proceedings Editors. Reviews were conducted by expert referees to the professional and scientific standards expected of a proceedings journal published by IOP Publishing.

Differential-Drive Mobile Robot Control Design based-on Linear Feedback Control Law

Siti Nurmaini[1], **Kemala Dewi**[2], **Bambang Tutuko**[3]

[1,2,3] Robotic and Control Research Lab. Computer Engineering Department, Universitas Sriwijaya

[1] sitinurmaini@gmail.com, [2] kemaladewi09@gmail.com, [3] tutukocn235@gmail.com,

Abstract. This paper deals with the problem of how to control differential driven mobile robot with simple control law. When mobile robot moves from one position to another to achieve a position destination, it always produce some errors. Therefore, a mobile robot requires a certain control law to drive the robot's movement to the position destination with a smallest possible error. In this paper, in order to reduce position error, a linear feedback control is proposed with pole placement approach to regulate the polynoms desired. The presented work leads to an improved understanding of differential-drive mobile robot (DDMR)-based kinematics equation, which will assist to design of suitable controllers for DDMR movement. The result show by using the linier feedback control method with pole placement approach the position error is reduced and fast convergence is achieved.

1. Introduction

According to various studies that have been conducted, mobile robot control system can be classified into three categories. The first category is namely the sensor-based control-based approach. Such control system is emphasized on how to model the motion of a robot in a dynamic environment [1]. The control process to produce estimation and predictions of the mobile robot movement is based on information from sensor detection [2]. The intelligent control scheme is an approach that is most widely used [3][4][5]. However, its results produce sub-optimal response, because the motion is only around the trajectory detection [5]. The second category is the approach of decomposed execution process using a path planning [6][7]. The control system regulates the movement of the mobile robot through the planned path, therefore it can move according to the target that has been set up. The environmental mapping is created for producing collision-free path. Such scheme of control is based on minimal distance, energy and time. The third category presenting the optimization algorithm is developed for controlling the mobile robot with accurate trajectory. The controller design is based on mathematical model of mobile robot. The approach is for tracking the mobile robot errors between reference and actual trajectory [8][9].

However, the whole categories of the control system only operate when the condition of linear velocity is not zero. Therefore, a mobile robot is difficult to control, especially in the case of following the reference of trajectory in a short time with minimal errors. Nonlinear control approaches have been employed to solve this problem [10][11][12][13]. Although the regulation problem is solved to track the mobile robot move to desired trajectory, but it found to yield slow asymptotic convergence [13]. In order to obtain faster convergence, an alternative approach must be proposed. This paper proposes the linier feedback control law to overcome such limitation with pole placement approach to regulate the polynom desired. By using the scheme, the controller performance can be regulated and fast asymptotic convergence is achieved.

2. Differential Drive Mobile Robot Kinematic Model

The differential-drive mobile robot (DDMR) is a class of computer-controlled vehicles whose motion can be described or transformed into the following model with constrained movement in a plane. Figure 1 describes the DDMR on 2D-cartesian plane, it means the mobile robot move only in the *x y* axis, while the contour and elevation of the z-axis is ignored [13].

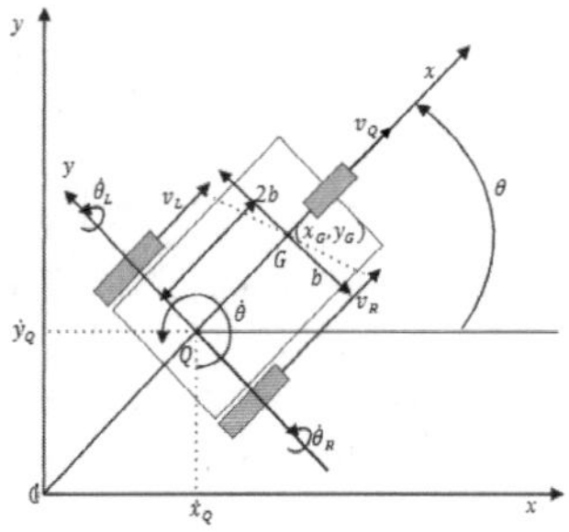

Figure 1. DDMR on 2D cartesian plane

From Figure 1, the representation of $\dot{x}_Q, \dot{y}_Q$ are x, y coordinate relation to center of the robot Q, v_Q is a linier velocity, v_L are linier velocity of left wheels and v_R linier velocity of right wheel, r is a radius of wheel, $2b$ is a distance from right to left wheel, Q is a center point of the robot, G is a center of gravitation and b is a distance between 2 wheels with G. By ignoring the analysis of the castor-free, the configuration of the mobile robot can be described into three general variables $q(t)$ to describe the position of the robot and the control input $u(t)$. It can be defined in matrix:

$$q(t) = (x(t), y(t), \theta(t))^T, \text{ and } u(t) = \left(\dot{\theta}_R(t),\ \dot{\theta}_L(t)\right)^T \quad (1)$$

The angular velocity equations for the right wheel and left wheel such as,

$$\dot{\theta}_R(t) = \frac{1}{2\pi r} \cdot v_R(t)\ ;\ \dot{\theta}_L(t) = \frac{1}{2\pi r} \cdot v_L(t) \quad (2)$$
$$v_R(t) = v(t) + b\dot{\theta}(t)\ ;\ v_L(t) = v(t) - b\dot{\theta}(t) \quad (3)$$

From equations (2) and (3) above, a linear velocity equation is $v_Q(t)$ and an angular velocity equation is $\dot{\theta}(t)$ as follow,

$$v_Q(t) = \frac{1}{2}(v_R(t) + v_L(t));\ \dot{\theta}(t) = \frac{v_R(t) - v_L(t)}{2b} \quad (4)$$

However, the DDMR has a limitation in terms of movement, due to the non-holonomik constraint. The constrained are obtained by two main assumptions. The first assumption no lateral slip motion. It's mean the robot can move only in a curved motion (forward and backward) but not sideward. The velocity equation at the midpoint of the both wheels (G) is obtained by using Figure 2(b), the equation is described as follows,

$$v(t) = \dot{x}(t)\cos\theta(t) + \dot{y}(t)\sin\theta(t) \quad (5)$$

In the robot frame, this condition means the velocity of the wheel in the centre point G is zero along the lateral axis and it's only move along x axis and y axis [7]. Therefore, the velocity equation gives,

$$\dot{x}(t)\sin\theta(t) - \dot{y}(t)\cos\theta(t) = 0 \tag{6}$$

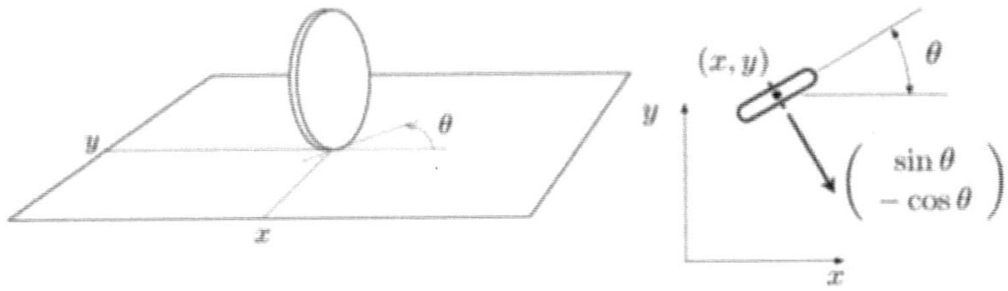

Figure 2. No lateral slip

The second assumption is the pure rolling constraint, due to the wheels do not slip on the floor and has limitations of pure rolling (Figure 3). The velocities of the contact points in the robot frame are related to the wheel velocities for the right wheel and left wheel gives,

$$v_R(t) = r\dot{\theta}_R(t) \text{ and } v_L(t) = r\dot{\theta}_L(t) \tag{7}$$

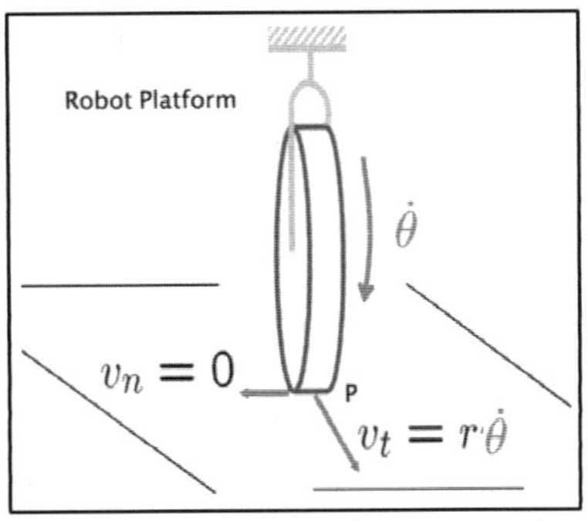

Figure 3. Pure rolling motion constraint

By looking at the starting position of the wheel to the Cartesian field xy can be obtained the equation of position wheeled robot, $x(t) = v(t)\sin\theta(t)$, $y(t) = v(t)\cos\theta(t)$, and $\theta(t) = v_\theta(t)$. Thus, the transformation matrix equations from the initial position can be defined as,

$$T_{NH}(q) = \begin{bmatrix} \frac{r}{2}\cos\theta\,(t) & \frac{r}{2}\cos\theta(t) \\ \frac{r}{2}\sin\theta\,(t) & \frac{r}{2}\sin\theta\,(t) \\ \frac{r}{2b} & -\frac{r}{2b} \end{bmatrix} \tag{8}$$

Furthermore, the angular velocity $u(t)$ as input to the robot have three parameters such as, $u_1(t) = \dot{\theta}_R(t)$, $u_2(t) = \dot{\theta}_L(t)$, and $\dot{\theta}(t) = \omega(t)$. From equation (8) and three parameters of input equation, the model of DDMR kinematic equation is, $\dot{q}(t) = T_{NH}(q)\,u(t)$ or,

$$\begin{bmatrix} \dot{x}(t) \\ \dot{y}(t) \\ \dot{\theta}(t) \end{bmatrix} = \begin{bmatrix} \cos\theta(t) & 0 \\ \sin\theta(t) & 0 \\ 0 & 1 \end{bmatrix} \begin{bmatrix} v(t) \\ \omega(t) \end{bmatrix} \tag{9}$$

3. The Error Model of DDMR Movement

The error model of DDMR is obtained by reducing the reference position $q_r(t) = (x_r(t), y_r(t), \theta_r(t))^T$ with the actual position of the robot $q_c(t) = (x_c(t), y_c(t), \theta_c(t))^T$. The visualisation of DDMR movement is presented in Figure 4. Thus, the error of mobile robot movement is represented by $q_e(t) = (e_1(t), e_2(t), e_3(t))^T$. $q_e(t)$ is the transformation result between the reference position $q_r(t)$ in a local coordinate system with the previous position $q_c(t)$ and X axis which is the direction toward the robot is $\theta_c(t)$ or $q_e(t) = T_e(q_r(t) - q_c(t))$. It can be realized in equation (10) as follows:

$$\begin{bmatrix} e_1(t) \\ e_2(t) \\ e_3(t) \end{bmatrix} = \begin{bmatrix} \cos\theta_c(t) & \sin\theta_c(t) & 0 \\ -sin\,\theta_c(t) & \cos\theta_c(t) & 0 \\ 0 & 0 & 1 \end{bmatrix} \begin{bmatrix} x_r(t) - x_c(t) \\ y_r(t) - y_c(t) \\ \theta_r(t) - \theta_c(t) \end{bmatrix} \tag{10}$$

By lowering the equation (10) and involving the equations that exist in the equation of non-holonomic DDMR constraints, the error equation is obtained as equation (11) below:

$$\begin{bmatrix} \dot{e}_1(t) \\ \dot{e}_2(t) \\ \dot{e}_3(t) \end{bmatrix} = \begin{bmatrix} \cos e_3(t) & 0 \\ \sin e_3(t) & 0 \\ 0 \quad 0 & 1 \end{bmatrix} \begin{bmatrix} v_r(t) \\ \omega_r(t) \end{bmatrix} + \begin{bmatrix} -1 & e_2(t) \\ 0 & -e_1(t) \\ 0 & -1 \end{bmatrix} \begin{bmatrix} v_c(t) \\ \omega_c(t) \end{bmatrix} \tag{11}$$

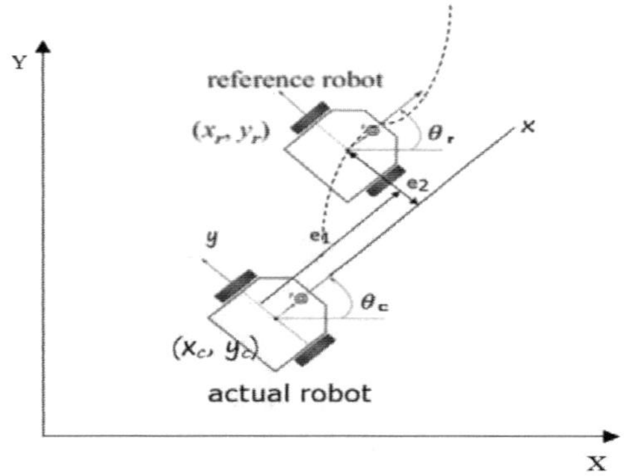

Figure 4. Transformation of error movement

From equation (11), $v_r(t)$ as a reference linear velocity, $\omega_r(t)$ as a reference angular velocity, $v_c(t)$ as an actual linear velocity, and $\omega_c(t)$ as an actual angular velocity. In this model the combination between feed forward control and linier feedback control based on pole placement approach is proposed, therefore a new input is created below,

$$v_c(t) = v_{ff}(t) - v_1(t)$$
$$v_c(t) = v_r(t)\cos e_3(t) - v_1(t) \tag{12}$$
$$\omega_c(t) = \omega_{ff}(t) - v_2(t)$$
$$\omega_c(t) = \omega_r(t) - v_2(t) \tag{13}$$

By substituting equation (12) and (13) into equation (11) nonlinear equation of error control velocity is obtained,

$$\begin{bmatrix} \dot{e}_1(t) \\ \dot{e}_2(t) \\ \dot{e}_3(t) \end{bmatrix} = \begin{bmatrix} 0 & \omega_c(t) & 0 \\ -\omega_c(t) & 0 & 0 \\ 0 & 0 & 0 \end{bmatrix} \begin{bmatrix} e_1(t) \\ e_2(t) \\ e_3(t) \end{bmatrix} + \begin{bmatrix} 0 \\ \sin e_3(t) \\ 0 \end{bmatrix} v_r(t) + \begin{bmatrix} 1 & 0 \\ 0 & 0 \\ 0 & 1 \end{bmatrix} \begin{bmatrix} v_1(t) \\ v_2(t) \end{bmatrix} \tag{14}$$

By using the linearization approach around its operating point (OP : $e_1 = e_2 = e_3 = 0, v_1 = v_2 = 0$). Thus, the linear model of error in equation (14) become,

$$\Delta\dot{e}(t) = \begin{bmatrix} 0 & \omega_r(t) & 0 \\ -\omega_r(t) & 0 & v_r(t) \\ 0 & 0 & 0 \end{bmatrix} . \Delta e(t) + \begin{bmatrix} 1 & 0 \\ 0 & 0 \\ 0 & 1 \end{bmatrix} . \Delta v(t) \tag{15}$$

4. Control System Design

In this research, the controller design has three states, namely $x(t), y(t),$ and $\theta(t)$ and two inputs, namely $v(t)$, and $\omega(t)$. The general equation linear state space controller. $v(t) = K.e(t)$, therefore the feedback control law becomes

$$\begin{bmatrix} v_1(t) \\ v_2(t) \end{bmatrix} = \begin{bmatrix} -k_1 & 0 & 0 \\ 0 & -sign\,(v_r(t))k_2 & -k_3 \end{bmatrix} \begin{bmatrix} e_1(t) \\ e_2(t) \\ e_3(t) \end{bmatrix} \tag{16}$$

From equation (16) a linear velocity of robot $v(t)$ replaced by $v_1(t)$, an angular velocity of robot $\omega(t)$ replaced by $v_2(t)$, some values of $k_1, k_2,$ and k_3 are a gain in the controller. The diagram block of the control system design can be seen in Figure 4. Moreover, the value of k control law corresponding will be determined. This problem is solved by optimizing some function value. In this research, the determination of the k value is determined by comparing the characteristics of the real polynoms and desired polynoms of control system characteristics [5]. These polynomials take the following form as follow,

$$(s + 2\zeta\omega_n)(s^2 + 2\zeta\omega_n s + \omega_n{}^2) \text{ and } s^3 + (4\zeta\omega_n)s^2 + \left(4\,\zeta^2\omega_n{}^2 + \omega_n{}^2\right)s + 2\zeta\omega_n{}^3 \tag{17}$$

Where, two parameters are selected such as, the damping coefficient $\zeta \in (0,1)$ and the natural frequency $\omega_n > 0$. At the additional pole $s = -2\zeta\omega_n$ which is useful to increases the rise time and to reduce the overshoot system. The polynomial characteristic of the closed loop control law in equation (15) with input to the state space controller in equation (16) are the two important parameters design. All equations from (8) to (16) can be visualized as a block diagram with linier feedback control law as shown in Figure 5.

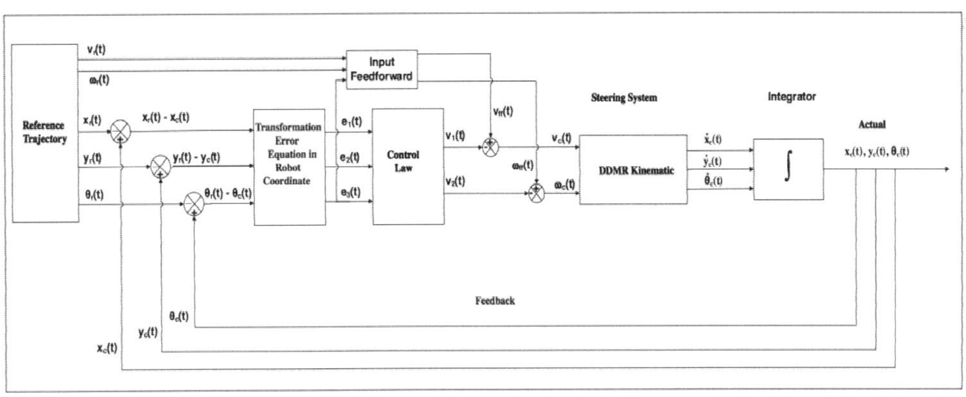

Figure 5. Block diagram control law of DDMR system

5. Simulation Results

The purpose of this research is to find the parameters of error value including *x(t), y(t)* and *θ(t)* based on polynomials approach to produce good control performance such as, rise time, and steady state error parameters. Some simulations are conducted to show the DDMR performance in terms of error and delta error values by using the proposed control law. The starting position of the robot at coordinates is set at (0; 0; 0.78) and initial linear velocity is set at 0,01 m/s and angular velocity is set at 0,02 rad/s. Therefore, it can be calculated to obtain the value of the average error and time to steady state. Initial performance of controller are the damping factor and natural frequency. The damping factor is set variable from $0,2 < \zeta < 0.8$ and natural frequency are set at $\omega_n = 5$, $\omega_n = 2$, and $\omega_n = 1$ respectivelly. The error position of DDMR based on selected parameter as shown in Figure 6.

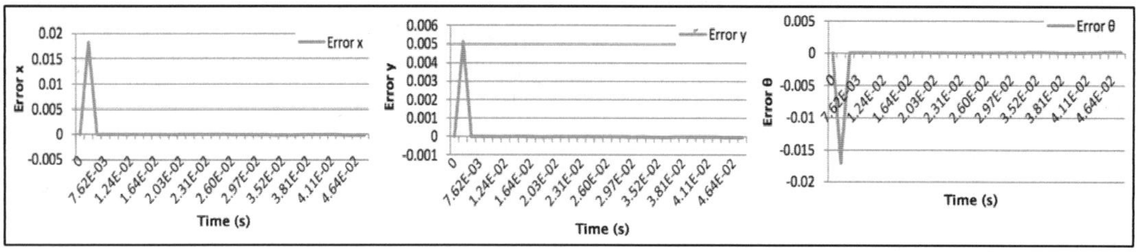

Figure 6. The DDMR error position control

The summary all experiments for DDMR movement in simulation process is described in Table 1 and Table 2.

Table 1. (a) Comparison data of average error value with $0,2 < \zeta < 0.8$ and ω_n is fixed
(b) Comparison data of the average interval time with $0,2 < \zeta < 0.8$ ω_n is fixed

(a)

No	ζ	ω_n	g	k_1	k_2	k_3	Average of error		
							Δx	Δy	$\Delta\theta$
1	0.2	2	1	0.8	0.8	0.8	$2.8x10^{-3}$	$1.4x10^{-3}$	$-3x10^{-3}$
2	0.3	2	1	1.2	0.8	1.2	$-4x10^{-3}$	$-2x10^{-3}$	$4.9x10^{-3}$
3	0.4	2	1	1.6	0.8	1.6	$-3x10^{-3}$	-10^{-3}	$4x10^{-3}$
4	0.5	2	1	2	0.8	2	$3x10^{-4}$	$7x10^{-5}$	$-2x10^{-4}$
5	0.6	2	1	2.4	0.8	2.4	$4x10^{-4}$	10^{-4}	$-4x10^{-4}$
6	0.7	2	1	2.8	0.8	2.8	$-7.3x10^{-4}$	$-2.1x10^{-4}$	$7.7x10^{-4}$
7	0.8	2	1	3.2	0.8	3.2	10^{-4}	$2x10^{-5}$	$-9.6x10^{-5}$

(b)

No	ζ	ω_n	g	k_1	k_2	k_3	Average of Δt		
							$e_1(t)$	$e_2(t)$	$e_3(t)$
1	0.2	2	1	0.8	0.8	0.8	$7x10^{-2}$	$7x10^{-2}$	$7x10^{-2}$
2	0.3	2	1	1.2	0.8	1.2	$2.8x10^{-1}$	$2.8x10^{-1}$	$2.8x10^{-1}$
3	0.4	2	1	1.6	0.8	1.6	$2.4x10^{-1}$	$2.4x10^{-1}$	$2.4x10^{-1}$
4	0.5	2	1	2	0.8	2	$3.3x10^{-2}$	$3.3x10^{-2}$	$3.3x10^{-2}$
5	0.6	2	1	2.4	0.8	2.4	$2.5x10^{-2}$	$2.5x10^{-2}$	$2.5x10^{-2}$
6	0.7	2	1	2.8	0.8	2.8	$8.7x10^{-2}$	$8.7x10^{-2}$	$8.7x10^{-2}$
7	0.8	2	1	3.2	0.8	3.2	$3.6x10^{-2}$	$3.6x10^{-2}$	$3.6x10^{-2}$

In table 1(a) and 1(b), it can be seen the value of the error and the smallest travel time obtained by setting the value of $\zeta = 0.6$. In Table 2, it compared with the test data in value $\zeta = 0.6$, $\omega_n = 2$, and $g = 1$ before, the delta error value and the smallest interval time are obtained by assigning a value $\omega_n = 2$.

Table 2. Comparison data of the average error value and time interval with $\omega_n = 5$, $\omega_n = 2$, and $\omega_n = 1$ and ζ is fixed

No.	ζ	ω_n	g	k_1	k_2	k_3	Average of Δe			Average of Error time		
							x	y	θ	x	y	θ
1	0.6	5	1	0.8	0.8	0.8	$1.2x10^{-3}$	$1.7x10^{-4}$	$-9.3x10^{-4}$	$3.5x10^{-2}$	$3.5x10^{-2}$	$3.5x10^{-2}$
2	0.6	2	1	2.4	0.8	2.4	$4x10^{-4}$	10^{-4}	$-4x10^{-4}$	$2.5x10^{-2}$	$2.5x10^{-2}$	$2.5x10^{-2}$
3	0.6	1	1	1.2	0.8	1.2	$2x10^{-4}$	$8x10^{-5}$	$-2x10^{-4}$	$5x10^{-2}$	$5x10^{-2}$	$5x10^{-2}$

6. Conclusion

From the experimental results, it is concluded that the method of linearization error with feed-forward and feedback control law with poles placement approach produces small value of the average error position, and it achieved in a short time to steady state condition. From Seven values of ζ and two values of ω_n, are selected $\zeta = 0.6$, $\omega_n = 2$, and $g = 1$, due to by using the values it can be obtained the average value of delta error is $4\text{x}10^{-4}$ m on *x* position; the average value of delta error is 10^{-5} m on *y* position and the average value of delta error of direction side of the robot is $-4\text{x}10^{-4}$ rad with an average time interval is $2.5\text{x}10^{-2}$ s. By regulating performance of controller based on pole placement approach, the small error, faster time to steady state and small overshoot is achieved.

Acknowledgments
The Authors thank The Ministry of Technology Research and Higher Education (Kemenristek-Dikti), Indonesia and Universitas Sriwijaya for their financial support in Excellence University Research Project or Penelitian Unggulan Perguruan Tinggi (PUPT)

References
[1] Fiorini, P., & Shiller, Z. (1998). Motion planning in dynamic environments using velocity obstacles. *The International Journal of Robotics Research*, *17*(7), 760-772.
[2] Elfes, A. (1989). Using occupancy grids for mobile robot perception and navigation. *Computer*, *22*(6), 46-57.
[3] Fierro, R., & Lewis, F. L. (1998). Control of a nonholonomic mobile robot using neural networks. *IEEE transactions on neural networks*, *9*(4), 589-600.
[4] Hu, H., & Brady, M. (1996). A parallel processing architecture for sensor-based control of intelligent mobile robots. *Robotics and autonomous systems*, *17*(4), 235-257.
[5] Ge, Shuzhi S., and Yun J. Cui. "Dynamic motion planning for mobile robots using potential field method." *Autonomous Robots* 13.3 (2002): 207-222.
[6] Salomon, B., Garber, M., Lin, M. C., & Manocha, D. (2003, April). Interactive navigation in complex environments using path planning. In *Proceedings of the 2003 symposium on Interactive 3D graphics* (pp. 41-50). ACM.
[7] Yang, J. M., & Kim, J. H. (1999). Sliding mode control for trajectory tracking of nonholonomic wheeled mobile robots. *IEEE Transactions on Robotics and Automation*, *15*(3), 578-587.
[8] Hoffmann, G. M., Tomlin, C. J., Montemerlo, M., & Thrun, S. (2007, July). Autonomous automobile trajectory tracking for off-road driving: Controller design, experimental validation and racing. In *American Control Conference* (pp. 2296-2301).
[9] Lee, T. C., Song, K. T., Lee, C. H., & Teng, C. C. (2001). Tracking control of unicycle-modeled mobile robots using a saturation feedback controller. *IEEE transactions on control systems technology*, *9*(2), 305-318.
[10] Fierro, R., & Lewis, F. L. (1995, December). Control of a nonholonomic mobile robot: backstepping kinematics into dynamics. In *Decision and Control, 1995., Proceedings of the 34th IEEE Conference on* (Vol. 4, pp. 3805-3810). IEEE.
[11] Yang, H., Fan, X., Shi, P., & Hua, C. (2016). Nonlinear Control for Tracking and Obstacle Avoidance of a Wheeled Mobile Robot With Nonholonomic Constraint. *IEEE Transactions on Control Systems Technology*, *24*(2), 741-746.
[12] Huang, J., Wen, C., Wang, W., & Jiang, Z. P. (2014). Adaptive output feedback tracking control of a nonholonomic mobile robot. *Automatica*, *50*(3), 821-831.
[13] Dhaouadi, Rached, and Ahmad Abu Hatab. "Dynamic modelling of differential-drive mobile robots using lagrange and newton-euler methodologies: A unified framework." *Advances in Robotics & Automation* 2013.

IOP Conf. Series: Materials Science and Engineering **190** (2017) 012002 doi:10.1088/1757-899X/190/1/012002

Conveyor Performance based on Motor DC 12 Volt Eg-530ad-2f using K-Means Clustering

Zaenal Arifin[1], Sri Artini DP[2], Imam Much Ibnu Subroto[3]

[1,2,3] Electrical Engineering Department, Universitas Islam Sultan Agung, Semarang, Indonesia

[1]zenalarif21@yahoo.com,[2]artini@unissula.ac.id, [3]imam@unissula.ac.id.

Abstract. To produce goods in industry, a controlled tool to improve production is required. Separation process has become a part of production process. Separation process is carried out based on certain criteria to get optimum result. By knowing the characteristics performance of a controlled tools in separation process the optimum results is also possible to be obtained. Clustering analysis is popular method for clustering data into smaller segments. Clustering analysis is useful to divide a group of object into a k-group in which the member value of the group is homogeny or similar. Similarity in the group is set based on certain criteria. The work in this paper based on K-Means method to conduct clustering of loading in the performance of a conveyor driven by a dc motor 12 volt eg-530-2f. This technique gives a complete clustering data for a prototype of conveyor driven by dc motor to separate goods in term of height. The parameters involved are voltage, current, time of travelling. These parameters give two clusters namely optimal cluster with center of cluster 10.50 volt, 0.3 Ampere, 10.58 second, and unoptimal cluster with center of cluster 10.88 volt, 0.28 Ampere and 40.43 second.
Keywords: *optimization, clustering, k-means.*

1. Introduction

Separation process has become a part of production process. Separation process is carried out based on certain criteria to get optimum result. By knowing the characteristics performance of a controlled tools in separation process the optimum results is also possible to be obtained. Clustering analysis is popular method for clustering data into smaller segments. Clustering analysis is useful to divide a group of object into a k-group in which the member value of the group is homogen. Homogeneity in the group is set based on certain criteria.

The work in this paper based on K-Means method to conduct clustering of loading in the performance of a conveyor driven by a direct current motor 12 volt eg-530-2f. The prototype system separation is controlled by using Omron PLC CPM1A for separating cube height of 2 cm and 5 cm running on a conveyor driven 12 V_{dc} motor. The problems in this work can be defined is how the K-Means clustering method can optimize the loading on the performance of the conveyor 12volt dc motor-530ad-2f eg on a prototype system based on the height as separator items.

2. Workload Conveyors

Prototype system is separation equipment based on this height, using a base control PLC Omron Sysmac CPM1A with 30 Input–Output (I/O). Power Supply 12 V_{dc} is used to give electric supply to all motors, relays and sensors. All buttons on the prototype is connected to a PLC by using relay.

The function of this prototype to separate the goods is in cube form. The cube has a height difference. For the height of 2.5 cm, it is called short body and a height of 5 cm high, it is called objects. All the cubes have the same length and width in 5 cm. Based on the loading of data acquisition, measurements are carried out in the term of parameters of voltage, current and time.

3. Research Methodology

Stages of the research is started by collecting data on the results of the loading conveyor system. The next stage is data calculation by using the K-Means clustering method. Analysis is carried out on the results of calculation loading data to get optimal performance of conveyor.

First step in this work is collecting loading data conveyors. The second is determining the method optimization loading data on performance of conveyor. In this case employed researchers are using K-means clustering. The pre-processing or the integration of data is carried out and it is followed by data processing and testing to get optimization analysis of K-means loaded data for the performance of conveyor. The stage of evaluation of the test results is also conducted. If the desired results are not obtained, then the data is returned to the stage of processing the data. If the data evaluation of test results is in accordance with the desired result, then the conclusion is drawn.

3.1. K-Means

Data Clustering is one method of Data Mining that is without direction. Clustering is a technique in data mining is used to enter data into a corresponding group without a deep knowledge of the group. There are two types of data clustering often used in the process of grouping data. Those are is hierarchical and non-hierarchical data clustering. This method of partitioning makes the data into clusters or groups so that the data with the same characteristics are grouped into the same cluster and data with different characteristics are grouped into another group [1-3].

On Clustering algorithms, the data load will be grouped into clusters based on the similarity of the data. Clustering is the principle of maximizing the similarity between members of the cluster and minimizes the similarity between members of different clusters. K-Means algorithm is a non-hierarchical clustering method based on the distance that divides the data into clusters and this algorithm works on numerical attributes. At the initial stage, the K-Means algorithm randomly selects k pieces of data as a centroid. The distance between the data and the centroid are computed using the Euclidean distance. Data placed in the nearby clusters, is calculated from the midpoint of the cluster. Centroid will be determined when all the data has been placed in the nearby clusters. The process of determining the centroid and placement of data in the cluster is repeated until the value of convergent centroid. Convergent is obtained when the centroid of the entire clusters do not change anymore. Distance Space is the process of calculating the distance between points with another point. Distance space used is Euclidean distance i.e. finding the shortest distance between the centroid point data or objects. Data that has a short distance or closest to the centroid will form a cluster [1, 4-6].

$$d_{Euclidean}(i,j) = \sqrt{(x_{i1}-x_{i1})^2 + (x_{i2}-x_{j2})^2 + \ldots + (x_{in}-x_{jn})^2} \quad (1)$$

where $d_{(i,j)}$ is the distance between i^{th} and j^{th} data, x_{i1} is the first attribute value of the ith data, x_{j1} is the first attribute value of the jth data, and n is the number of attributes.

K-Means algorithm basically conducts two processes i.e. clustering the centre location detection process and the process of finding members of each cluster.

$$C_k = \left(\frac{1}{n_k}\right) \sum d_i \quad (2)$$

where n_k is the amount of objects in cluster-k and d_i is the ith object of cluster-k.

4. Result and Discussion

The test results on optimization performance of loading data conveyor driven by 12 volt dc motors eg-530ad-2f. By using the method of K-means clustering, data that has been prepared in data loading conveyor. Starting voltage, current and time will be integrated into one table. Data indicated loading data integration results are presented in V for voltage in Volt, I in current for Ampere) and t for time in second.

Table1. Data integration imposition

No	V (Volt)	I (Ampere)	t (second)
1	11.60	0.17	27.00
2	10.90	0.25	10.00
3	11.60	0.17	27.00
4	10.90	0.25	10.00
5	10.60	0.33	45.00
6	11.20	0.35	30.00
7	10.10	0.34	10.00
8	11.00	0.25	30.00
9	11.60	0.17	29.00
10	11.60	0.17	31.00
…	…	…	…
84	10.90	0.28	35.00

Data processing optimization using the K-means clustering has been done through several stages. For K-Means stage 1, the determination of the centre of the centroid is done by taking three pieces of data from the original data with reference to Equation (1) and (2) to produce a new centroid. The K-Means stage 2 determining the center of the centroid is done by taking the data from the new first centroid, new second centroid and new third centroid on first stage K-means clustering and it refers to equations (1) and (2). K-means clustering processing process stops on second stage of cluster and yields optimal and not optimal condition by taking 10 data K-means to end result. Each member of the cluster has a characteristic resemblance as it is seen in Table 2.

Table 2. Result of Optimization cluster K-Means final stage

No	Cluster 1			Note
	V (Volt)	**I** (Ampere)	**t** (second)	
1	10.90	0.25	10.00	OPTIMAL
2	10.10	0.34	10.00	OPTIMAL
3	10.90	0.25	10.00	OPTIMAL
4	10.10	0.34	10.00	OPTIMAL
5	10.90	0.25	10.00	OPTIMAL
6	10.10	0.34	10.00	OPTIMAL
7	10.90	0.26	10.00	OPTIMAL
8	10.10	0.35	11.00	OPTIMAL
9	10.89	0.27	12.00	OPTIMAL
10	10.09	0.37	12.00	OPTIMAL
…	…	…	…	…
24	10.09	0.37	12.00	OPTIMAL

No	Cluster 2			Note
	V (Volt)	**I** (Ampere)	**t** (second)	
1	11.60	0.17	27.00	NOT OPTIMAL
2	10.60	0.32	40.00	NOT OPTIMAL
3	11.20	0.35	30.00	NOT OPTIMAL
4	11.00	0.25	30.00	NOT OPTIMAL
5	11.00	0.25	30.00	NOT OPTIMAL
6	10.60	0.37	80.00	NOT OPTIMAL
7	10.99	0.28	35.00	NOT OPTIMAL
8	11.59	0.20	42.00	NOT OPTIMAL
9	10.99	0.28	35.00	NOT OPTIMAL
10	10.99	0.28	35.00	NOT OPTIMAL
…	…	…	…	…
60	10.99	0.28	35.00	NOT OPTIMAL

The final result of optimization loading data on performance of the dc motor DC12 volt conveyors with method of K-means clustering is displayed in 3-dimensional maps of data clustering and it is classified by optimal and not optimal results with centroid position as it is shown in Figure 1.

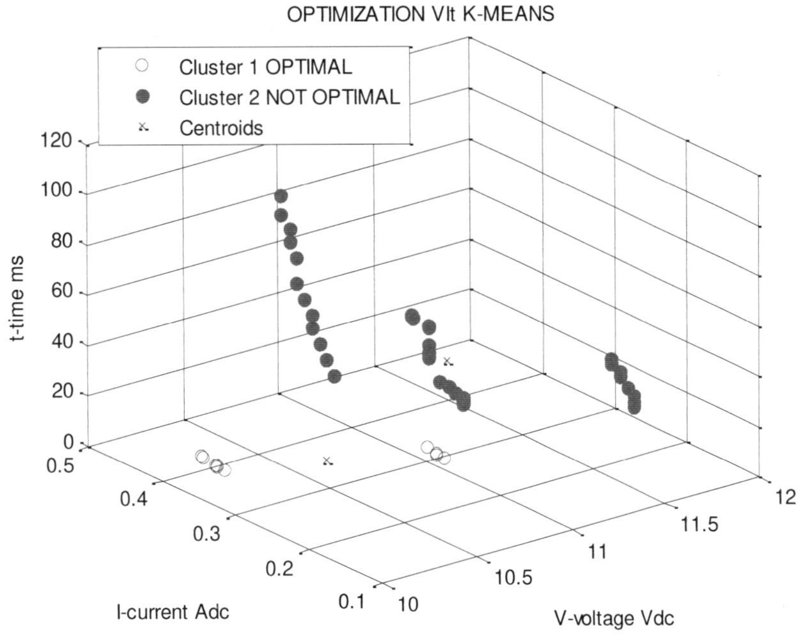

Figure 1. Optimization of loading 3D curve by the method of K-means clustering

Optimization Vit K-Means indicates that the optimal cluster is in cluster 1 where centroid position right in the centre of cluster 1 and cluster 2 is not optimal. Finally on each cluster member that has the characteristics of similarity. This character is the same as groups of clusters as a result of optimizing loading data conveyor 12 volt dc motor.

5. Conclusion

The implementation of the K-Means method to optimize loading data on the performance of the conveyor 12volt dc motor-530ad-2f eg on prototype separator system can produce the goods grouping or clustering of data. Data with similar characteristics similar will be grouped into one cluster. The optimization loading data by using the K-Means clustering has analyzed 84 data. The used data is divided into two clusters that marked as optimal cluster for 24 data and non optimal cluster for 60 data. The optimal data with central cluster consists of 10.50 Volt, 0.30 Ampere, and travel time in 10.58 second. Non optimal cluster with central cluster consists of 10.88 Volt, current 0.28 Ampere, and travel time in 40.43 second.

References

[1] S. Sharma and G.N.Purohit, 2012 ”Analysis of spectral clustering approach for tracking community formation in social network”, *International of Social Networking and Virtual Communities*, vol. 1, no. 1, pp. 31-37.

[2] Parthajit Roy and J.K.Mandal, 2015, “A novel spectral clustering based on local distribution”, *International of Electrical and Computing Engineering* , vol. 5, no. 2, pp. 361-370.

[3] B. Sai Chandana, K. Srinivas and R. Kiran Kumar, 2014 “Clustering Algorithm Combined with Hill Climbing for Classification of Remote Sensing Image”, *International of Electrical and Computing Engineering* , vol. 4, no. 6.

[4] Se-Hoon Jung, Jong-Chan Kim and Chun-Bo Sim, 2016, "Prediction Data Processing Scheme using an Artificial Neural Network and Data Clustering for Big Data,"*International of Electrical and Computing Engineering*, vol. 6, no. 1.

[5] M. Goyal and S. Kumar, 2014, "Improving the Initial Centroids of k-means Clustering Algorithm to Generalize its Applicability," *Journal of The Institution of Engineers (India)*, vol 95, pp 345-350.

[6] Janmenjoy Nayak and Bighnaraj Naik, 2015 "An improved swarm based hybrid k-means clustering for optimal cluster centers," *Springer Information System Design and Inteeligen*t.

A Position Controller Model on Color-Based Object Tracking using Fuzzy Logic

Budi Cahyo Wibowo[1], Imam Much Ibnu Subroto[2], Bustanul Arifin[3]

[1,2,3] Electrical Engineering Dept, Sultan Agung Islamic University, Semarang, Indonesia

[1] wibowo.budicahyo@gmail.com

Abstract. Robotics vision is applying technology on the camera to view the environmental conditions as well as the function of the human eye. Colour object tracking system is one application of robotics vision technology with the ability to follow the object being detected. Several methods have been used to generate a good response position control, but most are still using conventional control approach. Fuzzy logic which includes several step of which is to determine the value of crisp input must be fuzzification. The output of fuzzification is forwarded to the process of inference in which there are some fuzzy logic rules. The inference output forwarded to the process of defuzzification to be transformed into outputs (crisp output) to drive the servo motors on the X-axis and Y-axis. Fuzzy logic control is applied to the color-based object tracking system, the system is successful to follow a moving object with average speed of 7.35 cm/s in environments with 117 lux light intensity.
Keywords: *Vision Robotics, Colour Object Tracking, Fuzzy Logic.*

1. Introduction

The technology of robotics requires information vision to decide the action to be performed. Application of vision on a robot as a navigation tool, the search for the desired objects, etc. Vision information on the robots become very important as information received more details compared with the proximity sensor or other sensors. Robotics vision is a technology that combines the technology of robotics and computer vision.

Computer vision is a combination of image processing, pattern recognition and decision making. The main purpose of computer vision is to translate an image. Many benefits can be gained from the fields of computer vision, one of which can be applied to the color object tracking system. By using computer vision technology on the robot, it allows the robot to be able to move autonomous in accordance with the information obtained from the imaging camera mounted on the robot. The system's ability to follow a moving object became an obstacle in the process of positioning the camera to always follow the position of the object, so we need a reliable position control so that the camera is able to follow an object moving at a certain speed. Generally, the control of the position that has been applied are using conventional controls, such as PID control or a combination of them. In this paper we apply fuzzy logic method for controlling servo motors on the X axis and Y axis, so that the camera can follow the movement of the object and evaluate speed of response on a fuzzy logic control applied in the control of servo motor to position the camera always follows the object being detected.

2. Image Processing

One application of image processing is to separate the image in parts, e.g. certain parts that are wanted for feature extraction and pattern and get the desired image [1]. Image processing operations are categorized into: image enhancement, image segmentation, restoration, and image compression, by using a computer algorithm to perform digital image processing [2]. Image segmentation is the first step in extracting the information contained in an image. Image segmentation is a technique to divide the image into two parts so as not to overlap [3].

3. Fuzzy Logic

Fuzzy logic is used to declare the operation of a system of law with language expression and not with the mathematical equation. Many systems are too complex to be modeled accurately, even with complex mathematical equations. Fuzzy approach involves the rules expressed in words and does not require high precision and tolerance for less precise data[4].

Four main factors are needed in implementing FLC, namely the number of fuzzy sets in linguistic form, rule base for a process of inference, decision-making to determine the behavior of the controller and the shape of the membership function[5]. Fuzzy control does not require complex mathematical models. Control rules stated on the basis of the logic of language variation. Fuzzy control adaptive control techniques so far is good.[6]. Generally, the fuzzy control system consists of four basic factors: fuzzification, rule base, inference and defuzzification[7].

The membership function is a curve that describes the mapping input data into membership values, which have a range between 0 and 1. The membership function has several different types, such as a triangular shape, a trapezoidal shape, etc. For systems - systems that require significant dynamic variation, triangular or trapezoidal shape can be used. For the control system is very high, gaussian shape or S-curves can be utilized[8].

Position control using fuzzy logic in this paper using a triangle curve and Trapezoidal Curve, as shown in Figures 1 and 2.

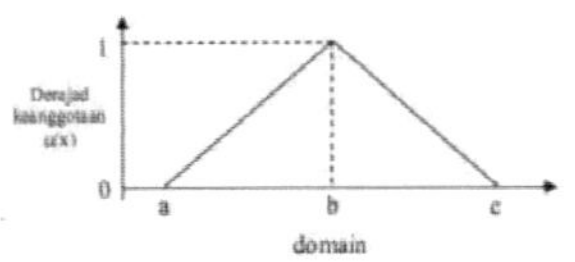

Figure 1. Triangle Curve

Membership function:

$$\mu[x] = \begin{cases} 0; & x \le a \text{ or } x \ge c \\ (x-a)/(b-a); & a \le x \le b \\ (c-x)/(c-b); & b \le x \le c \end{cases} \quad (1)$$

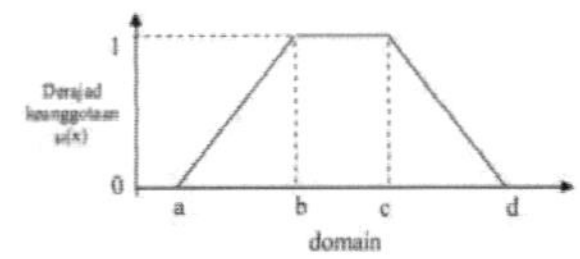

Figure 2. Trapezoidal Curve

Membership function:

$$\mu[x] = \begin{cases} 0; & x \le a \text{ or } x \ge d \\ (x-a)/(b-a); & a \le x \le b \\ 1; & b \le x \le c \\ (d-x)/(d-c); & x \ge d \end{cases} \quad (2)$$

4. Result and Discussion

Realization colour object tracking system using fuzzy logic as position control, done in two phases: the design and manufacture which includes the design and manufacture of hardware and software.

4.1. Design of Fuzzy Logic Controller

The system in this research is designed to implement the fuzzy logic that consists of two inputs: input data (X-axis data and Y-axis data) and input del_data_prev. Input data is data read out position of the object captured by the camera while del_data_prev input is the input data is reduced to previous data. And consists of one output of the PWM value (Pulse Width Modulation) for controlling DC servo motors for movement on the X axis and Y axis. Design of position control using fuzzy logic of this approach is to have a set of fuzzy triangular shape (a, b, c) according to the equation 1.

Input data and input del_data_prev then used as crisp for fuzzification process in determining the direction of rotary motors by setting the PWM (Pulse Width Modulation) to position the camera to always follow the position of the ball-shaped object. The analysis of fuzzification process for both inputs are as follows:

a. Input data

Input data is represented by the shoulders and triangular membership functions as in Figure 3.

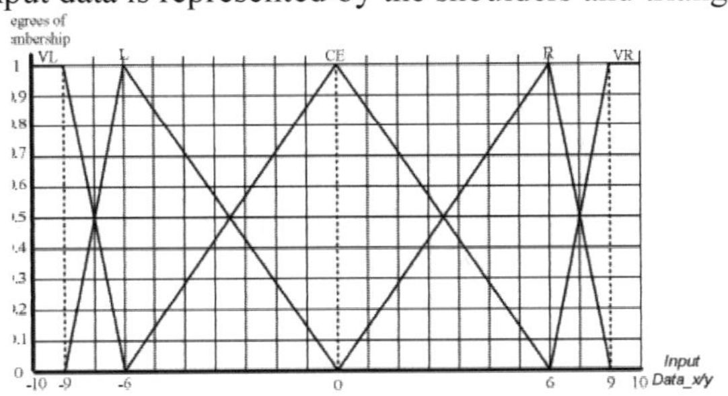

Description:
VL = Very Left
L = Left
CE = Center
R = Right
VR = Very Right

Figure 3. Fuzzy set for input data

b. Input del_data_prev

Input del_data_previous crisp second input derived from the input data is reduced to previous data.input del_data_prev presented by the shoulders and triangular membership functions such as figure 4.

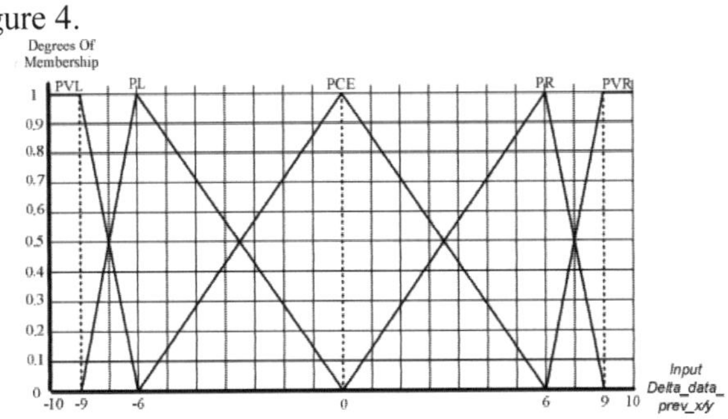

Description:
PVL = Previous Very Left
PL = Previous Left
PCE = Previous Center
PR = Previous Right
PVR = Previous Very Right

Figure 4. Fuzzy Set for Input del_data_prev

Both input fuzzification is used interchangeably to control the X-axis and Y-axis motors. After the fuzzification process is finished and then proceed with the next stage of the process of inference including the design Fuzzy Logic Rule. Here isthe design of Fuzzy Logic Rule applied in position control.

Table 1. Fuzzy Logic Rule

	VL	**L**	**CE**	**R**	**VR**
PVL	LF	LS	S	S	RS
PL	LS	LS	S	S	RS
PCE	LS	LS	S	RS	RS
PR	LS	S	S	RS	RS
PVR	LS	S	S	RS	RF

Description:
LF = Left Fast
LS = Left Slow
S = Stop
RS = Right Slow
RF = Right Fast

Then the next stage is the defuzzification process, process defuzzification used methods singleton weight average of Sugeno, while the design of defuzzification process is described as follows:

IOP Conf. Series: Materials Science and Engineering **190** (2017) 012003 doi:10.1088/1757-899X/190/1/012003

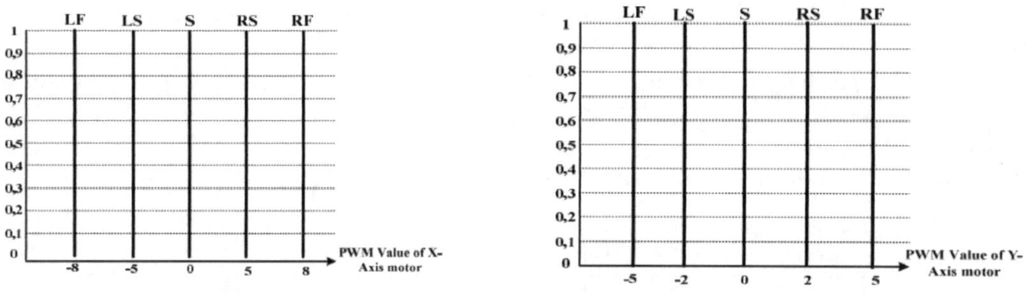

Figure 5. (a) Fuzzy Set for Output Position Control of X-axis Servo Motor (b) Fuzzy Set for Output Position Control of Y-axis Servo Motor

4.2. Hardware Design of Colour Object Tracking System

Hardware manufacture color object tracking system using some of the equipment and materials needed and make the design, while the block diagram of a color object tracking system can be seen in Figure 6.

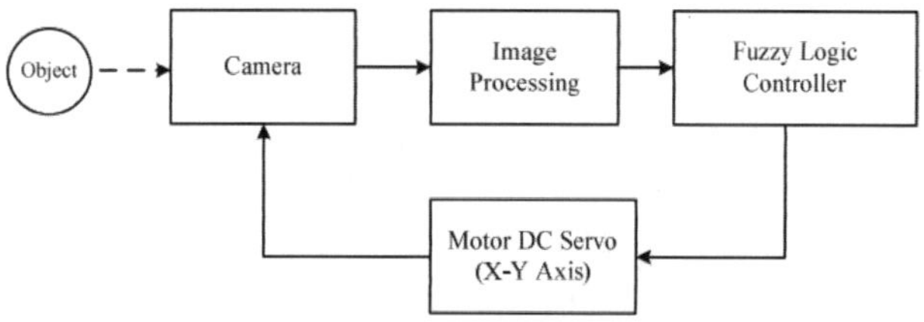

Figure 6. Block diagram of colour object tracking system

Table 2. Testing Results of object detecting with variations in the size of the object

Object size (mm)	Light intensity (lux)	Sightings camera detected objects	
		RGB *image*	HSV *image*
16.9	117	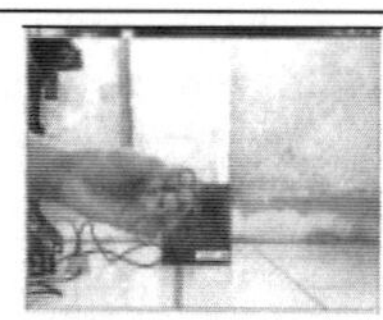	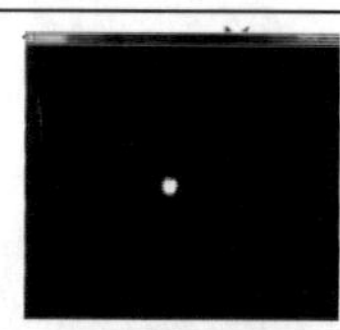
31	117	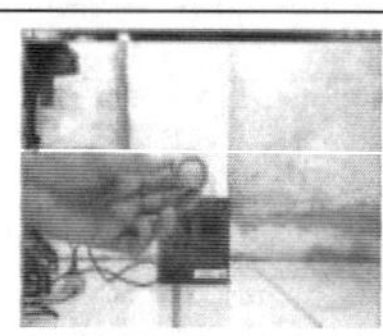	
63.7	117	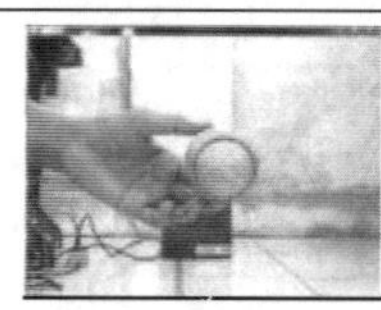	

Block diagram on Figure 6 describe show the color object tracking system follows the position of the ball-shaped object based on the pixel value of each axis into a microcontroller (Fuzzy Logic Controller), the results of data processing performed by the microcontroller are then sent to the X-axis and Y-axis servo motor to move the camera to follow the changing position of the object detected. The test results detecting a ball shaped object with a diameter of 16.9mm, 31mm, and 63.7mm. The test results are presented in Table 2.

Table 2 reveals that the camera is able to detecting objects green balls with a diameter of 16.9 mm, 31 mm, and 63.7 mm, it is marked with a red circle around the objects and the pixel value of the object's position will be read. Results of testing the response of the camera position control using fuzzy logic to change the position of objects, are presented in Table 3.

Table 3. Results of testing the response of control position in following moving objects

Average rate Speed Objects (cm/s)	Average rate Speed Cameras (cm/s)	Description
2.87	2.45	The object is able to follow
4.23	4.13	The object is able to follow
5.33	5.33	The object is able to follow
6.53	6.19	The object is able to follow
7.35	7.12	The object is able to follow
16.33	3.36	The object is not able to follow

Table 3 shows that the speed of the object of 2.87 cm/s - 7.35 cm/s, camera is able to follow the movement of the object. This is indicated by the rate of which has a speed camera near the speed of the object. While, at a rate of an object 16.33cm/s camera is not able to keep pace with the object.

5. Conclusion

Fuzzy logic controller was implemented using the C language on this system, which includes the process of fuzzification, inference and defuzzification able to control two servo motors that move on the X axis and Y axis to follow the change in position of an object well. Position control with fuzzy logic applied is able to follow an object moving horizontally at an average rate of 7.35 cm/s.

References

[1] P. Gupta, vandana Malik, and mallika gandhi, "Implementation of Multilevel Threshold Method for Digital Images Used In Medical Image Processing," vol. 2, no. 2, pp. 2–4, 2012.
[2] S. Sharma and R. Sharma, "Comparison Between Circular Hough Transform And Modified Canny Edge Detection Algorithm For Circle Detection," *(IJERT)*, vol. Vol. 1, no. Issue 3, May - 2012, pp. 1–5, 2012.
[3] R. Passarella, "Implementation of Image Segmentation Using RGB Color Detection on Virtual Reality Robot as Spy Tool," *Proceeding of EECSI*, no. August, pp. 20–21, 2014.
[4] S. K. Dewi, "Analisis Desain Sistem Fuzzy Menggunakan Tool Box Matlab." jogjakarta, 2002.
[5] M. Namazov, "DC motor position control using fuzzy proportional-derivative controllers with different defuzzification methods," *Turkish J. Fuzzy Syst.*, vol. 1, no. 1, pp. 36–54, 2010.
[6] K. Tahir, C. Belfedal, T. Allaoui, and G. Champenois, "Power Control of Wind Turbine Based on Fuzzy Sliding_Mode Control," *Int. J. Power Electron. Drive Syst.*, vol. 5, p. 502~511, 2015.
[7] M. R. Rahimi, S. Hajighasemi, and D. Sanaei, "Designing and Simulation for Vertical Moving Control of UAV System using PID , LQR and Fuzzy logic," *IJECE*, vol. 3, no. 5, 2013.
[8] Y. Bai and D. Wang, "Fundamentals of Fuzzy Logic Control – Fuzzy Sets , Fuzzy Rules and Defuzzifications," 1982.

Investigation of Diesel's Residual Noise on Predictive Vehicles Noise Cancelling using LMS Adaptive Algorithm

Sri Arttini Dwi Prasetyowati[1], Adhi Susanto[2], Ida Widihastuti[3]

[1,3] Electrical Engineering Dept, Islamic Sultan Agung University, Semarang, Indonesia
[2] Departement.of Electrical Engineering, Technology Faculty, Gadjah Mada University, Yogyakarta, Indonesia

[1] arttini@unissula.ac.id

Abstract. Every noise problems require different solution. In this research, the noise that must be cancelled comes from roadway. Least Mean Square (LMS) adaptive is one of the algorithm that can be used to cancel that noise. Residual noise always appears and could not be erased completely. This research aims to know the characteristic of residual noise from vehicle's noise and analysis so that it is no longer appearing as a problem. LMS algorithm was used to predict the vehicle's noise and minimize the error. The distribution of the residual noise could be observed to determine the specificity of the residual noise. The statistic of the residual noise close to normal distribution with = 0,0435, = 1,13 and the autocorrelation of the residual noise forming impulse. As a conclusion the residual noise is insignificant.

1. Introduction

Residual noise always appear after the execution of cancelling noise process. The correlation between residual noise and White Noise Gaussian (WNG) is important to be investigated, because WNG cannot be eliminated further. The previous proposal investigated that the noise cancelling process was still leaving the residual noise[1] while Ref. [2] examine about time-frequency domain methods for noise estimation and speech enhancement. The aims of this research are to analyze the statistic of residual noise and observe the distribution curve.

LMS Adaptive algorithm is the robust and simple algorithm so that it can be applied to the complex signal. The noise from the vehicles were quite complicated, because the noise change fluctuations or not static [1]. The simplest structure, Linear Combiner, can be executed simply. The Equation can be shown on the Equation (1).

$$y_k = w_{ok}x_k + w_{1k}x_{k-1} + w_{2k}x_{k-2} + \ldots + w_{Lk}x_{k-L} = \sum_{l=0}^{L} w_{lk}x_{k-l} \qquad (1)$$

LMS Adaptive is one of the simplest adaptive algorithm that can solve the complex problem of vehicle's noise. The LMS Adaptive Algorithm can be shown on the equation (2).

$$\mathbf{W_{k+1}} = \mathbf{W_k} + 2\mu\mathbf{X}_k\,\varepsilon_k \qquad (2)$$

Equation (2) is used to find the correct weight which used on equation (1). Besides two equation, needs to find the correlation of the error which is obtained from difference of noise that want to cancel and the output of the system [3].

The configuration in Figure 1 illustrates the most appropriate scheme for vehicle's noise cancelling and equation (1) and (2) are used in the block of that configuration [4].

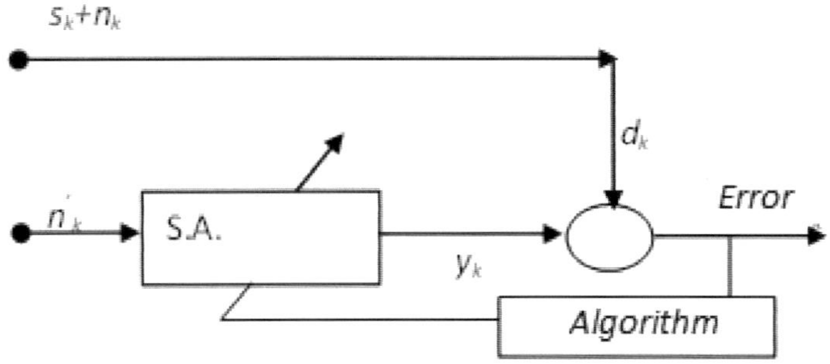

Figure 1. Configuration of Noise Cancelling

2. Method

Data from diesel vehicles were recorded. The collection of that data was processed by LMS predictive algorithm for minimizing the noise. Error that occur on the end of the process is called residual noise. The statistic of the residual noise was count. The grafic of the residual noise's statistic is shown in Figure 2. The horizontal axis shows the value of the signal, the vertical axis shows the amount of the each value.

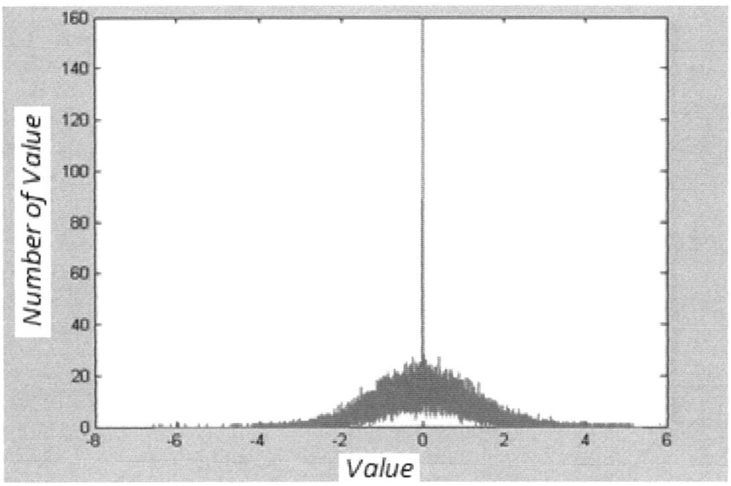

Figure 2. the statistic of the residual diesel's noise

It was called the statistic of the residual noise. The fitting curve of Figure 2 after taken the logarithm would be compared with the logarithm of the normal distribution. The analysis is as follows:

At the first time, scrutinized the autocorrelation. If there was still forming impuls, its shows that there was a noise. The autocorrelation of the residual noise shows in Figure 3.

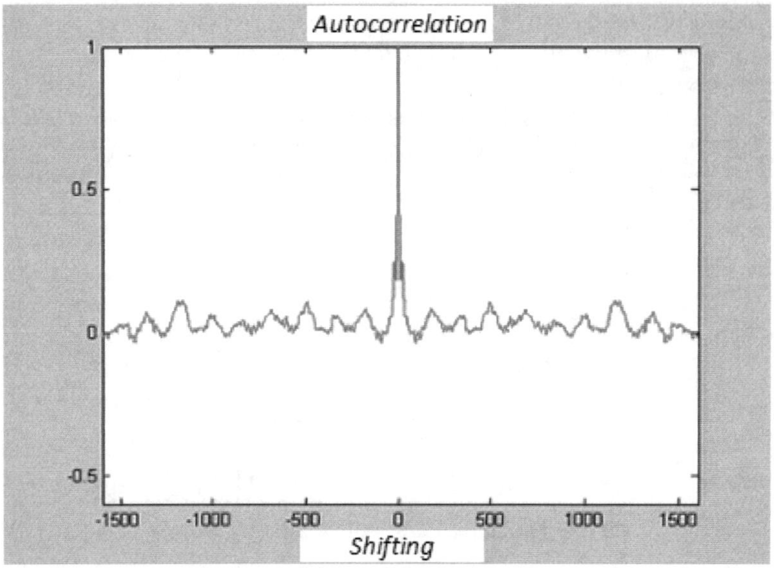

Figure 3. Autocorrelation of the residual Diesel's noise forming impuls

The condition that autocorrelation of the residual noise forming impuls confirm that there was a noise.

The next process is to analyse the distribution function of diesel's residual noise. It will be investigated if the distribution close to normal distribution or not.

The Normal distribution: $N(x,\mu,\sigma) = \frac{1}{\sigma\sqrt{2\pi}}\exp\left[-\frac{1}{2}(\frac{x-\mu}{\sigma})^2\right]$

$$\ln[N(x,\mu,\sigma)] = \ln\ \frac{1}{\sigma\sqrt{2\pi}}\exp\left[-\frac{1}{2}(\frac{x-\mu}{\sigma})^2\right]$$

$$= \ln\ (\sigma^2 2\pi)^{-1/2}\exp\left[-\frac{1}{2}(\frac{x-\mu}{\sigma})^2\right]$$

$$= -\frac{1}{2}\ln(2\pi\sigma^2) - \frac{1}{2}\left[\frac{x-\mu}{\sigma}\right]^2 \quad \text{.....................................} \quad (3)$$

$$\text{y\_diesel} = -0{,}39x^2 - 0{,}034x + 2{,}6 = -\frac{1}{2}\frac{(x+0{,}0435)^2}{1/0{,}78} - (0{,}39)(-6{,}67) \quad \text{................} \quad (4)$$

From Eqs. (3) and (4) the result were

$$\sigma^2 = \frac{1}{0{,}78}$$

$$\sigma = \sqrt{\frac{1}{0{,}78}}$$

$$\sigma = 1{,}13$$

So, the first result is $\mu = -0{,}0435$ and $\sigma = 1{,}13$.

The value of σ also can be taken from

$$-\frac{1}{2}\ln(2\pi\sigma^2) = (0{,}39)(-6{,}67)$$

$$-\frac{1}{2}\ln(2\pi\sigma^2) = -2{,}6013$$

IOP Conf. Series: Materials Science and Engineering **190** (2017) 012004 doi:10.1088/1757-899X/190/1/012004

$$\ln(2\pi\sigma^2) = 5{,}2$$
$$2\pi\sigma^2 = e^{5,2}$$
$$\sigma^2 = \frac{e^{5,2}}{2\pi}$$
$$\sigma = \sqrt{\frac{e^{5,2}}{2\pi}}$$
$$\sigma = 5.373$$

There were two result for σ: $\sigma = 1{,}13$ and $\sigma = 5{,}373$. Because σ or standard deviation, measured how the data values are spread. Standard deviation (σ) could also be defined as the average of the difference of data measured by the mean of data. In addition to mathematical calculation, the fitting curve of statistic of the residual noise is in Figure 4 and be enlarged at Figure 5.

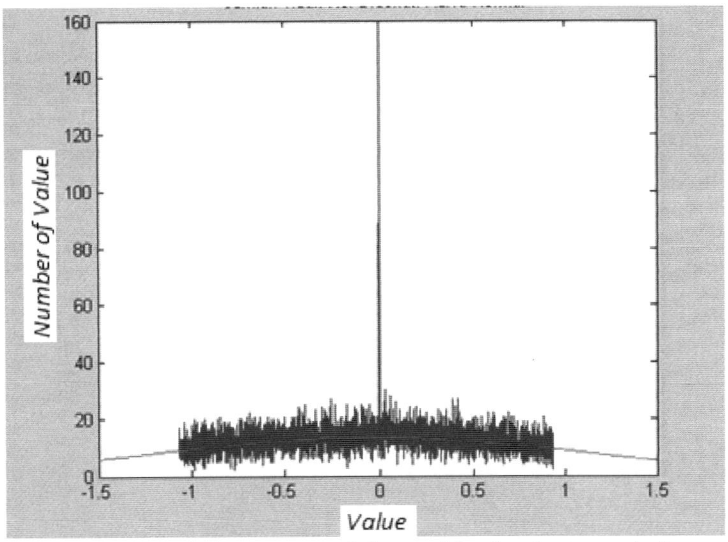

Figure 4. Fitting curve of the residual noise's statistic

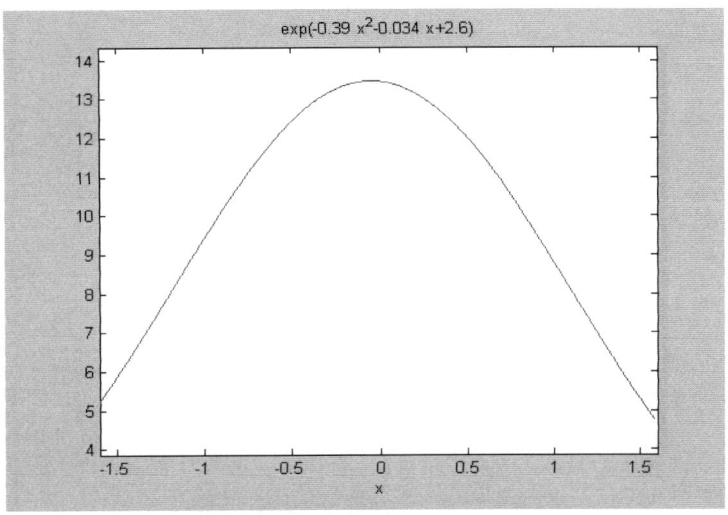

Figure 5. Enlarge of Figure 4

Figure 5 is close to the normal distribution (with $\mu = 0$ and $\sigma = 1$). Therefore, the diesel's residual noise was a white noise Gaussian. Based on the mathematical calculation and Figure 5 it is found that $\mu = -0{,}0435$ and σ close to1,13 than 5,373.

3. Conclusion

From the description above, it can be concluded that the statistic of the residual noise close to normal curve with $\mu = -0{,}0435$, $\sigma = 1{,}13$ and the autocorrelation of the residual noise forming impulse. That confirms the noise is no longer appear as a problem or no longer disturb.

References

[1] Prasetyowati, S.A.D, Adhi, Susanto," Multiple Processes for Least Mean Square Adaptive Algorithm on Roadway Noise Cancelling," International Journal of Electrical and Computer Engineering (IJECE), Vol. 5, No. 2, April 2015, pp. 355~360,ISSN: 2088-8708
[2] Sorensen, K.V; S.V. Anderson, "Speech Enhancement with Natural Sounding Residual Noise Based on Connected Time-Frequency Speech Presence Regions", EURASIP Journal on Applied Signal Processing, 2005: 18, 2954-2964
[3] Prasetyowati, S.A.D., Bustanul, A., Eka N.B.S., "Solution for Vehicles Noise Cancellation With Modification of LMS Adaptive Algorithm", International Journal on Computer Science and Engineering, Vol: 4 Issue: 5, Engg Journal Publications, May 2012
[4] Widrow, B., and S.D. Stearns, "Adaptive Signal Processing", 1985, Prentice-Hall, Inc., Englewood Clifts, New Jersey.

IOP Conf. Series: Materials Science and Engineering **190** (2017) 012005 doi:10.1088/1757-899X/190/1/012005

Quadrotor Proportional-Derivative Regulation for Nonzero Set Point on SO(3) with Disturbance Compensation

S Herdjunanto[1], A P Sandiwan[2], A I Cahyadi[3]

[1,2,3] Department of Electrical Engineering and Information Technology, Universitas Gadjah Mada, Bulaksumur, Yogyakarta 55281, Indonesia (telp: +62274-555225; fax: +62274-4321982)

Email: [1]samiadji@ugm.ac.id, [2]andreas.prasetyanto.s@mail.ugm.ac.id, [3]adha.imam@ugm.ac.id

Abstract. Disturbance compensation is a challenging problem in quadrotor control, especially in nonzero set point regulation. This paper presents proportional-derivative regulation for nonzero set point on SO(3) with disturbance compensation for quadrotor UAV. Quadrotor nonlinear kinematics and dynamics model in SO(3) are used to design the control law. Disturbance compensation is added to the control law by using the upper bound of the disturbance. The numerical simulation shows that the disturbance compensation is able to counter the disturbance effect and improve the bound of the state variables.

1. Introduction

Unmanned Aerial Vehicle (UAV) is popular flying machine for commercial, military, and academic purposes. Quadrotor is a famous UAV type because its mechanical structure is simple. In contrast to common single-rotor helicopter, quadrotor does not need complex swashplate to change the orientation rotor plane. Quadrotor controls its orientation in the air by manipulating the rotational speed of its four rotors. Because of this, quadrotor needs special control algorithm to control its rotors [1].

Generally, quadrotor attitude control can be divided into two approaches. The first approach, namely separate control, controls roll, pitch, and yaw (RPY) angles separately, while the second approach uses SO(3) control [2]. A drawback of the first approach (the separate control) is that it can not evade singularities of attitude representation. In the case of RPY angles, singularity occur when the pitch angle of UAV is 90 degrees [3].

The second approach considers quadrotor's motion in SO(3). Simply stated, SO(3) group is a mathematical group consisting of rotation matrices. In contrast to RPY representation, which represents a rotation by 3 consecutive rotations, SO(3) group uses single arbitrary axis in 3-dimension space. Some works that used this approach were the work of Yun Yu [2] and Yushu Yu [3]. Different from RPY representation that possesses singularity, SO(3) representation can avoid singularity.

Creating a control law that can neutralize disturbance is a challenging problem [4], especially in nonzero set point regulation. Some works modeled disturbance by calculating wind effects on quadrotor. The example of works that used this approach were conducted by Sydney [5] and Tran [6]. Sydney conducted wind effect estimation on quadrotor and used it to design a control law capable of countering disturbance. Tran modelled the studyied propeller-wind interaction, but made no disturbance compensation. Huang [7] considered aerodynamic effect in making a control law aggressively maneuvering quadrotor. He approached this problem by carefully considering blade flapping. However, all of them used separate roll, pitch, yaw control. Recent work that used control on SO(3) and taking disturbance into account was conducted by Fernando [8]. He used the upper bound of the disturbance to design disturbance compensation in the control law.

This research uses the form of disturbance compensation used in [8] and modifies it for regulation using nonzero setpoint. Using similar approach, the upper bound of disturbance magnitude is used to compensate the disturbance. The remainder of this paper is organized as follows. Section II explains quadrotor model in SO(3). This is continued by section III that describes proportional-derivative (PD) control with disturbance compensation. Next, the numerical simulation and analysis are presented in section IV. This paper ends with conclusion in section V.

2. Quadrotor Equations in SO(3)

This section explains the derivation of state-space equation based on [9], [10], and [11]. In exponential coordinates representation, the attitude representation is $\boldsymbol{\zeta} = [\zeta_1 \;\; \zeta_2 \;\; \zeta_3]^T$, where $\boldsymbol{\zeta}$ is the logarithm of the rotation matrix . The angular velocity is described in body frame axes, denoted as $\boldsymbol{\omega}_b = [p \;\; q \;\; r]^T$. The inertia matrix of the quadrotor is diagonal. The quadrotor's dynamics is defined by Euler equation:

$$\mathbf{J}\dot{\boldsymbol{\omega}}_b = -\boldsymbol{\omega}_b \times \mathbf{J}\boldsymbol{\omega}_b + \boldsymbol{\tau} \tag{1}$$

The quadrotor kinematics is written in terms of the angle $\boldsymbol{\zeta} = \log(\mathbf{R})^\vee$ [11]:

$$\dot{\boldsymbol{\zeta}} = \left(\mathbf{I} + \frac{1}{2}\hat{\boldsymbol{\zeta}} + [1 - \alpha(\|\boldsymbol{\zeta}\|)]\frac{\hat{\boldsymbol{\zeta}}^2}{\|\boldsymbol{\zeta}\|^2} \right)\boldsymbol{\omega}_b \tag{2}$$

where $\alpha(\|\boldsymbol{\zeta}\|) = (\|\boldsymbol{\zeta}\|/2)\cot(\|\boldsymbol{\zeta}\|/2)$. Therefore, the second order system of quadrotor UAV in SO(3) is defined as

$$\begin{cases} \dot{\boldsymbol{\zeta}} = \left(\mathbf{I} + \frac{1}{2}\hat{\boldsymbol{\zeta}} + [1 - \alpha(\|\boldsymbol{\zeta}\|)]\frac{\hat{\boldsymbol{\zeta}}^2}{\|\boldsymbol{\zeta}\|^2} \right)\boldsymbol{\omega}_b, \\ \mathbf{J}\dot{\boldsymbol{\omega}}_b = -\boldsymbol{\omega}_b \times \mathbf{J}\boldsymbol{\omega}_b + \boldsymbol{\tau}. \end{cases} \tag{3}$$

Equation 2 shows the benefit of using SO(3) for regulation since onversion from $\boldsymbol{\omega}_b$, which is obtained from gyroscope sensor, to $\dot{\boldsymbol{\zeta}}$ does not suffer from singularity. This is due to $\lim_{\boldsymbol{\zeta}\to\mathbf{0}} \alpha(\|\boldsymbol{\zeta}\|) = 0$ and $\lim_{\boldsymbol{\zeta}\to\mathbf{0}} \left(\hat{\boldsymbol{\zeta}}^2 \|\boldsymbol{\zeta}\|^{-2}\right) = 0$. On the other hand, in RPY representation, conversion from $\boldsymbol{\omega}_b$ to $\dot{\boldsymbol{\Theta}}$ suffers from singularity when $\theta = \pi/2$ (see [10] or [9] for details).

3. PD Control with Disturbance Compensation

Before disturbance compensation is taken into consideration, the simpler proportional-derivative (PD) control under no disturbance is explained. This control law uses the the result made by Bullo [11]. The control law to regulate the quadrotor to a nonzero set point $\boldsymbol{\zeta}_d$ is

$$\boldsymbol{\tau} = \boldsymbol{\omega}_b \times \mathbf{J}\boldsymbol{\omega}_b - \mathbf{K}_p(\boldsymbol{\zeta} - \boldsymbol{\zeta}_d) - \mathbf{K}_d\boldsymbol{\omega}_b. \tag{4}$$

where $\mathbf{K}_p$ and $\mathbf{K}_d$ are positive definite matrices. This control law is a modificatioin from [11]; the original control law uses zero set point. The proof of the control law's stabilizing ability, which invokes Lyapunov stability theorem, can be read at [11].

In reality, quadrotor flight encounters disturbance, which may be caused by parameter uncertainties, approximation or inaccuracy in aerodynamic effects modelling, and wind. The disturbance can be modelled by adding **g** into dynamics equation in the system:

$$\mathbf{J}\dot{\boldsymbol{\omega}}_b = -\boldsymbol{\omega}_b \times \mathbf{J}\boldsymbol{\omega}_b + \boldsymbol{\tau} + \mathbf{g}. \tag{5}$$

The magnitude of **g** has an upper bound of δ, where δ is a nonzero positive scalar. Under the influence of the disturbance, the closed-loop system dynamics becomes

$$\dot{\boldsymbol{\omega}}_b = -\mathbf{J}^{-1}\mathbf{K}_p(\boldsymbol{\zeta}-\boldsymbol{\zeta}_d) - \mathbf{J}^{-1}\mathbf{K}_d + \mathbf{J}^{-1}\mathbf{g}. \tag{6}$$

The disturbance compensation **u** is added to the control law:

$$\boldsymbol{\tau} = \boldsymbol{\omega}_b \times \mathbf{J}\boldsymbol{\omega}_b - \mathbf{K}_p\boldsymbol{\zeta} - \mathbf{K}_d\boldsymbol{\omega}_b + \mathbf{u}. \tag{7}$$

With the compensation, the closed-loop system dynamics becomes

$$\dot{\boldsymbol{\omega}}_b = -\mathbf{J}^{-1}\mathbf{K}_p\boldsymbol{\zeta} - \mathbf{J}^{-1}\mathbf{K}_d + \mathbf{J}^{-1}\mathbf{g} + \mathbf{J}^{-1}\mathbf{u}. \tag{8}$$

The disturbance compensation **u** is

$$\mathbf{u} = -\frac{\delta\left(k_1(\boldsymbol{\zeta}-\boldsymbol{\zeta}_d) + k_2\boldsymbol{\omega}_b\right)}{\left\|k_1(\boldsymbol{\zeta}-\boldsymbol{\zeta}_d) + k_2\boldsymbol{\omega}_b\right\| + \kappa}. \tag{9}$$

where k_1 and k_2 are positive constants. This compensation is modifed from [8].

4. Numerical Simulation and Analysis

The simulation is conducted by adding random disturbance to the system. The quadrotor model is based on [9] and [12]. The elements of the inertia matrix are J_{xx} = 0.082 kg m^2, J_{yy} = 0.0845 kg m^2, and J_{zz} = 0.1377 kg m^2. The initial condition of angle $\boldsymbol{\zeta}(0)$ and body angular velocity $\boldsymbol{\omega}_b(0)$ are, respectively, $\boldsymbol{\zeta}(0) = [\pi/6 \;\; 0 \;\; 0]^T$ and $\boldsymbol{\omega}_b(0) = [p \;\; q \;\; r]^T = [0 \;\; 0 \;\; 0]^T$. The set point is $\boldsymbol{\zeta}_d = (\pi/18)[1 \;\; 1 \;\; 1]^T$, which is equivalent to 10°. The proportionality and derivative constant matrix are written in form of $\mathbf{K}_p = k_p\mathbf{I}_{3\times3}$ and $\mathbf{K}_p = k_d\mathbf{I}_{3\times3}$, respectively, with $k_p = 8$ and $k_d = 5$. The upper bound of disturbance is δ = 2.18.

Figure 1 shows the angle $\boldsymbol{\zeta}$ with the control law without compensation, while Figure 2 shows the angle $\boldsymbol{\zeta}$ under compensation effect. In Figure 1, after about 2 seconds, the angle values' divergence from the set point is close to 5°. In figure 2, the angle values after approximately 2 second does not diverge close to 5° (the divergence is well below 5°). The fluctuation of angle values in Figure 2 is smoother than it is in figure 2. It is apparent that, under disturbance influence, the disturbance compensation reduces the bound of state variables around the set point $\boldsymbol{\zeta}_d$, whose all elements equal 10°.

Possible future works following this research can be conducted in some areas. For example, it will be necessary to test the control algorithm in physical experiment. It may be also possible to make adaptive control law or design backstepping control law on SO(3) that can encounter disturbance.

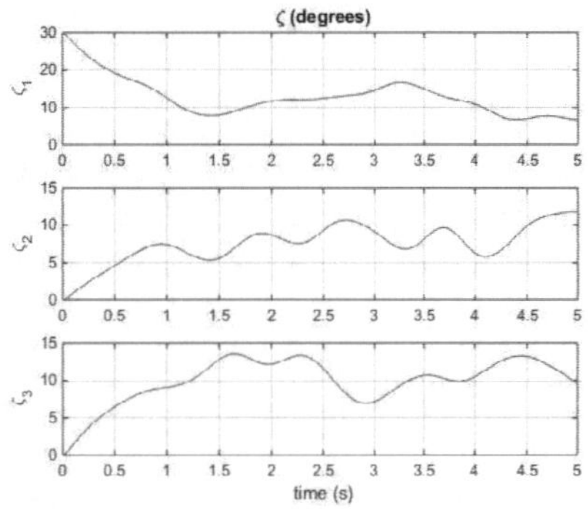

Figure 1. ζ without compensation. Set point for ζ_1, ζ_2, ζ_3 is 10°.

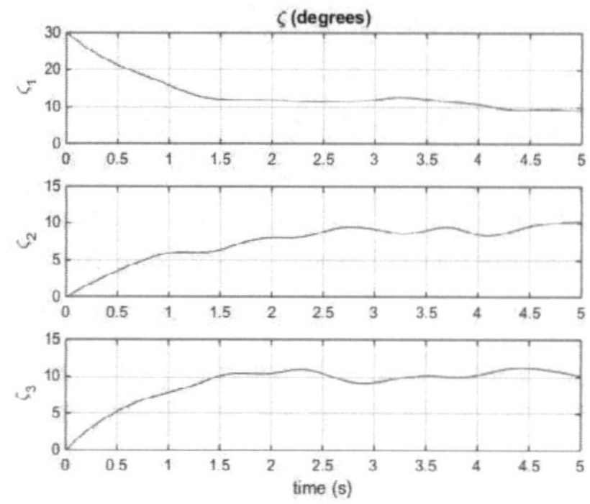

Figure 2. ζ with compensation. Set point for ζ_1, ζ_2, ζ_3 is 10°.

5. Conclusion

The disturbance compensation reduces the bound of state variables. The bound of state variables is smaller with the compensation than it is without the compensation. The compensated control law can handle constant disturbance better than the uncompensated control law can.

References

[1] Dikmen I C, Arisoy A, and Temeltas H 2009 Attitude control of a quadrotor *Proc. of 4th International Conference on Recent Advances in Space Technologies* Istanbul p 722–27.

[2] Yu Y, Yang S, Wang M, Li C, and Li Z 2015 High performance full attitude control of a quadrotor on SO(3) *Proc. of the IEEE Int. Conf. on Robotics and Automation* Seattle p 1698–1703.

[3] Yu Y, Ding X, and Zhu J J 2013 Attitude tracking control of a quadrotor UAV in the exponential coordinates *J. Frankl. Inst.* **350** 8, p 2044–2068, Oct. 2013.

[4] Alexis K, Nikolakopoulos G, and Tzes A 2010 Experimental model predictive attitude tracking control of a quadrotor helicopter subject to wind-gusts *Proc. of the 2010 18th Mediterranean Conf. on Control and Automation* Marrakech p 1461–66.

[5] Sydney N, Smyth B, and Paley D A 2013 Dynamic control of autonomous quadrotor flight in an estimated wind field *Proc. of 52nd IEEE Conf. on Decision and Control* Firenze p 3609–16.

[6] Tran N K, Bulka E, and Nahon M 2015 Quadrotor control in a wind field *Proc. of 2015 International Conference on Unmanned Aircraft Systems* Denver p 320–28.

[7] Huang H, Hoffmann GM, Waslander S L, and Tomlin C J 2009 Aerodynamics and control of autonomous quadrotor helicopters in aggressive maneuvering *Proc. of the IEEE Int. Conf. on Robotics and Automation* Kobe p 3277–82.

[8] Fernando T, Chandiramani J, Lee T, and Gutierrez H 2011 Robust adaptive geometric tracking controls on SO(3) with an application to the attitude dynamics of a quadrotor UAV *Proc. of 50th IEEE Conf. on Decision and Control and European Control Conference* Orlando p 7380–7385.

[9] Corke P I 2011 *Robotics, vision and control: fundamental algorithms in MATLAB* Berlin, Springer.

[10] Ataka A, Tnunay H, Inovan R, Abdurrohman M, Preastianto H, Cahyadi A I, and Yamamoto Y 2013 Controllability and observability analysis of the gain scheduling based linearization for UAV quadrotor Jogjakarta p 212–218.

[11] Bullo F and Murray R M 1995 Proportional Derivative (PD) Control on The Euclidean Group *Proceedings of 3rd European Control Conf* Rome p 1091-1097.

[12] P. Pounds, R. Mahony, and P. Corke, "Modelling and control of a quad-rotor robot," presented at the Australasian Conference on Robotics and Automation 2006, Auckland, New Zealand, 2006.

IOP Conf. Series: Materials Science and Engineering **190** (2017) 012006 doi:10.1088/1757-899X/190/1/012006

Comparison of PID and Fuzzy Controller for Position Control of AR.Drone

A Prayitno[1], V Indrawati[2] and I I Trusulaw[3]

[1,2,3] Electrical Engineering Department, University of Surabaya (UBAYA)
Raya Kalirungkut, Surabaya 60293, East Java, Indonesia, Tel.+62-31-2981157

[1] prayitno_agung@staff.ubaya.ac.id

Abstract. This paper describes the implementation of the PID Controller to control the position of the AR.Drone in the x-y-z. This position control scheme uses three PID controllers to maintain the position of x, y and z using the signal control pitch, roll and vertical rate. PID Controller implemented on AR.Drone 2.0 and then tested in an indoor space. The performance of the controller will be compared with Fuzzy Logic Controller schemes that have been implemented previously. The results show that the PID Controller generate a response with rise time less than 3 seconds at the x and y position with around 25% overshoot. The result for z position give better result without overshoot. The comparison between fuzzy logic and PID Controller indicates that the results of the PID controller is better although there is overshoot.

1. Introduction

Currently, the AR.Drone is widely used by researchers for its research platform. AR.Drone was chosen because it is relatively inexpensive but has onboard controller and internal software. AR.Drone body constructs from frame which made of carbon pipe with 4 brushless motors complete with propellers and the main body is made of plastics fiber in which there are onboard electronics and Lithium-Polymer (Li-Po) battery. User can access controller via Wi-Fi communication. The onboard controller consists of motherboard and navigation board. On the motherboard, there are the processor, front camera, bottom camera, and Wi-Fi chips. On the navigation board there is a microcontroller board that use as an interface to the sensors. AR.Drone equipped with accelerometer, gyroscope and ultrasonic altimeter [1].

AR.Drone maneuver can be controlled using 4 control signals (values between -1 and 1): pitch (positive values for backward, negative values for forward), roll (positive values for move to right, negative values for move to left), yawrate (positive values for the movement pivot clockwise, negative values for the movement pivot counterclockwise) and vertical rate (positive values for vertical rate to up, negative values for vertical rate to down). Moreover, there are 3 toggle switchs for fly, hover and emergency command. Navigation data can be accessed from the AR.Drone includes the actual pitch, forward speed, actual roll, sideward speed, actual yaw, vertical speed, altitude, and time stamp. Using this data we can estimate the x,y,z position of the drone. Figure 1 show the inputs and outputs of the AR.Drone system.

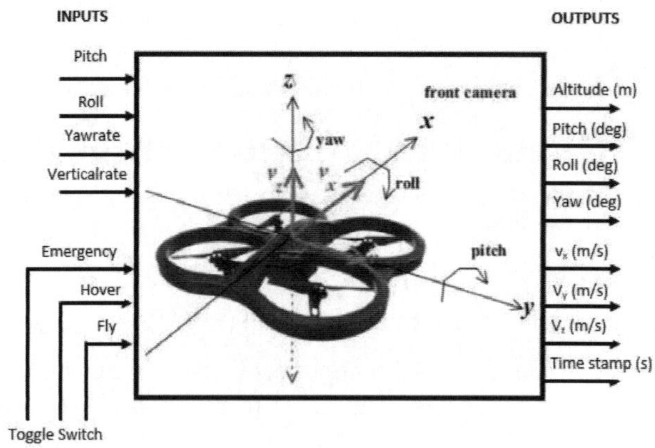

Figure 1. AR.Drone Inputs – Outputs.

Many studies have described a variety of development and application using the AR.Drone. Michael [2] created LabVIEW Toolkits consisting of Main VI, Video VI, NavData VI, and Additional Supporting VI to control the AR.Drone. This software has the ability to transmit control commands and keep the communication channel running, read UDP packets that contain video frame sent from the AR.Drone, sends UDP packets that contain sensor's navigation data outport to the IP address of the computer and estimate x, y, z position of the navigation data. Krajnik [3] formulated four AR.Drone structure models which has input output relations: setpoint pitch - actual pitch - forward speed; setpoint roll - actual roll - sideward speed, setpoint yaw rate - actual yaw rate - yaw, and setpoint vertical rate –actual rate vertical - altitude. Agung [4] calculated model of roll, pitch, yaw rate and critical rate with a step response test on the AR.Drone 1.0. Sarah [5] identified the model parameters of the internal and dynamic model of AR.Drone using second order system approach. Some control scheme are designed and implemented the the model obtained. The control schemes are waypoint navigation controller, leader-follower behavior, trajectory-following behavior. Agung [6] designed a fuzzy scheme trajectory tracking control for applications in x-y coordinates. This fuzzy scheme has two inputs that distance, angle and has two outputs, namely pitch, yawrate. Fuzzy control scheme tested on straight and curved trajectory. Veronica [7] implemented fuzzy logic control for waypoint navigation applications. This fuzzy scheme uses three fuzzy logic to control pitch, roll and vertical rate respectively with each input in the form of desired position and the actual position. Three scenarios were designed and implemented for the waypoint problem. The test results indicate that the fuzzy scheme can be used for waypoint applications. Veronica [8] implemented position control of ar.drone by using two fuzzy control. The first fuzzy used to control pitch and yawrate simultaneously with the input of distance and angle, while the fuzzy second is used to control the vertical rate using the input of desired and actual position z.

This paper reports the development of the PID controller scheme to control the position of the AR.Drone in the x-y-z. The result then compares with fuzzy control schemes that have been implemented in [8].

2. PID Controller Design

The design of PID Controller that will be implemented in this study is shown in figure 2. It can be seen that each position x, y and z of the drone will be controlled by individual PID. Input of each PID is error x, error y and error z, and the output is pitch, roll and vertical rate respectively. In this paper, PID controller design procedure is as follows:

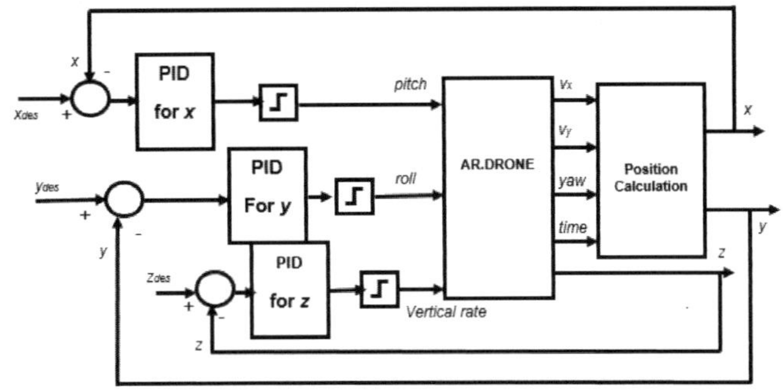

Figure 2. Block diagram of control system.

- Understanding the dynamics of the system AR.Drone. To understand the dynamics of the system, do the step response test on the system pitch, roll and vertical rate of AR.Drone by giving a value ranging from -1 up to 1. Navigation data such as actual pitch, forward speed, x, actual roll, sideward speed, y, vertical rate, and z are recorded and analyzed.
- From the data obtained, considering the limitations of indoor space there should be restrictions on the control signals. Pitch in the range of -0.115 to 0.115, roll in range of -0.125 to 0.125
- Formulate each subsystem PID controller, figure 3, into a computer program. For easy programming and implementation, the PID equation converted as follow.

$$PID = K_P e(t) + K_I \int e(t) + K_D \frac{de(t)}{dt} \tag{1}$$

$$PID = K_P * error + (K_I * integral * dt) + (K_D * \frac{derr}{dt}) \tag{2}$$

where:

K_P : Proportional Constant

K_I : Integral Constant

K_D : Derivative Constant

$error = desired\ point - actual\ point$

$integral = integral_{previous} + error$

$derr = error - error_{previous}$

dt : execution time

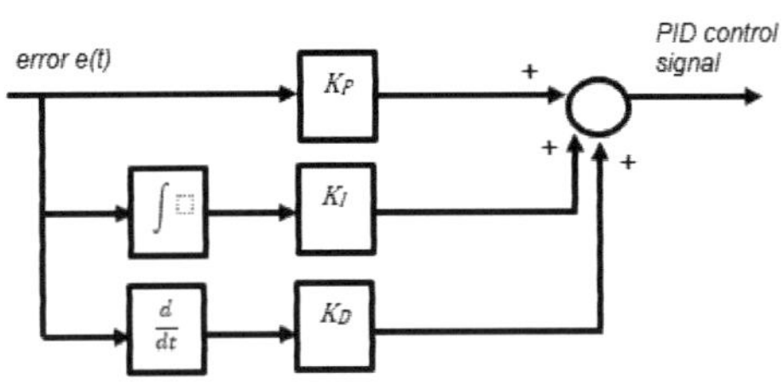

Figure 3. Block diagram of PID.

- Realize the system into computer program, block diagram and front panel.
- By trial and error, tune the PID parameter until obtain the best response. The obtained parameter is tabulated in table 1.

Table 1. PID Parameters.

Parameters	Values		
	PID for pitch	PID for roll	PID for vertical rate
Kp	1.6	1.6	1
Ki	0.005	0.005	1
Kd	0.5	0.5	2

3. Results and Analysis

PID controller is implemented on the AR.Drone 2.0 and flown at indoor laboratory. The testing process is conducted as follows:

- From the front panel, specify the setpoint position *x*, *y* and *z*, which are trying to reach by the AR.Drone.
- With the push button on the front panel, AR.Drone fly hover as high as 1 m. The hover position is considered as the initial coordinat (0,0,0).
- Then switch OFF hover in front panel, AR.Drone will fly autonomous with PID Controller designed to the setpoint. During flew the flight data is recorded.
- When it reaches the intended position, indicated hover fly, AR.Drone is landed.

The results of four time testing at the set point (1,1,1) are shown in figure 4 and figure 5. Figure 4 are 3-D representation, while figure 5 is 2-D representation of the AR.Drone response.

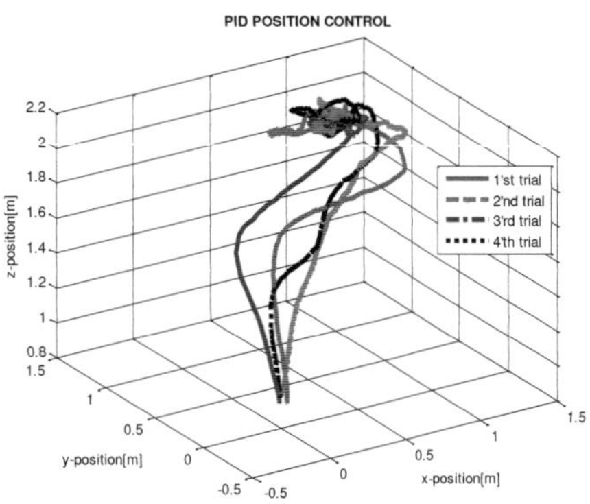

Figure 4. Experiment result of PID Position Control in 3-D.

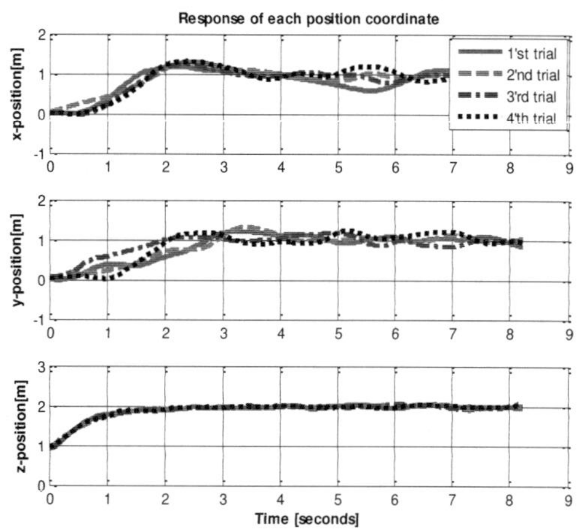

Figure 5. Experimental results of PID Position Control in 2-D.

From the test results, figure 5, it appears that the AR.Drone can reach the desired position with a rise time of about 2 seconds, has about 25% overshoot and settling time of about 4 seconds. From 3-D graphics, figue 4, it appears that the AR.Drone experienced oscillations around the setpoint.

Tests were also done on the set point (1,1,0) to see the response AR.Drone in flight the x-y plane.The results of four time testing are shown in figure 8.

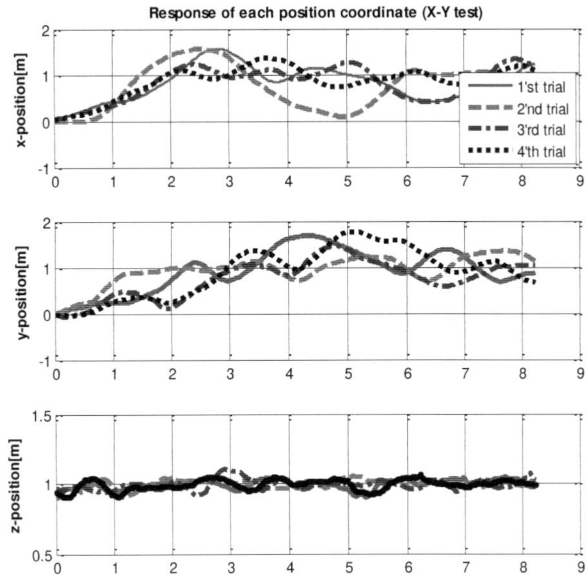

Figure 6. Experimental results of PID Position Control in 2-D.

4. Comparison with Fuzzy Control Scheme

Results PID position control is compared with the Fuzzy position control implemented by Veronica [8], where the fuzzy scheme is shown in figure 7 and figure 8:

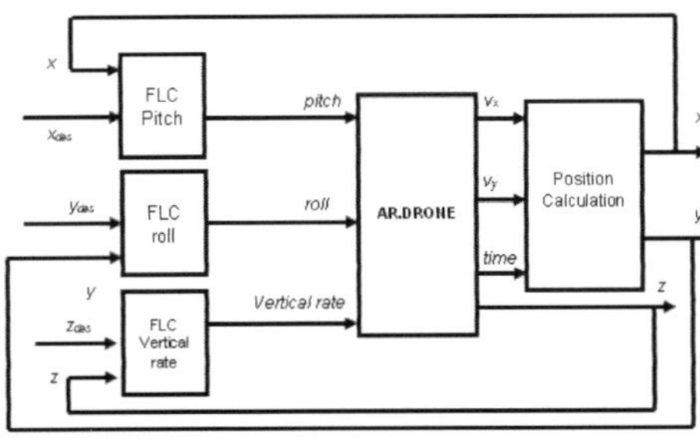

Figure 7. First fuzzy scheme implemented on [8].

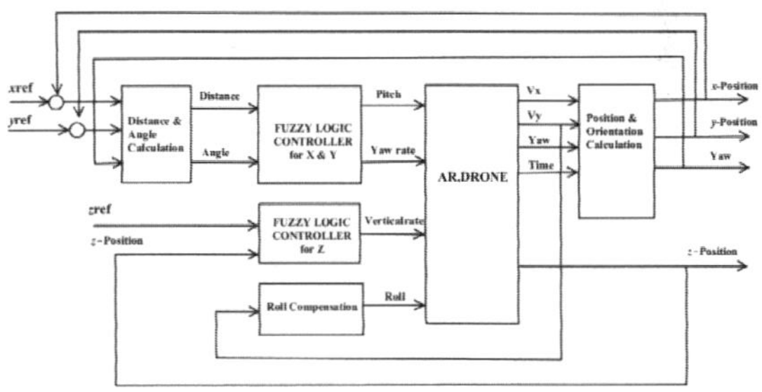

Figure 8. Second fuzzy scheme implemented on [8].

Similar to the PID Controller scheme, scheme in figure 7, uses three pieces of FLC controller to control pitch, roll and vertical rate simultaneously. Each fuzzy input is the desired position and the actual position. While scheme in figure 8 uses only two pieces of fuzzy control for pitch and yawrate simultaneously and fuzzy logic for the vertical rate.

The results of the comparison PID response with two fuzzy schemes that has been done is shown in figure 9. It is noticeable that the PID controller generates a better response both in the x, y and z, but has a slightly higher overshoot

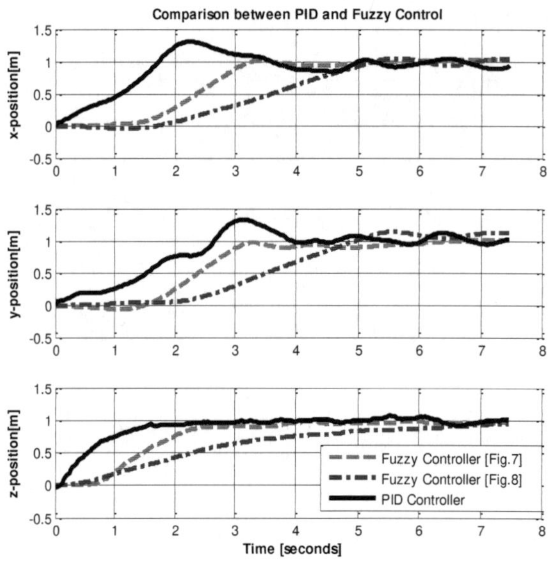

Figure 9. Comparation of two control schemes.

5. Conclusion

To control position of *x, y, z* of the AR.Drone can be realized with three PID controller to manipulate the pitch, roll and vertical rate. The implementation results show that PID controlled system gives response with a rise time of about 2 seconds, 25% overshoot and settling time of about 4 seconds. The results of the comparison between PID controlled system and fuzzy controlled system, the PID controlled system gives better result in this experiment.

References

[1] Pierre-Jean B, Francois C, David V and Nicolas P 2011 The navigation and control technology inside the AR drone micro UAV 18th IFAC World Congress Milano Italy

[2] Michael M 2012 The AR drone LabVIEW toolkit: A software framework for the control of low cost quadrotor aerial robots (Master of Science Thesis - TUFTS University)

[3] Krajnik T, Vonasek V, Fiser D and Faigl J 2011 AR-drone as a platform for robotic research and education. Research and Eductation in Robotics EUROBOT Heidelberg

[4] Agung P and Veronica I 2014 Model AR.drone dengan indoor dan outdoor Hull Proceeding CITEE Yogyakarta Indonesia.

[5] Sarah Y Tang 2013 Vision-based control for autonomous quadrotor (Final Report :Undergraduated Senior Thesis. Department of Mechanical and Aerospace Engineering. Princeton University)

[6] Agung P, Veronica I and Gabriel U 2014 Trajectory tracking of AR.drone quadrotor using fuzzy logic controller J. Telkomnika 12(4) 819

[7] Veronica I, Agung P and Thomas A Kurniawan 2015 Waypoint navigation of AR.drone quadrotor using fuzzy logic controller J. Telkomnika 13(3) 930

[8] Veronica I, Agung P and Gabriel U 2015 Comparison of two fuzzy logic controller schemes for position control of AR.drone *Proceeding of 7th International Conference on Information Technology and Electrical Engineering (ICITEE)* Chiangmai Thailand

Control of Liquid Sloshing Container using Active Force Control Method

Didik Setyo Purnomo[1], **Adnan Rachmad Anom Besari**[2] **and Zaqiatud Darojah**[1]

[1,3] Mechatronic Engineering Department, Politeknik Elektronika Negeri Surabaya, Indonesia
[2] Computer Engineering Department, Politeknik Elektronika Negeri Surabaya, Indonesia

E-mail: didiksp@pens.ac.id; zaqiah@pens.ac.id; anom@pens.ac.id

Abstract. This paper presents a robust control method to relieve the sloshing of liquid container transport using Active Force Control (AFC) method. A model of two degree-of-freedom (2-DOF) liquid container transfer was implemented in this research as the main dynamical system to be controlled. The surface of liquid is maintained in a flat position, so that changes the slope of liquid surface countered by changing the acceleration of container. The focus of this research is how to use AFC method being applied to the system, so that it can suppress liquid sloshing. The control scheme were simulated, compare between PID-AFC and pure PID. Simulations has been conducted, the results show that the PID-AFC have superior performance to suppress the sloshing compared with pure PID, especially if disturbance occurred.

1. Introduction

Sloshing or oscillation of the free surface of the liquid may occur during transfer of container filled with liquid. The sloshing happens due to the acceleration or deceleration when the container moves. Sloshing is undesirable condition, too many sloshes may cause spill-over, temperature changes, vibration and induce contamination of the molten metal is transferred. Therefore, the slosh should be suppressed while maintaining a high transfer speeds container for reasons of productivity and safety, for example transfer the liquid metal in the casting process. One way to reduce liquid sloshing low as controlling the acceleration or deceleration in accordance with the speed of the container, so as not to cause a socking. This research considers the fluid transfer machine can move the slide horizontally from one side to the other. An Inertial Measurement Unit (IMU) sensor fitted to the system for measuring acceleration and socks liquid container caused a stir. The purpose of this study was to develop a transfer container spilled liquid that can be suppressed using active force control.

2. Previous Works

Several methods to suppress the liquid slosh have done intensive research. In the tracking and sloshing suppression control to the moving object using a 2-DOF control system, the filter gain a different time using polynomial time can dampen spilled liquid container, and a fast track to get the target. The experimental results obtained indicate that the proposed control system is useful for both of the tracking control to the moving object and the sloshing suppression control [1] [2]. Additional work is presented using a flexible container as absorbent spill. Numerical predictions presented in which the flexible container is set to benefit more rigid container for effective control [3]. Numerical predictions presented in which the flexible container is set to benefit more rigid container for effective control [3]. Some researchers have been developed input shaping method to absorb sloshing on transfer of liquid in container using multiple degree-of-freedom robot arms [4]. They have also developed a vibration parameter identification algorithm integrated with design input shaper. Its applications for vibration absorber control of the transfer of the manipulator robot arm have been done. Simulation and experimental results have shown good accuracy in estimating the natural frequency and damping ratio. The final performances of forming an input have been also verified [5]. Other researchers presented analysis of dynamic effects and control of liquid stirring for an infinite number of modes spill as well as theoretically and experimentally. There are two control methods were proposed to suppress spill infinite mode in away minimize the slosh. The first is creating smooth input, while the second is a combination of input-shaping and smoothing command scheme. Both smooth and a controller coupled to produce good control effect that reduces unwanted transients and stir remainder. However, the combined controller can be shorter in time than a smooth ride. Simulations and experiments results were used to prove the dynamic character and the efficacy of the proposed method [6]. The both robustness and perform ability of the proposed scheme is guaranteed the proposed solution based on sliding mode control. The controller is shown to direct the robot to restricted environment of origin.

Another researchers have been created robust control such as slide mode control, adaptive control, quantitative feedback theory, passivity based control, etc. Yet another type of powerful method known as a method of active power control (AFC) was first proposed by Hewitt and Burdess (1981) has proven to be very robust against interference and the uncertainty of the dynamic system. However, no research or studies that have not been that incorporate the scheme into the liquid spilled oppressive and investigated its effects. Other researcher has been also developed of hybrid control schemes for vibration suppression applied to suspension system. The control scheme is developed on the basis of the AFC collaborate with *input shaping* techniques and it has been simulated, implemented and tested in an experimental suspension system. The characteristic of resonance modes of vibration suppression is evaluated to obtain performance of the control scheme. Using the both control strategies, the vibration suppression results had been accepted. A comparative appraisal of the control techniques applied to suppression of the suspension has shown that AFCIS with 4 impulses scheme results is better performance than the pure AFC in respect of vibration [7][8].

Other people use active force control (AFC) strategy to absorb the vibration and squealing noise phenomenon in disc brake model using a robust feedback control method. A model of a two-degree-of-freedom (2-DOF) *Wagner* considered in this research as the main dynamic element that must be controlled. The durability and effectiveness in suppressing vibration and screaming was tested, by considering account a number of disorders that work on the system. There are three control schemes involving simulated and compared to classical controller proportional-integral-derivative (PID), pure AFC with crude approximation (AFCCA) and combination of AFC with fuzzy logic (AFCFL) scheme. AFC-based schemes shown excellent performance compared with the pure PID counterpart with the AFCFL method generate the best performance [9].

3. Control Design

When someone brings a glass of water on his hand while walk, automatically the hand will suppresses the sloshing of the water surface. The hand will control the speed or acceleration of the moving. Although the hand moves in high speed, however the hand will keep the acceleration low as possible. There are two censors is used by human to measure the water sloshing, namely eyes and nerve. Eyes see value of the sloshing by the tilt angle of water surface. While the nerve detect water sloshing by the change of inertia or shocking. When eyes see water is sloshing and almost spilled or nerve sense change of inertia, the hand will react by do change the acceleration. If the sloshing caused by high acceleration, the hand will decrease the acceleration. In what follows, the system consist of a tank, mounted on two parallel bars. The tank can slide horizontal as reciprocally on the parallel bars. The movement of the tank is driven by a DC motor, which slides as long as rig gear. The surface of liquid is flat, when the tank at idle condition. But, when the tank move will occur slosh being caused by acceleration or deceleration. A laboratory scale experimental has been built to simulate the real operation and all progress is presented as shown in Fig. 1.

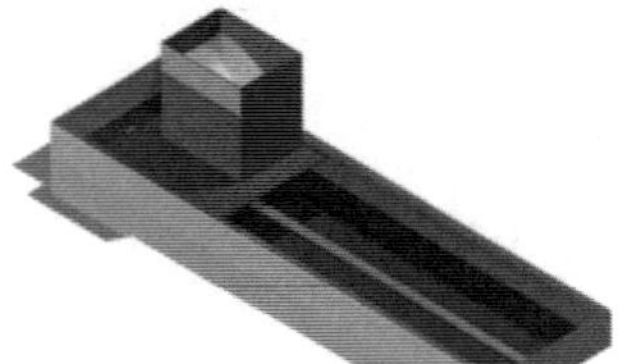

Figure 1. Design of liquid sloshing in container transfer

3.1. Sloshing Model

Sloshing in an open container as general can be well described by equivalent mechanical models, as well as a pendulum model or a spring-mass model [1]. One way is using a simple pendulum model, where one pendulum system is one mode. Slope magnitude of liquid surfaces similar with slope magnitude of vertical pendulum. The angle of liquid surface is similar to the angle of pendulum bar. The pendulum model is added a damper system to reflect the viscosity and friction of fluid with container walls. Given the fundamental mode only spill, which is dominant in the transfer container spilled, and next ignore other small modes, model of pendulum is shown in Figure 2 can adequately represent the dynamics spilling laterally.

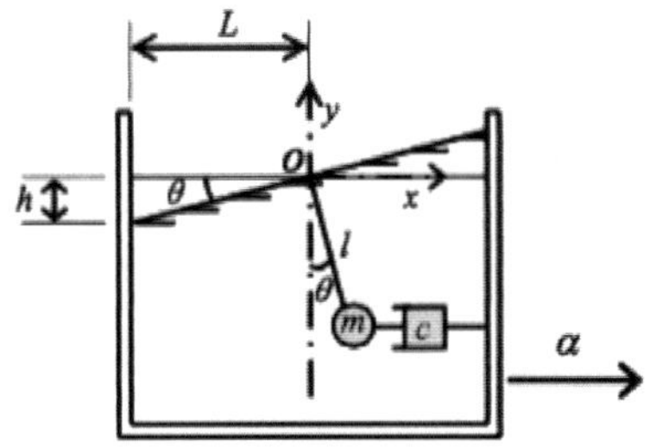

Figure 2. Pendulum model in liquid container transfer

The pendulum bar is perpendicular to the liquid surface, which tilts as the container accelerates or decelerates represent by α, configuring angle θ, which is angle of the horizontal lines and liquid surface. The container may slide to left or right direction, so that the angle θ has sign + or –, depend on the motion direction of container. The coefficient c is the constant damp of fluid viscosity and fluid friction with the walls of the container. The value of damper depends on the kind of liquid. Taking into account the balance of currents around the fulcrum of the pendulum, the dynamics of the model can be depicted by the equation:

$$J\frac{d^2}{dt^2} = c\frac{d(l\ \ )}{dt}l.\cos^2 \quad mgl\sin\ \ +m\ \ l\cos \tag{1}$$

Where $J\ (ml^2)$ is the moment of inertia, m is mass of liquid, and l equivalent pendulum length (m) and coefficient of viscosity c (N.s/m). The high level of liquid h on the wall equals $l{\cdot}tan(\theta)$. In view of only small value of θ, linear approximation of the above non-linear model is described in 2nd order equation as follows:

$$\ddot{\theta} = -\frac{c}{m}\dot{\theta} - \frac{g}{l}\theta + \frac{\alpha}{l} \tag{2}$$

$$\mathrm{h} = \mathrm{L}\,\alpha \tag{3}$$

The angular of acceleration will be integrated to the angular velocity, then integrate again will be obtained the value of angle θ. The transfer function relation between liquid high level h and translation acceleration α is described by the following equation:

$$\frac{h(s)}{(s)} = \frac{L/l}{s^2 + \frac{c}{m}s + \frac{g}{l}} \tag{4}$$

Sloshing occurs is caused from oscillation motion, so that its can be solved using vibration equation. The natural frequency ω_n and damping ratio ζ can be obtained by comparing the equation (4) to second-order damped linear oscillation:

$$_n = \sqrt{\frac{g}{l}} \tag{5}$$

$$= \frac{c}{2m}\sqrt{\frac{l}{g}} \tag{6}$$

3.2. Active Force Control

AFC method has been applied to many kinds of experimental rig, such as robot arm and mobile robot. The aim of use AFC method is to suppress any known or unknown disturbances that may present in the system, so that the system effectively remains robust and steady during operation. The method has convenient application that is readily realized and implemented since it is primarily based on measured or estimated quantities. The advantage of AFC method is its ability to suppress some disturbances effectively and reliably, without use complex mathematical computation. Actually, AFC is a simple methods, the mechanism largely depends on the accurate measurements of two parameters related to the torque and acceleration of the physical system with an appropriate estimation of the inertia matrix of the dynamical system. The AFC principle can be simply traced back from the Newton's second law of motion for a rotating mass. The sum of all torque (T) acting on a body is the product of the mass moment of inertia (I) and the angular acceleration ( ) of the body in the direction of torque applied, which can be expressed as:

$$T = I \tag{7}$$

Assuming a DC motor being mounted under water tank system that has only rotary motion, the equation governing the applied AFC method is as follow as:

$$T + Q = I(\;) \tag{8}$$

Equation (8) is the most significant expression that describes the strategy AFC and thus should be critically and carefully treated, especially during the implementation phase as well as in simulation and experimental studies. The successful implementation of the strategy lies in the AFC precise estimate of the matrix of inertia of a dynamic system. The estimated of inertia matrix can be obtained from calculation. The estimated disturbance vector Q^* is traverse the transfer function *H(s)*, before the vector summed with the command *c* in summing junction. An appropriate choice of *G(s)* associated to an actuator device can cause the controlled output *q* to be made invariant with respect to the introduced disturbances *Q'*. For more detail of a schematic of the AFC scheme can be seen in Figure 3.

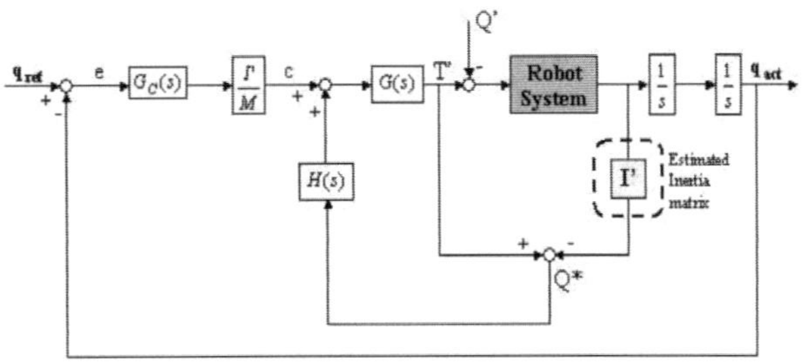

Figure 3. AFC schematic

Base on the schematic in Figure 3, outside the control loop transfer function $G_c(s)$ can be arbitrarily selected, the most common of which is the classical proportional-integral-derivative (PID) controller. The available of the inner control loop from *G(s)* to *H(s)* will localize the effect of the disturbances *Q'*, thus doing action necessary compensation. Choice of the appropriate *estimated inertia matrix I'* absolutely crucial and essential to ensure the effectiveness of the strategy of AFC. From the previous literature, a rough estimate of the approximate matrix of inertia, *IN* for a two-link robot arm and the Wheeled Mobile Robot (WMR) are expressed as:

$$\text{IN}=K_i\,\text{H} \tag{9}$$

Where K_i is a coefficient of the inertia matrix such that $0.5 < K_i < 1.5$, while H is the modelled inertia matrix that can be easily and directly obtained from the system equation of motion. This value obtained from trials, where the procedure has been also applied to the robot arm and mobile robot system. The main objective of using the intelligent means is to enable the computation of the required inertia matrix be made automatically, continuously and on-line.

4. Simulation Results

Simulations of sloshing control were conducted using MATLAB Simulink. For this simulation water thank can move slide along 120 cm, powered dc motor with constant motor torque (*Ktn*:0.263 N.m/s^2). Height of water tank 10 cm, and the mass of liquid about 1 kg. Maximum acceleration is set 2 m/s^2, whereas the initial speed is 0.5 m/s. The whole of simulation parameter is shown in Table 1.

Table 1. Parameter Values

Parameter	Value	Unit
Mass of liquid, *m*	1	kg
Initial velocity, *Vo*	0.5	m /s
Acceleration, *a*	0.8	m/s^2
High water level, *h*	10	cm
Coefficient of viscosity, *c*	1.02	N.s/m
Gravity acceleration, *g*	9.8	m/s^2

Firstly, simulation is conducted using PID controller, where Kp=10, Ki=0.4 and Kd=0.01. Disturbance is set 0, its mean no disturbance. The block diagram of suppress liquid control using PID controller is shown Figure 4.

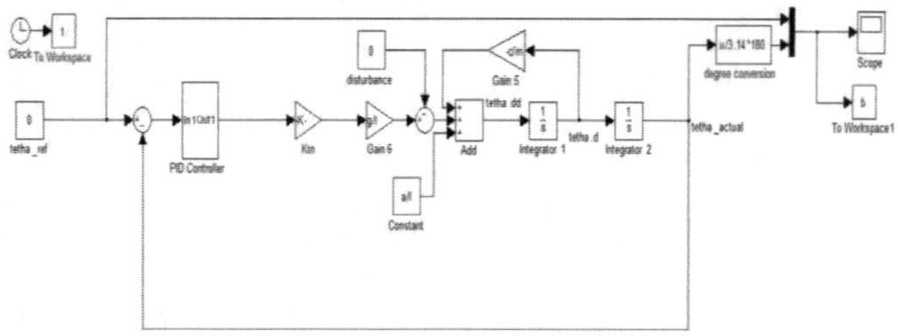

Figure 4. Simulation of suppress liquid control using PID controller

Secondly, simulation is conducted using PID-AFC. The gain controller is set same with the pure PID controller, and also without disturbance. The block diagram of suppresses liquid control using PID-AFC controller can be seen in Figure 5.

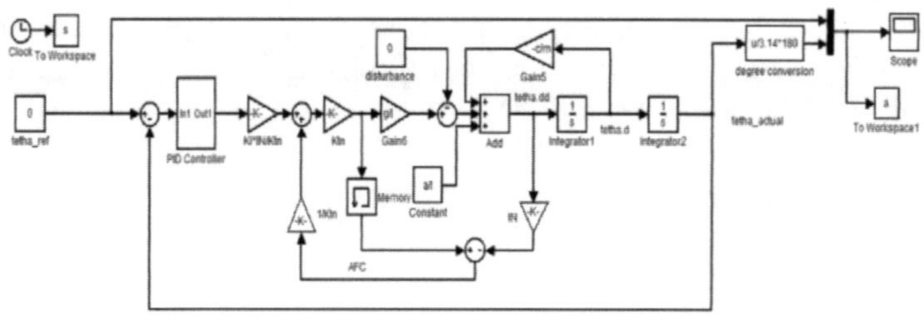

Figure 5. Simulation of suppress liquid control using PID-AFC controller

The result of simulation of PID and PID-AFC controller for 10 seconds can be seen in Figure 6. It can be seen that PID-AFC method more superior than pure PID method. The PID method occurs transient in the beginning and reachs steady state after 8 seconds, meanwhile the PID-AFC method reachs steady state from the strarting time.

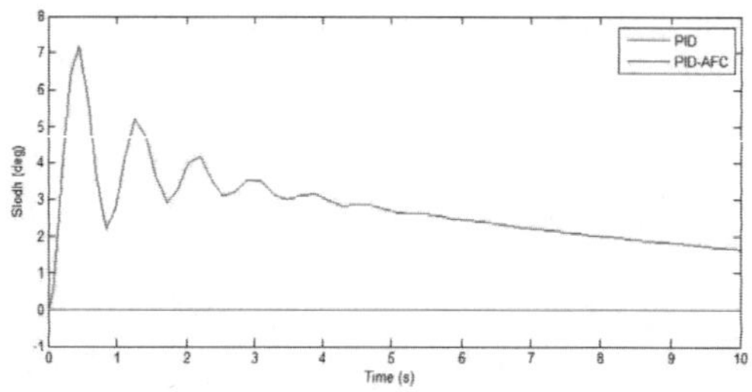

Figure 6. Simulation of suppress liquid control using PID-AFC and PID controller

The response of PID-AFC method in Figure 6 looks able to suppress slosh until zero. Actually, the slosh of PID-AFC method is not perfect zero, but vary in very small value such as shown in Figure 7.

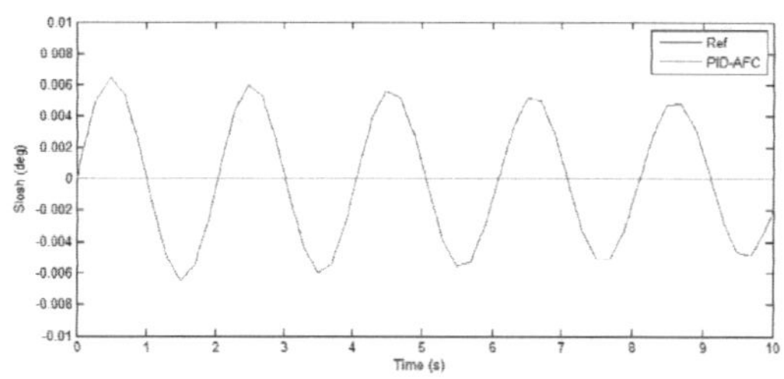

Figure 7. The result of simulation using PID-AFC controller

Third, simulation is conducted by giving constant disturbance. Using PID controller, disturbance give effect to the stability of system. Disturbance causes big slosh rising in early movement. The comparison results between no disturbance and using disturbance as shown in Figure 8.

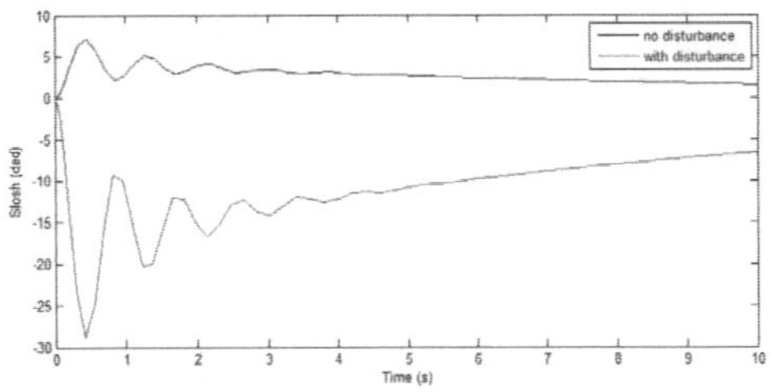

Figure 8. Simulation result using PID controller with disturbance

The next simulation is conducted using PID-AFC controller by giving disturbances. Either there are disturbance or no, the results is stable, the graph is almost same. While, the simulation time is extended to 100 seconds, precisely the sloshing toward to zero. The comparison results between no disturbance and with disturbance as shown in Figure 9.

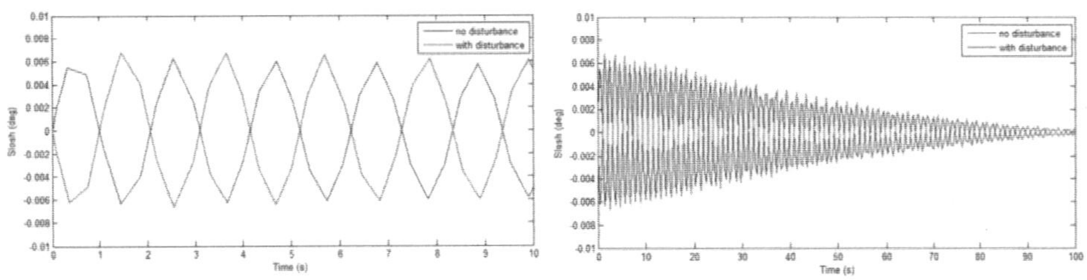

Figure 9. Simulation result using PID-AFC controller with disturbance

Conclusion

Simulations of liquid sloshing control have conducted. The simulation is conducted comparing between PID with PID-AFC control and also by giving constant disturbance. From the simulation result shows that using PID-AFC control be able to suppress the sloshing effectively than the PID control. The PID-AFC method reach steady state from the starting time. The simulation result clearly indicates that the proposed PID-AFC method performs more robust than PID method in the presence of disturbances. In the future work, we will focus in the experimental work.

References

[1] K. Yano and K. Terashima, "Sloshing suppression control of liquid transfer systems considering 3D transfer path", *IEEE/ASME Transaction on Mechatronics*, vol. 10-1, pp. 8-16, 2005.

[2] Y. Noda, K. Yano and K. Terashima, "Tracking to Moving Object and Sloshing Suppression Control Using Time Varying Filter Gain in Liquid Container Transfer" *SICE Annual Conference* in Fukui, 2003.

[3] M. Gradinscak , S.E. Semercigil and Ö.F. Turan, "Liquid Sloshing in Flexible Containers, Part2: Using A Sloshing Absorber with A Flexible Container for Structural Control" *Fifth International Conference on CFD in the Process Industries CSIRO*, Melbourne, Australia 13-15 December 2006.

[4] Wisnu Aribowo, Takahito Yamashita, Kazuhiko Terashima, and Hideo Kitagawa, "Input Shaping Control to Suppress Sloshing on Liquid Container Transfer Using Multi-Joint Robot Arm", *The 2010 IEEE/RSJ International Conference on Intelligent Robots and Systems,* October 18-22, 2010, Taipei, Taiwan.

[5] Wisnu Aribowo, Takahito Yamashita, Kazuhiko Terashima, Yoji Masui, Toru Saeki, Toshio

Kamigaki, Hirotoshi Kawamura, "Vibration Control of Semiconductor Wafer Transfer Robot by Building an Integrated Tool of Parameter Identification and Input Shaping". *The 18th IFAC World Congress Milano* (Italy) August 28 - September 2, 2011.

[6] Qiang Zang, Jie Huang, and Zan Liang, "Slosh Suppression for Infinite Modes in a Moving Liquid Container, *IEEE/ASME Transaction on Mechatronics*, Vol. 20, No. 1, pp. 217-225, Feb 2015.

[7] Mohd Zarhamdy Md Zain, Musa Mailah, G. Priyandoko, "Active Force Control with Input Shaping Technique for A Suspension System " *Jurnal Mekanikal December 2008.*

[8] Didik Setyo Purnomo and Endra Pitowarno, "Disturbance Cancellation of a Nonholonomic Wheeled Mobile Robot using Active Force Control Method", *Asian Control Conference*, 18-21 July 2016, Bali.

[9] Mohammed H. Al-Mola, M. Mailah, S. Kazi, A.H. Muhaimin, M.Y. Abdullah, "Robust Active Force Controller for an Automotive Brake System", *Third International Conference on Intelligent Systems Modelling and Simulation*, 2012.

IOP Conf. Series: Materials Science and Engineering **190** (2017) 012008 doi:10.1088/1757-899X/190/1/012008

Two-Link Flexible Manipulator Control Using Sliding Mode Control Based Linear Matrix Inequality

Zulfatman, Mohammad Marzuki, Nur Alif Mardiyah
Electrical Engineering Department
Faculty of Engineering, Univ. of Muhammadiyah Malang
Jl. Tlogomas No. 246, Malang, 65144, Indonesia

E-mail: zulfatman@umm.ac.id

Abstract. Two-link flexible manipulator is a manipulator robot which at least one of its arms is made of lightweight material and not rigid. Flexible robot manipulator has some advantages over the rigid robot manipulator, such as lighter, requires less power and costs, and to result greater payload. However, suitable control algorithm to maintain the two-link flexible robot manipulator in accurate positioning is very challenging. In this study, sliding mode control (SMC) was employed as robust control algorithm due to its insensitivity on the system parameter variations and the presence of disturbances when the system states are sliding on a sliding surface. SMC algorithm was combined with linear matrix inequality (LMI), which aims to reduce the effects of chattering coming from the oscillation of the state during sliding on the sliding surface. Stability of the control algorithm is guaranteed by Lyapunov function candidate. Based on simulation works, SMC based LMI resulted in better performance improvements despite the disturbances with significant chattering reduction. This was evident from the decline of the sum of squared tracking error (SSTE) and the sum of squared of control input (SSCI) indexes respectively 25.4% and 19.4%.

1. Introduction

Flexible robot manipulator is a robot manipulator, in which at least one of the links is made of a lightweight and flexible material. In general, flexible robot manipulator has several advantages than robot manipulator with rigid link. Flexible robot manipulator is lighter, require less power, low cost, and capable to result a larger payload [1].

Due to a lot of flexibility in its application, the two-link flexible manipulator is more preferable. However, to design a suitable control method for two-link flexible manipulator to maintain its accurate position is never easy. Two-link flexible manipulator has more complex problems than the manipulator with rigid link, because the flexible manipulator is a multi-input multi-output (MIMO) system. Other problems should be considered in the flexible manipulator are coupling problem in second link and vibration in between the two links [2]. Such that, an accurate model and efficient control method for the two-link flexible manipulator system must be developed.

A proper modelling is the main requirement for an optimum control method. In the previous studies, non-linear model was the model that frequently used in the system modelling. This type of model require a process that called as parameters identification and estimation. Implementation of this approach is based on the cutting of Taylor series until the first power degree around the certain point. Those modelling system has a drawback, in which the control systems cannot be synthesized to improve the performance of the system that is actually non-linear. Consequently, when the dynamics of the system is moving away from the operating point, then the performance of the controller can be reduced and the system can easily become unstable. It's due to availability of the control system only works properly in the nearest area of the operating point. The control of Proportional Integral Derivative (PID) is one of control methods that commonly used in industrial applications. However, PID controller becomes unstable when the system parameters changes and external disturbances exist. Moreover, it's not capable to drive states of the system to turn into its steady state with fast action [3].

In the class of nonlinear control, Variable Control Structure (VCS) via Sliding Mode Control (SMC) with specific characteristics is one of suitable answers for the problems of nonlinear systems. The main benefit of a control system using SMC is its insensitivity (robust) to parameter changes and disturbances. Moreover, SMC also be able to stabilize nonlinear system that can not be stabilized by using continuous state feedback control by using stability analysis [4]. The use of SMC method is expected to improve quality of the control algorithm with fast dynamic response, insensitive to parameter variations and disturbances, simple in control method, and can be implemented quite easy. Main factor affecting the performance of SMC are parameters of the sliding surface [5].

Regarding to the basic theory of SMC in [6], gain of the discontinuous control starts to result in discontinuous control action when the system states that are moving from somewhere reach a defined sliding surface and sliding on the surface toward their equilibrium point. However, the discontinuous control action displays a high frequency oscillation that is called as chattering. A higher gain will result a higher chattering amplitude. It is significantly undesirable and reduces the control performance [7]. In order to solve this drawback, several methods have been designed to reduce the chattering problem such as the use of a thin or varying boundary layer, reaching law method, low fast filtering, high-order SMC, adaptive robust control with disturbance observer, and any other methods.

In this study, SMC methods will be combined with linear matrix inequality (LMI) to obtain more robust control action and for chattering reduction [8-10]. The next section describes the flexible manipulator model in a nonlinear system platform. The proposed control technique is developed in section three. Then, the system and control parameters and set-up of the simulation are displayed in fourth section. The fifth section presents several discussions on results of the simulation works on the proposed technique in comparison with the conventional one. Conclusion of the study is appeared in the last section.

2. Model Formulation

Model of the two-link flexible manipulator expresses characteristics and behaviors of the system. The model was taken from the previous research [3], which created the system transfer function using system identification method from a set of input-output data taken from experiment. The complete mathematical model of the system is written as follow equation

$$G(s) = \frac{b_5 s^5 + b_4 s^4 + b_3 s^3 + b_2 s^2 + b_1 s + b_0}{s^6 + a_5 s^5 + a_4 s^4 + a_3 s^3 + a_2 s^2 + a_1 s + a_0} \tag{1}$$

where,

$a_0 = 6.84,\ a_1 = 47.46,\ a_2 = 112.12,\ a_3 = 113.17,\ a_4 = 50.95,\quad a_5 = 10.90$

$b_0 = 91.2, b_1 = 1080.3, b_2 = 2782.5.17, b_3 = 2571.1, b_4 = 840.9,\ b_5 = 84.1$

Mathematical model of the two-link flexible manipulator system in (1) can be changed into canonical form equation below

$$\dot{x} = (A_0 + \Delta A)x + B_0 u + d \tag{2}$$

Equation (2) can be rewritten by combining $\Delta A + d = d$

$$\dot{x} = A_0(x) + B_o u + d \tag{3}$$

While, output of the system

$$y = Cx \tag{4}$$

where,

$$A_0 = \begin{bmatrix} -a_5 & -a_4 & -a_3 & -a_2 & -a_1 & -a_0 \\ 1 & 0 & 0 & 0 & 0 & 0 \\ 0 & 1 & 0 & 0 & 0 & 0 \\ 0 & 0 & 1 & 0 & 0 & 0 \\ 0 & 0 & 0 & 1 & 0 & 0 \\ 0 & 0 & 0 & 0 & 1 & 0 \end{bmatrix} \tag{5}$$

$$B_0 = [1 \quad 0 \quad 0 \quad 0 \quad 0 \quad 0]^T \tag{6}$$

$$x = [x_1 \quad x_2 \quad x_3 \quad x_4 \quad x_5 \quad x_6]^T \tag{7}$$

$$Cx = [b_5\ b_4\ b_3\ b_2\ b_1\ b_0]\ x \tag{8}$$

Substitute (5) to (7) into (3), model of the system can be represented in state-space canonical form:

$$\dot{x} = \begin{bmatrix} -a_5 & -a_4 & -a_3 & -a_2 & -a_1 & -a_0 \\ 1 & 0 & 0 & 0 & 0 & 0 \\ 0 & 1 & 0 & 0 & 0 & 0 \\ 0 & 0 & 1 & 0 & 0 & 0 \\ 0 & 0 & 0 & 1 & 0 & 0 \\ 0 & 0 & 0 & 0 & 1 & 0 \end{bmatrix} \begin{bmatrix} x_1 \\ x_2 \\ x_3 \\ x_4 \\ x_5 \\ x_6 \end{bmatrix} + \begin{bmatrix} 1 \\ 0 \\ 0 \\ 0 \\ 0 \\ 0 \end{bmatrix} [u + d] \tag{9}$$

$$\boldsymbol{y = Cx = [b_5\ b_4\ b_3\ b_2\ b_1\ b_0]x} \tag{10}$$

In conversion of transfer function to state-space, substitute all parameters in (1) into (9) and (10), by ignorance disturbance.

3. Control Design

This control design is proposed to achieve a continuous control action in sliding mode control platform, such that, output of the system, x_i tracks the desired trajectories of the system states x_{id} as precise as possible in spite of parameters variation and the existence of disturbances [11].

3.1. Design of Sliding Mode Control

The proposed SMC is designed through several steps such as tracking error definition, sliding surface design, control law algorithm, stability analysis as follow:

3.1.1. Tracking Error. In order to design desired trajectories for the system states, the states errors of the system are defined as

$$e_i = x_i - x_{id} \quad \text{where, } i = (1, 2, , .. , 6) \tag{11}$$

where,

$$x_1 = \theta_1, x_2 = \theta_2, x_3 = \alpha_1, x_4 = \dot{\theta}_1,\ x_5 = \dot{\theta}_2, x_6 = \dot{\alpha}_1 \tag{12}$$

Referring to the formula of Euler-Lagrange, model in (9) can be rewritten in state equation as below

$$\begin{aligned} \dot{x_1} &= x_4 \\ \dot{x_2} &= x_5 \\ \dot{x_3} &= x_6 \\ \dot{x_4} &= -b_1x_4 + b_2x_3 + b_3u \end{aligned}$$

$$\dot{x_5} = -b_4x_5 - b_5x_3 + b_6u$$
$$\dot{x_6} = b_1x_4 - b_2x_3 - b_3u$$
$$\ddot{x}_1 = \dot{x}_4$$
$$\ddot{x}_2 = \dot{x}_5$$
$$\ddot{x}_3 = \dot{x}_6 \quad (13)$$

Such that tracking error for all states in the system as below

$$e = [e_1\ e_2\ e_3 e_4 e_5 e_6]^T \quad (14)$$

where,

$$e_1 = x_1 - x_{1d}$$
$$e_2 = x_2 - x_{2d} = \dot{x}_1 - \dot{x}_{1d}$$
$$e_3 = x_3 - x_{3d}$$
$$e_4 = x_4 - x_{4d} = \dot{x}_3 - \dot{x}_{3d}$$
$$e_5 = x_5 - x_{5d}$$
$$e_6 = x_6 - x_{6d} = \dot{x}_5 - \dot{x}_{5d} \quad (15)$$

3.1.2. Sliding Surface. A function of sliding surface needs to define to get attain the system states as defined in (12) track their defined desired trajectories at the same time [5], as follows

$$S = s_1 + \lambda_1 s_2 + \lambda_2 s_3 \quad (16)$$

where,

$$s_i = \dot{e}_i + c_i \int_0^t e_i(t)dt \quad (17)$$

$$s_i = \dot{e}_i + c_i e_i \quad (18)$$

where $\lambda_i (i = 1,2)$ and $c_i\ (i = 1,2,3)$ is a positive constant which is determined in accordance with the desired dynamics and the closed-loop system. S parameter is called as the sliding surface. The stability of the system relies on the constant.

Sliding surface is given $S = 0$ for the desired dynamic response of the system. If S is driven to zero, the desired dynamics is reached and the tracking error will converge to zero.

*3.1.3. Reaching Law.*In order to boost states of the system to reach the sliding surface as fast as possible, the reaching time equation is needed. The dynamic equation of constant reaching law is given as in [5].

$$\dot{\boldsymbol{S}} = -\boldsymbol{\rho}\ \boldsymbol{sign}\ (\boldsymbol{S}) \quad (19)$$

where, $\rho > 0$. Smaller value of ρ is suggested to reduce chattering effect.

3.1.4. Control Law. Control law will bring the system states trajectories towards the sliding surface which has been designed in advance and kept it in sliding condition. In general, the control input is m-vector *u(t)*. In order to determine the control law, the sliding surface (16) is defined in derivative form

$$\dot{\boldsymbol{S}} = \dot{\boldsymbol{s}}_1 + \boldsymbol{\lambda}_1 \dot{\boldsymbol{s}}_2 + \boldsymbol{\lambda}_2 \dot{\boldsymbol{s}}_3 \quad (20)$$

where,

$$\dot{\boldsymbol{s}}_1 = \ddot{\boldsymbol{e}}_1 + \boldsymbol{c}_1 \dot{\boldsymbol{e}}_1$$

$$\dot{s}_2 = \ddot{e}_2 + c_2\dot{e}_2$$

$$\dot{s}_3 = \ddot{e}_3 + c_3\dot{e}_3$$

By substituting Equation (11), (13), and (19) into (20), then control law of the system is obtained

$$\boldsymbol{u} = \left(-\frac{\mathbf{1}}{\boldsymbol{k}}\right)[\boldsymbol{x}_6(\boldsymbol{\lambda}_2\boldsymbol{c}_3) + \boldsymbol{x}_5(-\boldsymbol{\lambda}_1\boldsymbol{b}_4 + \boldsymbol{\lambda}_1\boldsymbol{c}_2)\boldsymbol{\lambda}_2\boldsymbol{b}_1$$
$$+\boldsymbol{x}_4(-\boldsymbol{b}_1 + \boldsymbol{\lambda}_2\boldsymbol{b}_1 + \boldsymbol{c}_1) + \boldsymbol{x}_3(\boldsymbol{b}_2 - \boldsymbol{\lambda}_1\boldsymbol{b}_5 - \boldsymbol{\lambda}_2\boldsymbol{b}_2) - \boldsymbol{\lambda}_2$$
$$-(\boldsymbol{c}_3\boldsymbol{e}_3) - \boldsymbol{e}_1 + (\boldsymbol{\lambda}_1\boldsymbol{c}_2\boldsymbol{e}_2)] - \boldsymbol{\rho}\,\boldsymbol{sign}\,(\boldsymbol{S}) \tag{21}$$

where, $\boldsymbol{k} = \boldsymbol{b}_3 + \boldsymbol{\lambda}_1\boldsymbol{b}_6 - \boldsymbol{\lambda}_2\boldsymbol{b}_3$, and c_1, c_2, c_3 are given.

3.2. Stability Analysis

One of the main issues in control system design is how to guarantee stability of the control system. In this study, stability of the proposed control algorithm is assured by a Lyapunov function candidate.

$$V(S) = \frac{1}{2}S^2 \tag{22}$$

The Lyapunov stability criterion of the proposed control is defined as

$$\dot{V}(S) > 0 \tag{23}$$

Then, to reach the condition in (23), it follows that

$$\dot{V}(S) = \frac{1}{2}\frac{d}{dt}S^2 = S\dot{S} \ll -\eta - \rho\, sign\,(S) \tag{24}$$

where, $\eta > 0$.

Hence, ρ in Equation (19) can be determined by substituting Equations (16), (19) and (20) into (24). While, the control design was examined by using Matlab Simulink as in Figure 1 below

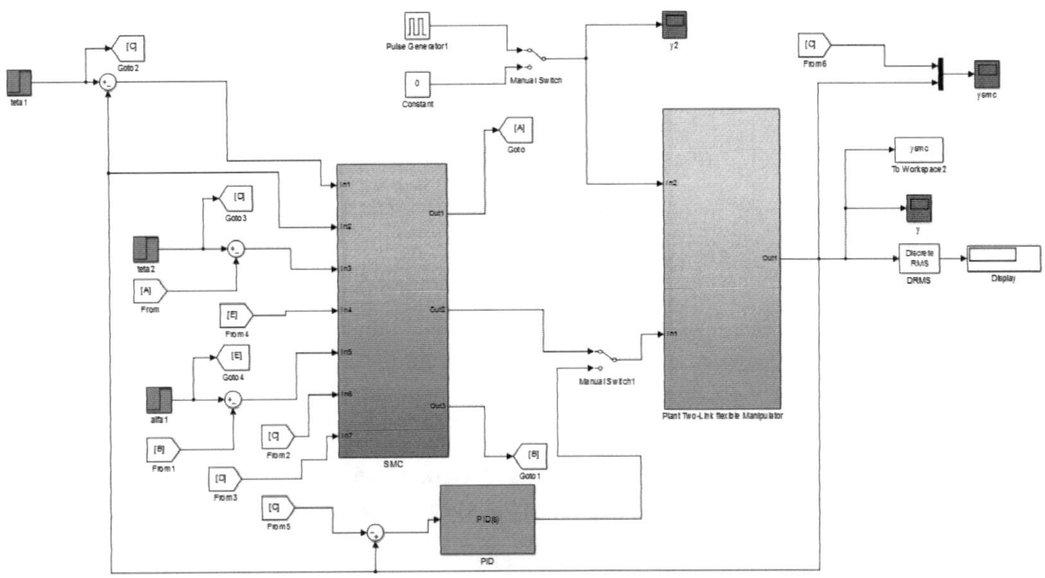

Figure 1. SMC design in Matlab Simulink

3.3. Design of Sliding Mode Control Based Linear Matrix Inequality

In this part, SMC based LMI is designed to complete the SMC design in Figure 1 above in order to decrease effect of chattering on control action of the SMC. The SMC based LMI designed is represented by block diagram in Figure 2.

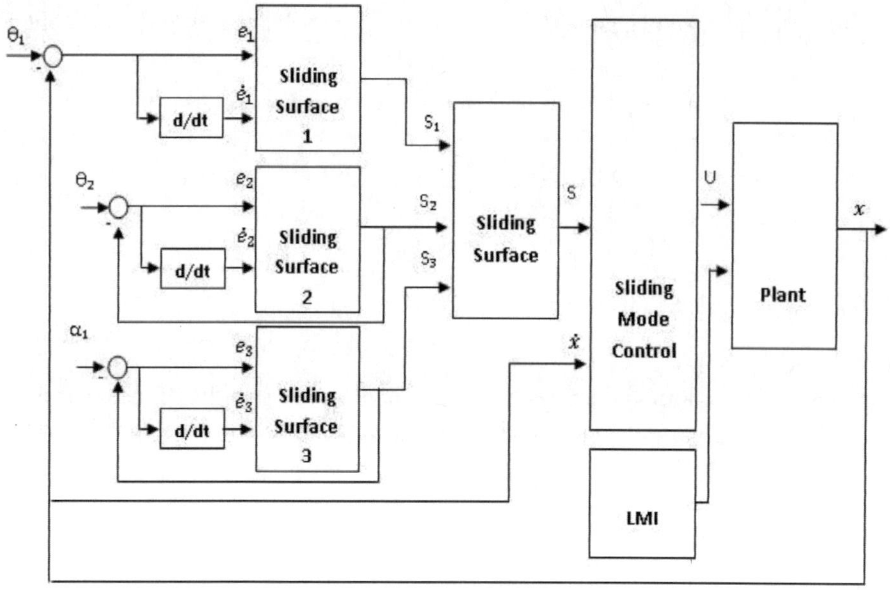

Figure 2. Block diagram of SMC based LMI

Then, the steps to design LMI are corresponded to flow chart in Figure 3 below

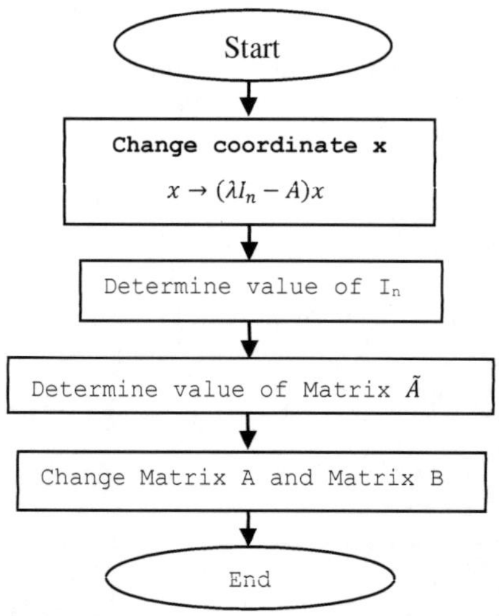

Figure 3. Flowchart of LMI Design

Before design the control algorithm, firstly, determine parameters of the control design. Control parameters are obtained from Equation (3), by assuming that A, B, C, D are constant matrices, which are parts of the state equation of the system model. Design of the optimized SMC with the LMI can be done by changing the initial coordinates of x.

$$x \to (\beta I_n - A)x \tag{25}$$

When β is not located along the spectrum A and a non-singular transformation, a matrix structure will be replaced with new coordinates.

$$\tilde{A} = (\beta I_n - A)A(\beta I_n - A)^{-1} \tag{26}$$

$$\tilde{B} = (\beta I_n - A)B \tag{27}$$

β is a positive constant. It is set in 9,887, which is gotten from trial and error. While I_n is the 6 x 6 identity matrix. Having obtained the value of $\tilde{A}$ and the value of$\tilde{B}$, then both parameters were included in the state-space block in the MATLAB Simulink.

Such that, matrices $\tilde{A}$ and $\tilde{B}$ are obtained as below

$$\tilde{A} = \begin{bmatrix} -988.182 & -6809.556 & -17833.656 & -18956.899 & -8297.994 & -1215.942 \\ 177.862 & 951.253 & 2252.238 & 2294.929 & 984.645 & 142.827 \\ -20.892 & -49.948 & -113.169 & -112.117 & -47.456 & -6.836 \\ 1 & -9.988 & 1 & 0 & 0 & 0 \\ 0 & 1 & -9.988 & 1 & 0 & 0 \\ 0 & 0 & 1 & -9.988 & 1 & 0 \end{bmatrix} \tag{27}$$

$$\tilde{B} = \begin{bmatrix} 6.836 & 0 & 0 & 0 & 0 & 9.887 \end{bmatrix}^T \tag{28}$$

4. Simulation Setup

Simulation is done in evaluating performance of the controller algorithm by delicately tune its parameters. In this study, the initial SMC and the proposed SMC based LMI abilities are assessed in minimizing chattering effect on the control signal of the SMC when the system parameters changes and unknown disturbances exist.

How to reduce chattering in the control action of the SMC is the main task of this work. In examining the smoothness of control signal, a step reference signal is employed in a transient performance assessment. The reference signal will facilitate the investigation on the chattering effect when the system reaches its steady state condition. The simulation works employ Runge-Kutta solver with a sampling time of 0.001s. Sum of squared tracking error (SSTE) and sum of squared control input (SSCI) are used to measure index of the control action performance. The SSTE is employed to measure the system tracking accuracy, and the SSCI is used to detect the control activities when LMI added to the initial SMC design.

$$SSTE = \sum_{t=0}^{N} (x_{1d}(t) - (x_1(t))^2 \tag{29}$$

$$SSCI = \sum_{t=0}^{N} u(t)^2 \tag{30}$$

5. Results and Discussion

Detail of observations through step input signal assessment on the transient response of the proposed design of SMC relates to its consideration on chattering reduction when system parameters changes and disturbances exist are carried out in this section. Figure 4 to Figure 6 presents the output responses of the system, tracking error, and control signals of the initial SMC in comparison with SMC based LMI. It can be detected by comparing the first and the second version of the control design in tracking error and control action performance. The displayed figures are supported by a comparison table of SSTE and SSCI as present in Table 1.

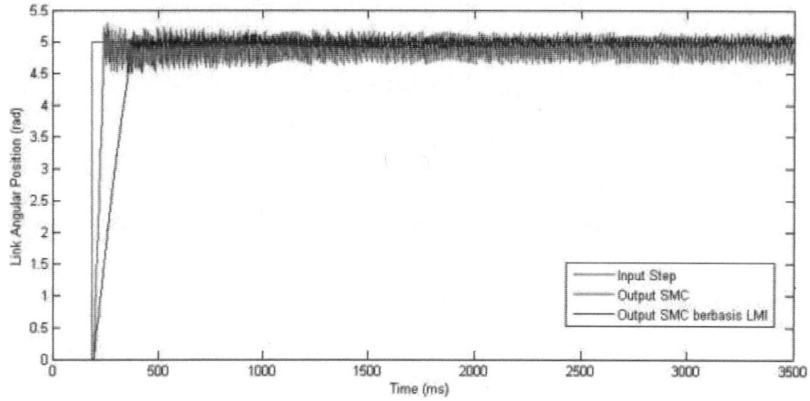

Figure 4. Output response of SMC and SMC based LMI

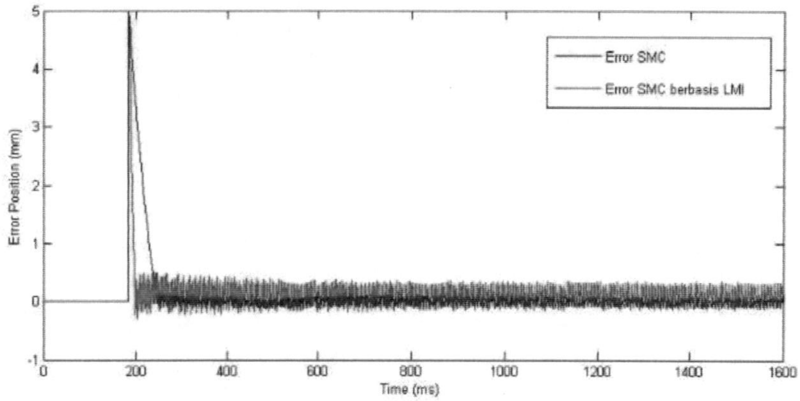

Figure 5. Tracking error of SMC and SMC based LMI

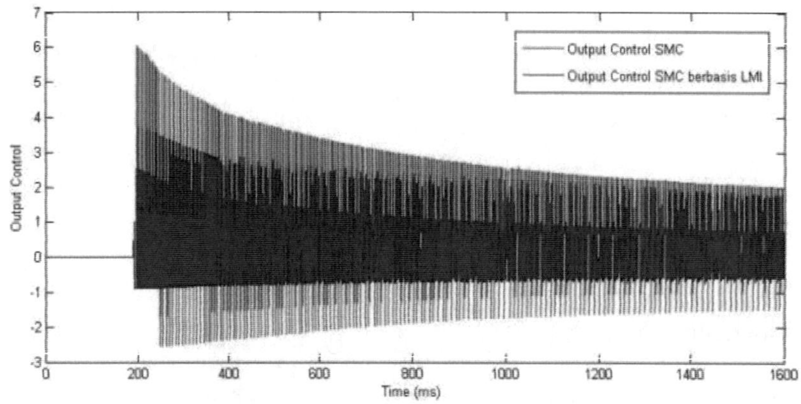

Figure 6. Control signals of SMC and SMC based LMI

As can be seen from Figure 4 to Figure 6, SMC based LMI was capable to minimize chattering compare to the conventional SMC. Transformation of the system model by using LMI guarantees the system become more robust and stable in the face of parameters changes and disturbances. Since the control gain of the SMC was calculated from the transformed model, it was possible to obtain a smaller control gain. In basic theory of SMC has already state that, a smaller control gain will produce smaller chattering in the control signal. This conclusion also was justified by comparison SSTE and SSCI indexes in Table 1, which display that SSTE and SSCI of SMC based LMI significantly smaller than SMC conventional.

Tabel 1. Control Performance Comparison

Controller/Index	SSTE	SSCI
SMC	0.106	0.237
SMC based LMI	0.079	0.191

6. Conclusion

Each Model of the two-link flexible manipulator has been successfully formulated from the existing transfer function form into state equations. This state equations model is required to develop SMC algorithm. The proposed SMC algorithm also has been developed successfully for the flexible manipulator system to be robust and insensitive on parameters variation and disturbances. Through step input signal test, it was seen that the development of SMC with LMI has increased capability of the SMC to be more robust and result smaller chattering in its control action. This was verified from the decline of the SSTE and SSCI indexes around 25.4% and 19.4%, respectively.

7. References

[1] Ishak M H 2005 "Dynamic modelling of a two-link flexible manipulator *UTM Thesis*

[2] Mohamed Z, Khairudin M, Husain A and Subudhi B 2014 Linear matrix inequality-based robust proportional derivative control of a two-link flexible manipulator *Journal of Vibration and Control* **22(5)** 1244-1256

[3] Khairudin M, Mohamed Z and Husain A 2014 System Identification and LMI Based Robust PID Control of a Two-Link Flexible Manipulator *TELKOMNIKA* **12(4)** 830-838

[4] Hassan K K 2002 *Nonlinear systems 3rd Ed* (Prentice Hall Inc., New Jersey)

[5] Gao W and Hung J C 1993 A Variable Structure Control of Nonlinear Systems: A New Approach *IEEE Trans. on Ind. Electronics* **40(1)** 45-55

[6] Utkin V 1978 Sliding modes and their applications in variable structure systems *Mir Publ.* Moscow

[7] Zulfatman, Rahmat M F, Husain A R, Ishaque K, Ghazali R, Ahmad M N, Sam Y M and Rozali S Md 2014 Robust position tracking control of an electro-hydraulic actuator *Arabian Journal for Science & Engineering* **39(4)** 2965-2978

[8] Edwards C 2004 A practical method for the design of sliding mode controllers using linear matrix inequalities *Automatica* **40(10)** 1761–1769

[9] Lochan K, and Roy B K 2015 Position control of two-link flexible manipulator using low chattering SMC technique *IJCTA* **8(3)** 1137-1145

[10] Lochan K, Roy B K and Subudhi B 2016 SMC controlled chaotic trajectory tracking of two-link flexible manipulator with PID sliding surface *4th IFAC Conference on ACODS 2016* **49(1)** 220-224

[11] Lochan K, Suklabaidya S and Roy B K 2015 Modelling and sliding mode control of flexible link flexible joint robot manipulator *Conference Paper* DOI: 10.1145/2783449.2783509

Acknowledgments

This research is supported by Direktorat Penelitian dan Pengabdian Masyarakat (DPPM), University of Muhammadiyah Malang (UMM) through Internal Research Grant Scheme (PKID-SI). Authors are grateful to the UMM for supporting the present work.

IOP Conf. Series: Materials Science and Engineering **190** (2017) 012009 doi:10.1088/1757-899X/190/1/012009

Lightweight UDP Pervasive Protocol in Smart Home Environment Based on Labview

Wijaya Kurniawan[1], Mochammad Hannats Hanafi Ichsan[2], Sabriansyah Rizqika Akbar[3] and Issa Arwani[4]

Computer Engineering, Computer Science Faculty, Brawijaya University, Indonesia

[1]wjaykurnia@ub.ac.id, [2]hanas.hanafi@ub.ac.id, [3]sabrian@ub.ac.id and [4]issa.arwani@ub.ac.id

Abstract. TCP (Transmission Control Protocol) technology in a reliable environment was not a problem, but not in an environment where the entire Smart Home network connected locally. Currently employing pervasive protocols using TCP technology, when data transmission is sent, it would be slower because they have to perform handshaking process in advance and could not broadcast the data. On smart home environment, it does not need large size and complex data transmission between monitoring site and monitoring center required in Smart home strain monitoring system. UDP (User Datagram Protocol) technology is quick and simple on data transmission process. UDP can broadcast messages because the UDP did not require handshaking and with more efficient memory usage. LabVIEW is a programming language software for processing and visualization of data in the field of data acquisition. This paper proposes to examine Pervasive UDP protocol implementations in smart home environment based on LabVIEW. UDP coded in LabVIEW and experiments were performed on a PC and can work properly.

Keywords. UDP, Pervasive Protocol, Smart Home, LabVIEW

1. Introduction

Internet nodes consist of not only computers or mobile devices but also with everyday appliances that are used such as household electronic devices that have connected to the Internet. Based on a report from the National Intelligence Council realm of IOT (Internet of Things) became one of the six technologies with the greatest risk of the potential impact until 2025 in the United States [1]. The main objective was internet nodes connect different devices inside communication of several different objects [2]. The object has a device, operating system and programming language that works independently. Household devices over time becomes devices that that has an internet device nodes and has a sensor that connected. This combination of technologies from household devices, computers, sensors and computer networks on IOT also emphasize the growing communication technology with ubiquitous [3]. The technology has the properties of low cost, pervasive connectivity and make "all" the wireless device connected and has four models consisting of a connectivity device to device, device to the cloud, the device to the gateway, and the back end of data sharing [4].

Pervasive computing was often referred as ubiquitous field of research that brings revolutionary paradigm for computing models. Pervasive computing offers a large area for computation of size, mobility and usability [5]. Ideally, pervasive model should be able to recognize and accommodate each connected device [6]. Suppose that a device which has a sensor that automatically collects information, transfer and take their own decisions can be categorized as pervasive devices. In a computer network, pervasive protocol was able to recognize a device and could provide communication services automatically [7].

TCP was a protocol that can be directed, could mean transmit data via the specified route. This would result in adding density of traffic on the network that is used so that the process will take longer [8]. In the smart home environment is not required sending large amounts of data [9]. UDP has an advantage compared with no use of sequence and acknowledgment field as was done by TCP and the most obvious advantage was extra bytes fewer [10]. In addition, UDP does not need to wait for the reception or storing data in memory until the data received [11]. This means, the UDP application is not slowed down by the admissions process and memory be released early.

LabVIEW was an acronym for Laboratory Virtual Instrumentation Engineering Workbench, computer software for processing and visualization of data in the field of data acquisition [12], control instrumentation and automation industry first developed by the company National Instruments. A previous study [13] built the same concept but using three devices, the agent, and the agent directory user agent that uses Salutation Discovery Protocol and Service Location Protocol for pervasive process. Another study [14] proposes for wireless sensor network environment on smart home using LabVIEW with HTTP-CoAP.

In the smart home environment, it does not need a large size and complex data transmission between monitoring sites and monitoring center is needed in the Smarthome tension monitoring system. Therefore, researchers wanted to examine how technology utilizes LabVIEW to configure the UDP protocol is pervasive in Smart Home Environment. In this research we discuss about how the design and configuration of UDP is pervasive in LabVIEW and the successful implementation of the UDP protocol in LabVIEW.

2. System Design

In this section, we design the pervasive protocol on the domain host and client, each using a pc. The design of pervasive protocol on the host and client is indicated in **Figure 1**. As the initial conditions, the hosts were in Listen condition. Then the client introduces himself on the host to provide such acknowledgment Client Name, its IP and Service on the host. In this process host indirectly receive the client, but to check duplication, whether the client is already there or not, even if the client is already on another host. The host then sends an ACK, Host Name, IP and Host Header which is then stored by the Client.

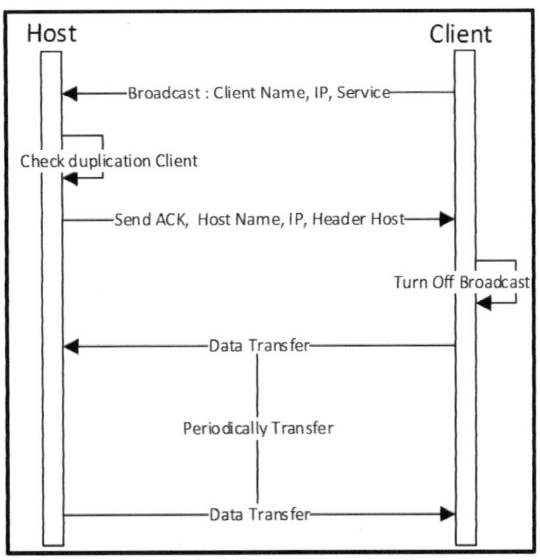

Figure 1. System design

Once the Client recognize the host, the next process was turning off the broadcast. Furthermore, Client and Host recognize each other and was able to transmit data. The service that can be delivered was sensor data transmission services, the provision of services control at client were boolean or numerically. Then it would be designed how to make or how to design a protocol that will be used on the host and the client.

2.1. *Protocol Design on The Host*

In the Host we designed in detail about what is done by the host from beginning. The details about what would be done by the hosts can be seen in **Figure 2**.

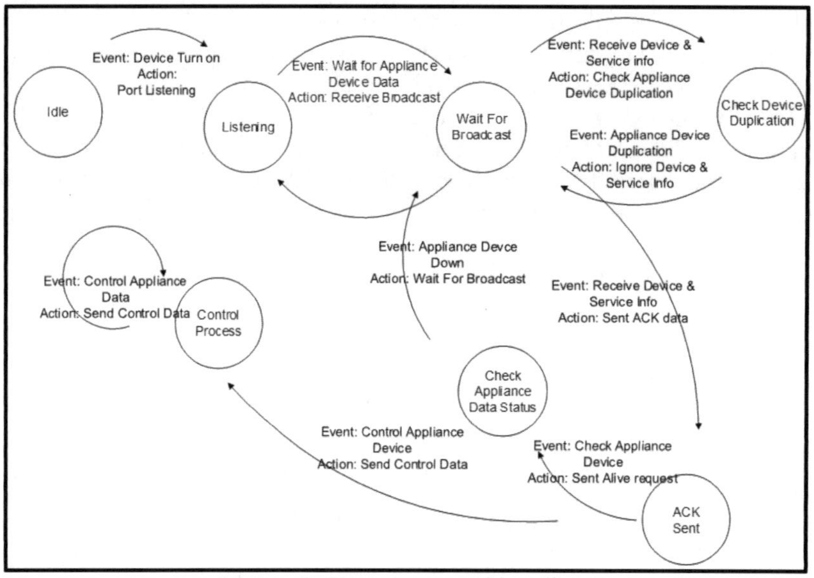

Figure 2. Host state machine diagram

Conditions on State Machine Diagrams in **Figure 2** each would be explained as follows:

1. Idle : The first condition to the host when the host first time run was idle, this condition states that the host does not perform any process but at idle only has the acting only listen.
2. Listening : In this condition, the host waiting for clients who are in the host area. Clients who process the data broadcast would be heard hosts on this process. This process would run continuously and this process explained in Wait For Broadcast Process.
3. Wait For Broadcast : This condition states that the hosts just waiting to receive broadcast data. If the host does not receive the data broadcast, the host simply repeating the process of listening and wait for broadcast. But if the host receives broadcast data then will proceed to the Check Device Duplication process.
4. Check Device Duplication : In the broadcast process performed by the client, the host receives a client name and client ip then this process will be checked whether the previous client had been in communication with the host or not. If you've ever it will be continued in the control process, but if it is not going to continue the process of ACK sent and check the status of the data appliance.
5. ACK Sent : After the hosts knew no data were duplicated, the host would send an ACK, in the sense that the host is ready to receive and transmit data. ACK sent by the host was to make requests about what services are owned by the client so that the host knew about anything that can be controlled by the host.
6. Check Appliance Data Status : After replying to client services that are owned, the host process Check Status Data Appliance to return at idle condition to wait if there is another client who wants to join the host..

7. Control Process : However, a client that has previously been identified will be stored and exchanged data. In this process service owned by the client will be able to take total control of the host. In the sense of total control over suppose lit LED, the data transmission gyroscope sensor or services that are owned by the client.

2.2. *Protocol Design on The Client*

After designing Host, will be designed Client. In this design will do the state machine design, the details about what will be done by the client, from the beginning until client being recognized by the Client Host well as in **Figure 3**.

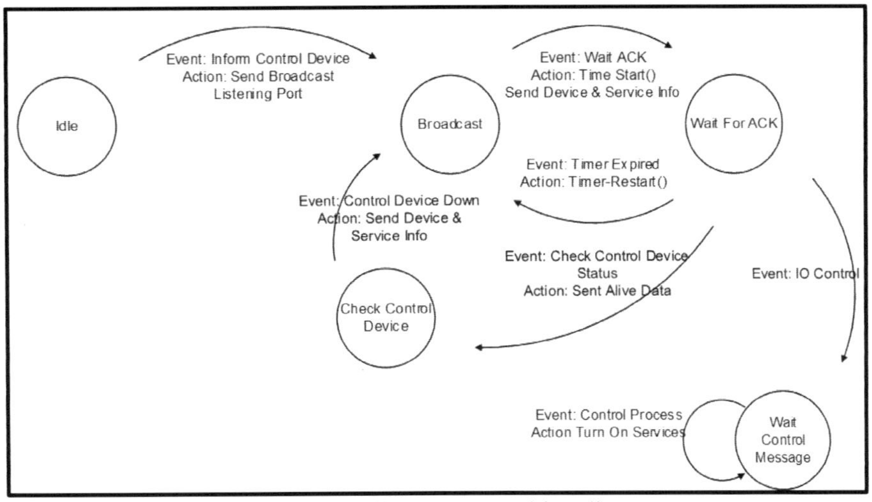

Figure 3. Client state machine diagram

Each condition in client state machine in **Figure 3** will explained as follows:

1. Idle : this condition is the initial condition of the client, the client sends the data broadcast was used to spread the data to find the host.
2. Broadcast : in this condition, data was broadcasted until the hosts detected on the "listen" condition. Once the data received by the host and the host sends an ACK, the condition of the client was no longer broadcasted the data, but to wait and count time until certain conditions were described in the Wait For ACK.
3. Wait For ACK : in this condition allows three further conditions. The first possibility when ACK was received from the host, it means the client and the host can communicate well and continued in "wait control message". The second condition of the host ACK waiting time is too long exceeded the specified time, so that in these conditions the client would have to broadcast back on the "broadcast" process. The third condition that if two conditions have been performed was entered in the process “check control device”.
4. Wait Control Message : in these conditions was a continuation of the conditions of the host "control process", the host and client were in condition to communicate. Client waits for commands from the host to operate in accordance the data transmitted from the host.
5. Check Control Device : This condition was the worst conditions, where the client does not receive the ACK sent by the host at the "ACK Sent", and this condition is also re-broadcast after data broadcasted where both did not work well at the time who determined. Client will check whether the package sent was correct or not. This process is repeated to the "broadcast" to the client find a host.

IOP Conf. Series: Materials Science and Engineering **190** (2017) 012009 doi:10.1088/1757-899X/190/1/012009

3. Implementation and Result

At this session discussed about the implementation on the client side and on the host side, each of which will be implemented in accordance with a design that have been made in the previous session. As well as the results will show the results of the processing system so that conclusions can be drawn.

2.3. Host and Client Implementation

Each state in the state machine implemented on a "case structure" on LabVIEW, each case illustrates the structure of each state on the design in Figure 2 that implemented as shown in **Figure 4**(a) for the host and Figure 4(b) for the client.

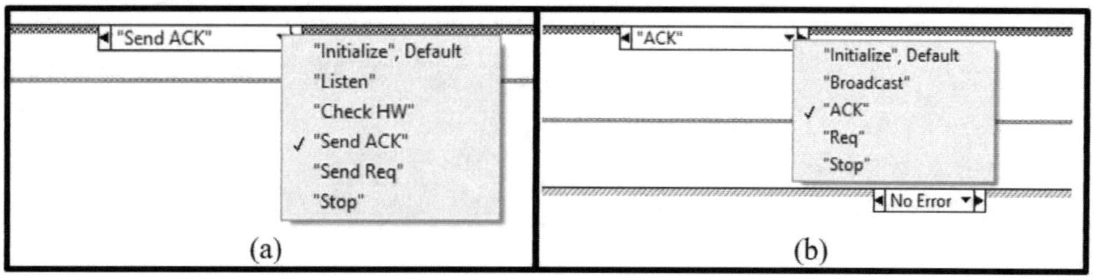

Figure 4. (a) Host case implementation. (b) Client case implementation.

Each process has a "case structure" itself, for one implementation of the delivery ACK as shown in **Figure 5**. In Figure ACK delivery process was implemented on the client. In case "Send ACK" There are some case for handling the "error" or "no error", as well as for handling if any of these conditions occurs. Data sent in the form of a "bundle" which contains the "client ID", "client service", "IP hardware" and "host name". "Bundle" in the sense of all that data into one package in which the packet sent to the client.

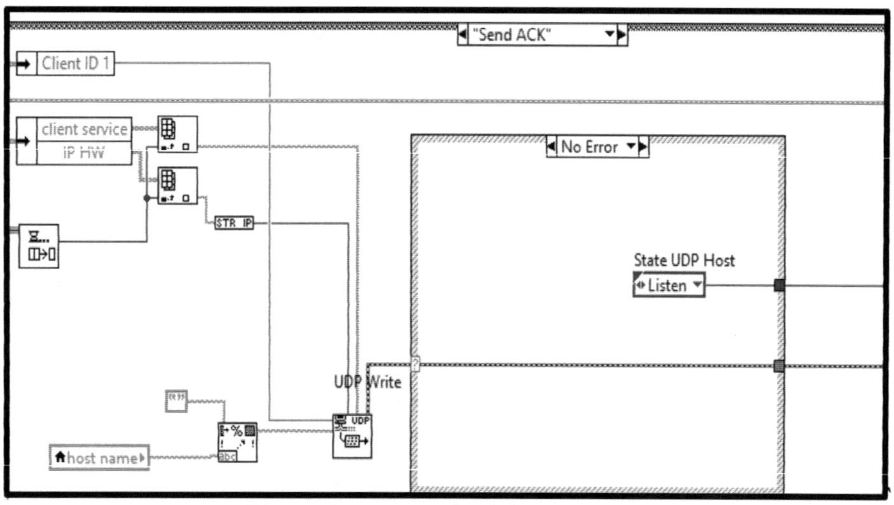

Figure 5. Host send ack implementation

Implementation sending ACK on the host will be received by ACK state on the client side. For the implementation of the client ACK in **Figure 6**. This figure shows that the client receives the data bundle that will be processed further.

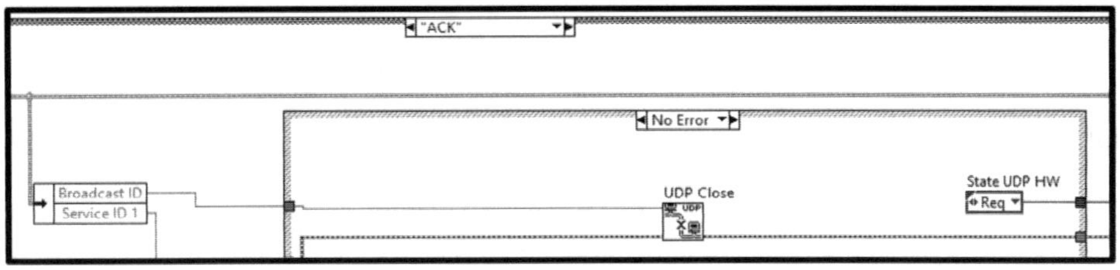

Figure 6. ACK client implementation

Just like on the host side, on the client side is also provided for handling the case of two conditions, the condition of "error" and "no error". Client receives an ACK to be processed In later steps. Implementation on both sides of the host and client successfully and completed for further testing processes that get results.

2.4. Result

After encoding, the result on each side of both the client side and the host side. For results from each side of the interface is in **Figure 7**.

Figure 7. Host interface

After running the system, in **Figure 7** there is a menu "UDP Host State" which claims the system is running in the state "Send ACK" as well if it is in another state then the column will declare it is running on the state. List UDP HW is a list of clients who are interacting with the host. Function "no hw" when running the simulation it.

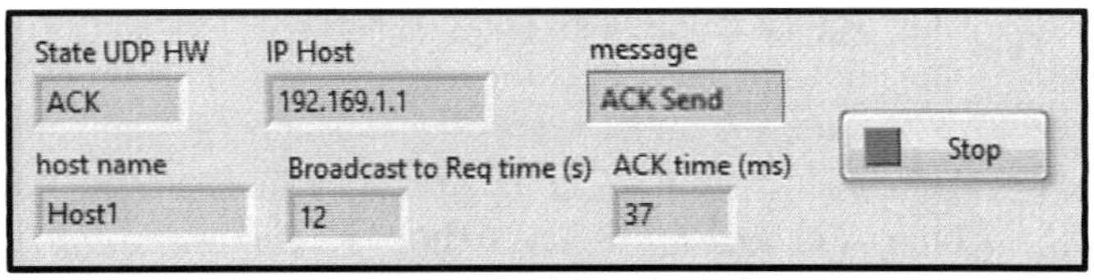

Figure 8. Client interface

For the client on **Figure 8**, it only displays "State UDP HW" that functions the same as the host, then the client can figure out "Host IP" and could exchange messages via the "message" to the host. In this trial Broadcast to produce as much as 12s Time Req. In terms of the host broadcast time to find a client takes the 12s, while the time required for delivery of ACK takes 37ms.

4. Conclusion and Future Works

We have built this system with mild or as a prototype or usually should called lightweight UDP in LabVIEW. Client can relate to Host with both pervasive. In this research service discovery was performed by the host can identified the client properly so that said construction of this study is successful. Host can control any services which provided by client. This research prototype was undertaken on two PCs with the assumption that the PC are a client and a host on a home device. So for future works this system can be implemented on an embedded system which has a high complexity with heterogeneous devices that can communicate well. We hope to implement this systems in Arduino, Xbee etc, or using National Instrument hardware such as MyRio etc.

Acknowledgments
Special thanks to the Laboratory of Computer System and Robotics Computer Engineering team, Department, Faculty of Computer Science University of Brawijaya. This research to support our research ini "Penelitian Unggulan 2016" scheme in BPP Program, Faculty of Computer Science, Brawijaya University.

5. References

[1] Atzori, L., Iera, A. and Morabito, G., "The Internet of Things: A survey. pp. 2787-2805," *Elsevier Computer Networks,* pp. 2787-2805, 2010.

[2] Vermesan, O. and Friess, P., Internet of Things - Global Technological and Societal Trends, Aalborg, Denmark: River Publisher, 2011.

[3] Munir, S.A., Ren, B., Jiao, W., Wang, B., Xie, D. and Jian, M., "Mobile Wireless Sensor Network: Architecture and Enabling Technologies for Ubiquitous Computing," in *International Conference on, Advanced Information Networking and Applications Workshops*, Niagara Falls, Ontario, Canada, 2007.

[4] Rose, K., Eldridge, S. and Chapin, L., "The Internet of Things: An Overview Understanding the Issues and Challenges of a More Connected World," *The Internet Society (ISOC),* 2015.

[5] Dourish, P. and Bell, G., Divining A Digital Future - Mess and Mythology in Ubiquitous Computing, London, England: The MIT Press, 2011.

[6] Banavar, G., Beck, J., Gluzberg, E., Munson, J., Sussman, J. and Zukowski, D., "Challenges: an application model for pervasive computing," in *MobiCom '00 Proceedings of the 6th annual international conference on Mobile computing and networking*, New York, USA, 2000.

[7] Saha, D. and Mukherjee, A.,, "Pervasive computing: a paradigm for the 21st century," *Computer,* vol. 36, pp. 25-31, 2003.

[8] M. Welzl, Network Congestion Control - Managing Internet Traffic, Chichester, England: John Wiley & Sons, Ltd, 2005.

[9] Petrovic, D., Shah, R.C., Ramchandran, K. and Rabaey, J., "Data funneling: routing with aggregation and compression for wireless sensor networks," in *Sensor Network Protocols and Applications, Proceedings of the First IEEE*, 2003.

[10] Snoeren, A.C. and Balakrishnan, H., "An End-to-End Approach to Host Mobility," in *MobiCom* '00 *Proceedings of the 6th annual international conference on Mobile Computing and Networking.*, New York, USA, 2000.

[11] Byers, W.J., Luby, M., Mitzenmacher, M. and Rege, A., "A Digital Fountain Approach to Reliable Distribution of Bulk Data," in *SIGCOMM,* New York, USA, 1998.

[12] Jimenez, F.J. and Frutos, D.J., "Virtual Instrument for Measurement, Processing Data, and Visualization of Vibration Patterns of Piezoelectric Devices," *Elsevier,* vol. 27, no. 6, pp. 653-663, 2005.

[13] Bhatti, N., Dhomeja, L.D. and Malkani, Y.A., "Service Discovery Protocols in Pervasive Computing: A Review," in *Multi-Topic Conference (INMIC), IEEE 17th International*, 2014.

[14] Serna, M.A., Sreenan, J.C. and Fedor, S., "A Visual Programming Framework for Wireless Sensor Networks in Smart Home Application," in *Sensor Networks and Information Processing (ISSNIP)*, Singapore, 2015.

Cooperative Learning for Distributed In-Network Traffic Classification

S.B.Joseph[1], H.R.Loo[1], I.Ismail [1],T. Andromeda[2], M.N.Marsono[1]

[1]Department of Electronics and Computer Engineering, Faculty of Electrical Engineering, Universiti Teknologi Malaysia, 81310 Johor Bahru, Malaysia
[2]Department of Electrical Engineering, Faculty of Engineering, Universitas Diponegoro, Semarang, Indonesia

sjbassi74@gmail. com, loohuiru@gmail.com, ismahani@fke .utm.my, trias1972@gmail.com; nadzir@fke . utm.my

Abstract. Inspired by the concept of autonomic distributed/decentralized network management schemes, we consider the issue of information exchange among distributed network nodes to network performance and promote scalability for in-network monitoring. In this paper, we propose a cooperative learning algorithm for propagation and synchronization of network information among autonomic distributed network nodes for online traffic classification. The results show that network nodes with sharing capability perform better with a higher average accuracy of 89.21% (sharing data) and 88.37% (sharing clusters) compared to 88.06% for nodes without cooperative learning capability. The overall performance indicates that cooperative learning is promising for distributed in-network traffic classification.

1. Introduction

Network traffic classification is a crucial network processing task for network traffic management. Traffic measurement and classification enable network administrators to be aware of the current network state. Data stream mining algorithms [1, 2] have been explored recently for online traffic classification to overcome the shortcoming of conventional data mining algorithms. They are designed to cope with concept drift, adapt to new knowledge and react to changes promptly.

Recently, distributed network management (DNM) strategies such as [3], have been introduced as a solution to the increasing management complexity of communication networks. These strategies are autonomic and decentralized in nature, towards improving network performance, scalability and to reduce human participation. Each DNM entity participates in a distributed management process, which requires cooperation among DNM entities to monitor, analyze and make decision to achieve global network objectives. Cooperative Learning (CL) enables network nodes to discover each other, exchange information, disseminate local decision, enhance self-adaptation of nodes, improve scalability and finally enforce management decision [4].

This paper proposes an analysis for online distributed in-network traffic classification based on our CL framework in [5], using incremental k-means network traffic classification [1]. This paper analyzes the effect of information exchange among nodes on the overall classification accuracy. We tested our proposed method on two schemes: sharing of training labeled data in the form of information and sharing of clusters in the form of knowledge. Our proposed system has been applied

IOP Conf. Series: Materials Science and Engineering **190** (2017) 012010 doi:10.1088/1757-899X/190/1/012010

to UPC [6] real network datasets in order to evaluate the performance of such system. This system is able to classify network traffic online with an average accuracy of 89.21% and 88.37% sharing clusters, compared to 88.06% over the non-sharing approach.

The remainder of this paper is structured as follows. Section 2 introduces related works on monitoring networks, DNM and CL. Section 3 presents online traffic classification and our proposed method. Section 4 analyzes the experimental results. Conclusion is in Section 5.

2. Related Works

The rapid growth of Internet traffic today practically demonstrates the concept of Big Data, characterized by having large data size with millions of network connections (variety, volume, veracity and velocity), which motivated the proposals of embedding network management functions into the network itself, e.g. [3] towards managing the network easier, automatically and with reduced complexity.

Several research works have also proposed different monitoring schemes for different network architecture over the years. Stadler et al. [7] has outlined principles for decentralised monitoring using spanning trees and gossiping methods. However, their techniques are targeted for network aggregates and not for online measurements. Raz et al [8] has presented an insight of network monitoring algorithms and also several monitoring algorithmics that utilize monitoring techniques. Although these techniques are proposed for distributed monitoring, they did not consider the issue of collaborative/cooperative learning among autonomic nodes.

In a distributed system where each node is semi or fully autonomic in solving a global problem, the role of interaction/cooperation among such distributed nodes can not be over emphasized. References [4, 9] proposed interaction/cooperative learning among distributed networks for improving accuracy and scalability. The issues of nodes discovery for interaction among nodes is presented in [9], network nodes discover one another in a network in other to cooperate with each other. In the same note, reference [4] proposed bootstrapping and discovery mechanisms to ensure the proper information dissemination in distributed network process. However, these methods are proposed for discovering neighbouring nodes, they did not consider the task of nodes interaction for information sharing. Finally, we consider the issue of information/knowledge sharing among distributed network nodes, through cooperative learning.

3. Online Network Traffic Classification With Cooperative Learning

Table 1: List of online features selected for online classification

ID	Name	Long Description
01	Source Port	Source port number
02	Destination	Destination port number
03	TL IP	Total bytes in IP packet
04	UL IP	Total bytes in IP packet (uplink)
05	DL IP	Total bytes in IP packet (downlink)
06	TL Eth	Total bytes in Ethernet packet
07	UL Eth	Total bytes in Ethernet packet
08	DL Eth	Total bytes in Ethernet packet
09	TL Ctr	Total control bytes in packet
10	UL Ctr	Total control bytes in packet
11	DL Ctr	Total control bytes in packet

We propose a network traffic classification technique with cooperative learning capabilities to support distributed network monitoring and to enhance the monitoring capabilities of the DNM nodes. The system framework is based on our proposed system in [5], which consists of three distinct layers, each performing a different successive task towards the achievement of global management objectives. In this paper, we only focus on classification layer and the effect of cooperative learning. Different from [5], we increase the number of nodes to show the scalability of the system framework. Figure 1 shows the framework of our proposed system. Network traffic are sensed at different network locations by autonomic nodes. These nodes share newly derived information in the form of information or knowledge among each other to improve their states.

In this paper, we propose two sharing methods for cooperative learning for network traffic classification. We consider the sender of information as host node, while the receiving node as client node.

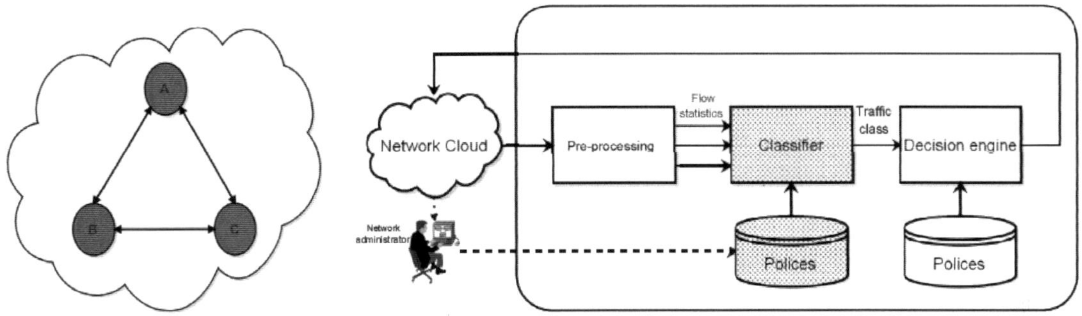

(a) (b)

Figure 1: System block diagram (a) Cooperating nodes, (b) Individual node structure

3.1. Scheme 1: Sharing of information

In this scheme cooperative learning is initiated by the host node upon receiving a labeled flow instance ($x_i y_i$) from the network administrator for retraining. The host retrains its model and at the same time shares the labeled instance with randomly selected neighboring node. Upon receiving shared labeled instance, a client node will update its state by retraining its model with the new instance. This sharing is done based on Algorithms 1.

3.2. Scheme 1: Sharing of information

This sharing scheme is activated by a node upon creating a new cluster after retraining with received labeled flow instance, the host shares newly created cluster (c_j) with randomly selected neighboring nodes. Each of the nodes will either merges or creates new cluster. This sharing is done based on Algorithms 2.

Let x_i, represents incoming data stream; M is classification model; $c_1 c_2$ are first and second nearest cluster from x_i; y_i and y'_i are the true and predicted labels for x_i; x_j and y'_j are labeled data from nearest neighboring node; c_j is newly created cluster. The classifications are initiated upon receiving incoming flow instances. In both schemes not all information received from network administrator will be shared, host will only share a percentage (%S) of the received information to a neighboring client. This is to minimize network resources overhead due to sharing.

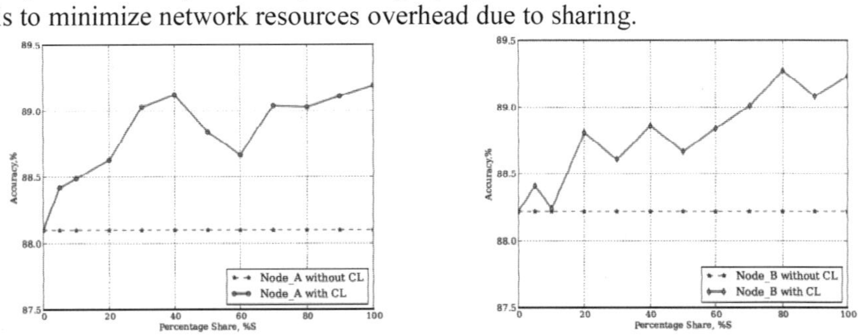

(a) (b)

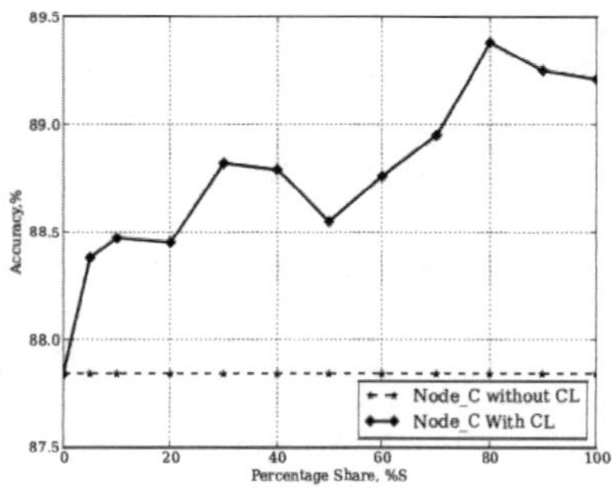

(c)

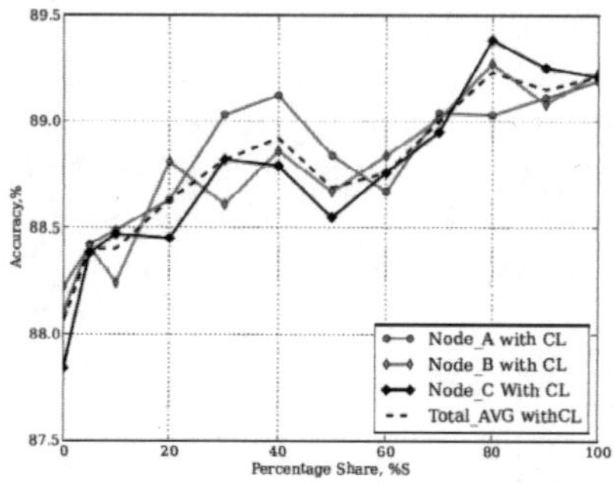

(d)

Figure 2: Comparison of average accuracy plots for scheme 1 in (a) Node A, (b) Node B, (c) Node C, (d) All Nodes

Algorithm 1 Proposed Cooperative Learning Algorithm (Case 1)

```
while new xi do
    yi i = Classify M,xi
    if (xi is labeled) then
    yi is known
    retrain(M,xi,yi)
    select random neighbor
    share (xi,yi)to selected neighbor
end if
end while
Re-Training
if (xj yj) received then
    retrain (M,xj,yj)
end if
```

Algorithm 2 Proposed Cooperativ Learning Algorithm (Case 2)

```
while new xi do
  y i = Classify M,xi
  if (xi is labeled) then
    yi is known
    retrain (M,xi,yi)
  end if
  if (ci)is created then
    select random neighbor
    share (ci)to selected neighbor
  end if
end while
Re-Training
if (cj)is received then
  find nearest cluster to (cj)
  if (cj)is close to nearest cluster then
    merge (cj)to nearest cluster
  else
    adapt(cj)as new cluster
  end if
end if
```

Table 2: Summary of results

Nodes	Scheme 1			Scheme 2		
	Accuracy (%) without CL	Accuracy (%) with CL	Improvement (%)	Accuracy (%) without CL	Accuracy (%) with CL	Improvement (%)
A	88.10	89.19	1.09	88.10	88.59	0.49
B	88.22	89.23	1.01	88.22	88.31	0.09
C	87.84	89.21	1.37	87.84	88.22	0.38

IOP Conf. Series: Materials Science and Engineering **190** (2017) 012010 doi:10.1088/1757-899X/190/1/012010

4. Results and Discussion

4.1. Dataset

We evaluated our proposed method using real network datasets. The UPC dataset [6] is used for the experiment; a total of 339061 flow instances was extracted. We represented each host address as an autonomic node representing different sensing segment in our experiment. We used eleven (11) attributes, and six (6) traffic classes as suggested in [1] as listed in Table 1.

4.2. Performance Evaluation

The average accuracy plots of individual nodes with and without cooperative learning for scheme 1 is presented in Figure 2. Figure 3 shows the cumulative accuracy plots for nodes with and without cooperative learning for scheme 2. Both plots show higher classification accuracy on each nodes with cooperative learning compared to nodes without cooperative learning. The comparison of average accuracy of each nodes with different percentage of sharing (%S) shows that with at least 5% of sharing we can increase the accuracy by 0.34% for scheme 1 and 0.15% for scheme 2.

Table 2 presents a summary of accuracy at each node for our proposed method. The experimental results indicate an improvement on the classification accuracy when cooperative learning is activated compared to independent accuracy without cooperative learning. An improvement in accuracy was observed for both classification models. The overall performance indicates that cooperative learning is promising for distributed network monitoring.

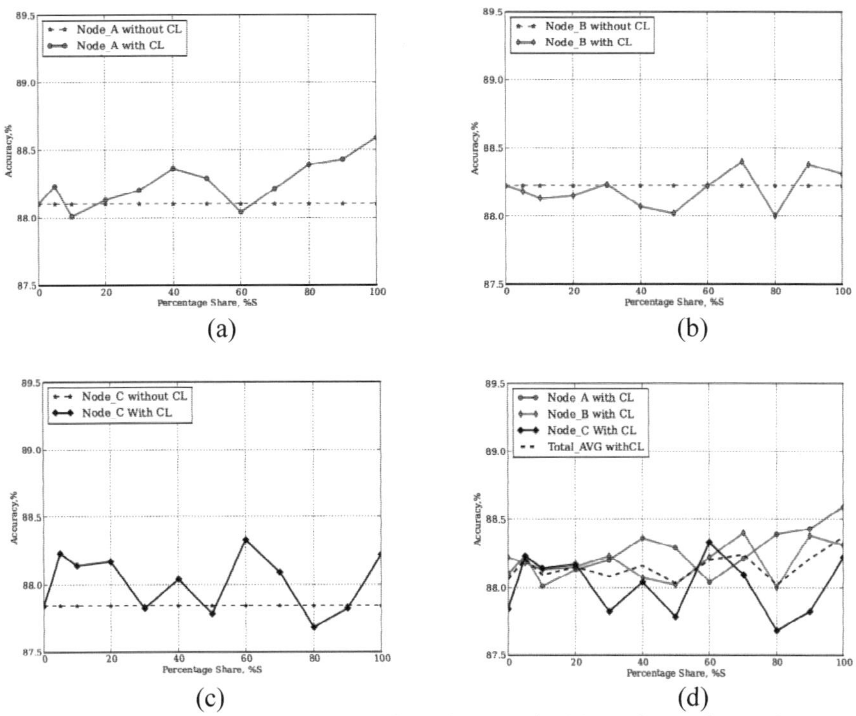

Figure 3: Comparison of average accuracy plots for scheme 2 in (a) Node A, (b) Node B, (c) Node C, (d) All Nodes

5. Conclusion

The proposed CL algorithm for online distributed network performance monitoring based on incremental k-means traffic classification algorithm has been presented. In difference to existing classification methods, the proposed method explores the capabilities of cooperative learning to improve the accuracy, reliability and scalability of classification accuracy. The performance of the proposed method was justified with real network traces. The obtained results have shown the significant increment in classification knowledge of nodes, and improvement in accuracy of classification in cooperative learning network nodes

Acknowledgment
This work is supported in part by CREST grant (UTM Vote No. 4B176) and Universiti Teknologi Malaysia Grant matching (UTM Vote No. 00M75)

References

[1] H. Loo, S. Joseph, and M. Marsono, "Online incremental learning for high bandwidth network traffic classification," Applied Computational Intelligence and Soft Computing, vol. 2016, 2016.
[2] G. Mingliang, H. Xiaohong, T. Xu, M. Yan, and W. Zhenhua, "Data stream mining based real-time highspeed traffic classification," in Broadband Network & Multimedia Technology, 2009. IC-BNMT'09. 2nd IEEE International Conference on, pp. 700–705, IEEE, 2009.
[3] D. Dudkowski, M. Brunner, G. Nunzi, C. Mingardi, C. Foley, M. P. de Leon, C. Meirosu, and S. Engberg, "Architectural principles and elements of in-network management," in Integrated Network Management, 2009. IM'09. IFIP/IEEE International Symposium on, pp. 529–536, IEEE, 2009.
[4] L. Guardalben, S. Sargento, P. Salvador, and V. Mirones, "A cooperative hide and seek discovery over in network management," in Network Operations and Management Symposium Workshops (NOMS Wksps), 2010 IEEE/IFIP, pp. 217–224, IEEE, 2010.
[5] S. Joseph, H. Loo, I. Ismail, and M. Marsono, "Cooperative learning for online in-network performance monitoring," in Communications (MICC), 2015 IEEE Malaysia International Conference on, pp. 219–224, Nov 2015.
[6] V. Carela-Español, T. Bujlow, and P. Barlet-Ros, "Is our ground-truth for traffic classification reliable?," in Passive and Active Measurement, pp. 98–108, Springer, 2014.
[7] R. Stadler, M. Dam, A. Gonzalez, and F. Wuhib, "Decentralized real-time monitoring of network-wide aggregates," in Proceedings of the 2nd Workshop on Large-Scale Distributed Systems and Middleware, p. 7, ACM, 2008.
[8] D. Raz, R. Stadler, C. Elster, and M. Dam, "In-network monitoring," in Algorithms for Next Generation Networks, pp. 287–317, Springer, 2010.
[9] K. M. Konwar, D. Kowalski, and A. A. Shvartsman, "Node discovery in networks," Journal of Parallel and Distributed Computing, vol. 69, no. 4, pp. 337–348, 2009.

IOP Conf. Series: Materials Science and Engineering **190** (2017) 012011 doi:10.1088/1757-899X/190/1/012011

Monitoring and Indentification Packet in Wireless With Deep Packet Inspection Method

Ahmad Fali Oklilas[1], Tasmi[2]

[1]Computer Engineering, Computer Science Department, Sriwijaya University
[2]Master of Engineering Informatics, Computer Science Department, Sriwijaya University

fali@ilkom.unsri.ac.id. faliunsri@gmail.com
tasmi@ilkom.unsri.ac.id, tasmisalim@gmail.com

Abstract. Layer 2 and Layer 3 are used to make a process of network monitoring, but with the development of applications on the network such as the p2p file sharing, VoIP, encrypted, and many applications that already use the same port, it would require a system that can classify network traffics, not only based on port number classification. This paper reports the implementation of the deep packet inspection method to analyse data packets based on the packet header and payload to be used in packet data classification. If each application can be grouped based on the application layer, then we can determine the pattern of internet users and also to perform network management of computer science department. In this study, a prototype wireless network and applications SSO were developed to detect the active user. The focus is on the ability of open DPI and nDPI in detecting the payload of an application and the results are elaborated in this paper.

1. Introduction

The various types of data traffic on the Internet often complicate the process of identification of traffic on the network monitoring system. Callado et al [1] stated that in many internet services there are applications that use the same port with used standard services.

There are many challenges in the classification of traffic. Running a network application can produce a series of packets which consists of header, payload and header information carrying layering. In the study conducted by [2], currently a mechanism of network monitoring system is desperately needed to monitor the status of infrastructure, LAN / WAN and ensure devices under normal conditions and active. The method used in the monitoring of the network in order to obtain a connection and good information is a way to process classification of traffic. In general, the methods used in the classification consist of a Port-Based, Payload-Based, or a Heuristic Behaviour Protocol Based Classification.

Based on previous research [3] systems on the network not only focus in process of identifying the data on the type of protocol and port number, but rather than identification and regulation of the signed application. Similarly for case of access restriction to certain web applications cannot be done simply by using a firewall by stopping access to the web IP address.

This paper focuses on the development of a prototype system to monitoring and identification data packets on a wireless network. This study was submitted to prove the real method of Deep Packet Inspection in the classification process data packets

2. Overview and Background Problems

Previous research [3] stated that systems in the network not only focus on the process of data identifying on the type of protocol and port number but rather than identification and regulation of the signed application. Similarly for case of restriction access to certain web applications, it cannot be done simply by using a firewall by stopping access to the web IP address.

Solutions for classifying data packets have been carried out both in the network online and offline by using machine learning, such as in [4] using the port number for the classification process data packets, as well as research conducted by [5] and [6] with combining methods of machine learning (ML) and artificial intelligent to classify IP traffic. Meanwhile, the research conducted by [7] and [8] explain that the classification of data packets on network online was by recognizing the payload. Moreover, network research conducted by [9] used Deep Packet Inspection with the signature on the payload to identify the protocol.

Deep Packet Inspection is a method used to perform thorough inspection process in the packet header of the TCP / IP and Payload. Some previous researches conducted by [10], [11] and [12] have been using the DPI as a method for packet classification process incoming and outgoing data

3. Proposed Design

In this research, a prototype of data packet monitoring system in wireless networks has been developed. Figure 1 is a schematic of experiments conducted.

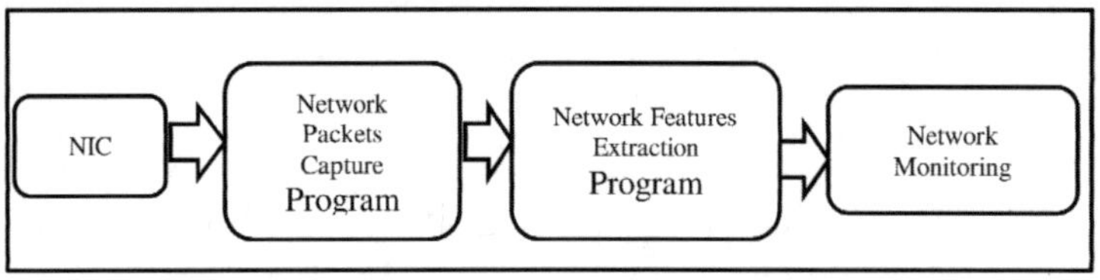

Figure 1. The procedure of monitoring data packets

At this research stage, it begins with Phase (i) to produce raw data obtained from sniffing packets of data using a sniffing application tool. IPv4 header structure is unique and has a hidden header depend on the protocol and encapsulated process that causes the raw data is difficult to read by human. Therefore, we need a new file type that can be processed and facilitate the training process. In this study, the type of files generated from the processing of raw data is .csv, where these file type easily processed and can be general received.

Phase (ii) is Feature extraction process to generate strong standards to serve as a benchmark as the archetype of the process. At this stage it made an algorithm to extract the layers to find the same patterns that mapping the characteristics possessed by a field. The third stages (iii) is the Feature Selection, as a field produces features as the unique cause difficulties in terms of identification of data packets, because not all features can be used the classification process data packets, therefore it created an algorithm to process the kind of selection data packets.

The last stage in this study (iv) is the training process (training) using Deep Packet Inspection methods and algorithms used are AC (Aho-Corasick). Aho-Corasick algorithm uses multi pattern that can solves the string matching problem relating to the timing and size of output. We uses Single-Sign-On (SSO) as the central authentication of users which are used in the Fasilkom Sriwijaya University (UNSRI)'s campus wireless network as shown in Figure 2.

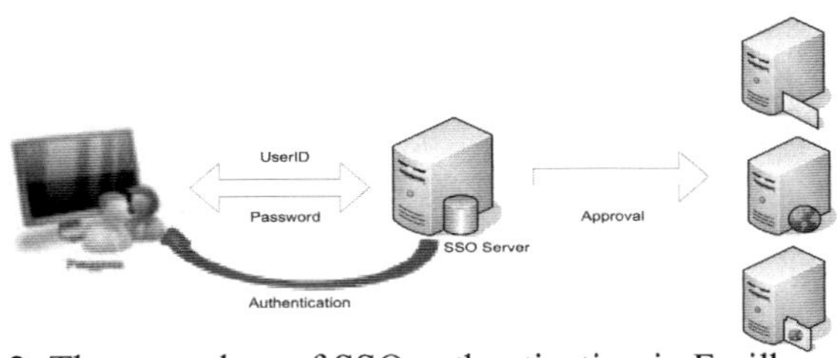

Figure 2. The procedure of SSO authentication in Fasilkom UNSRI

4. Implementation and Result

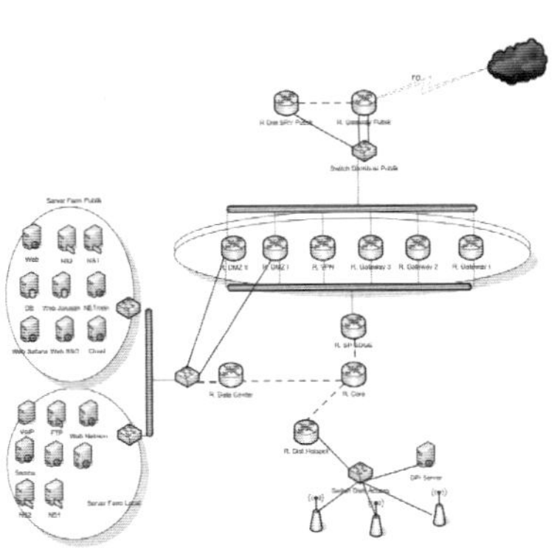

Figure 3. Topology research

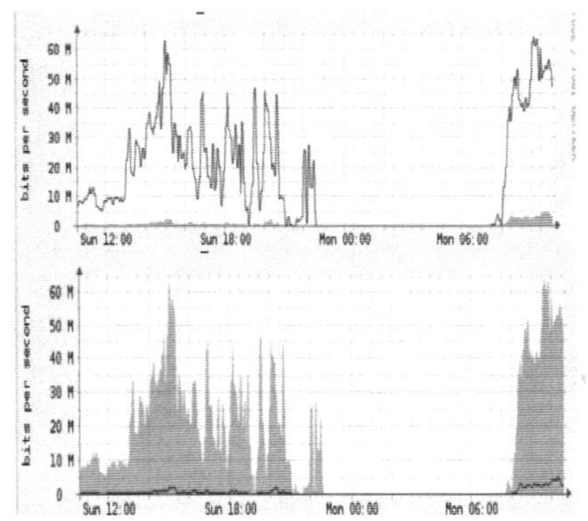

Figure 4. Traffic in the graph of router

Draft Design to create a depiction topology, router configuration, IP address allocation, configuration of multiple devices Wi-Fi, and election monitoring system is shown in Figure 3. In this case, a limited prototype was implemented in the Laboratory of Computer Science UNSRI network. The detail of works on this stage are: (i) Topology integrated into the Fasilkom Unsri campus network, (ii) selection of the device, (iii) Coding system applications and configuration monitoring and (iv) integrating the device into the application

Figure 4 shows the results of the traffic monitoring in which data obtained from users who conduct connection test. Results obtained shows heavy traffic between 40 -70 Mbps, this is because of the manual process of uploading and downloading for several applications. Figure 5 is the result of a user login which uses SSO.

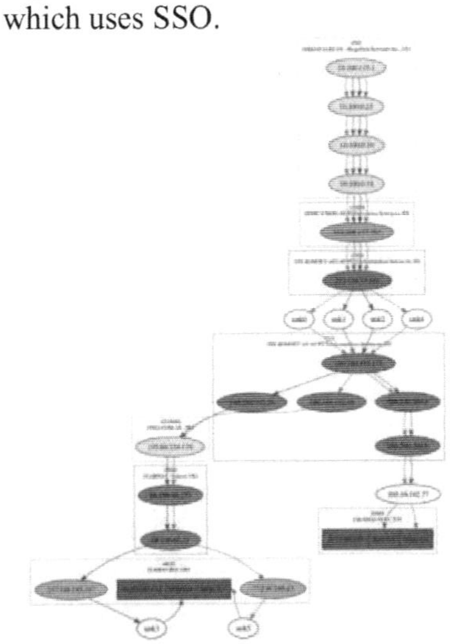

Figure 5. Routing packet data

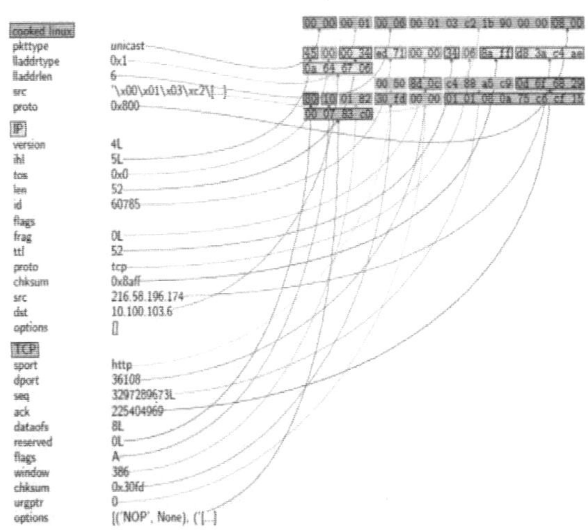

Figure 6. Result of extraction packet data

Table 1. Results of monitoring packet data

Protocol	∑ Packets	∑ bytes	∑ flows
DNS	292	64167	147
HTTP	44	8980	5
NETBIOS	48	4567	4
SSDP	51	21330	2
DHCP	4	1378	3
Unknown	22127	20765840	97

In the process of getting the results as shown in Table 1, first made the process sniffing packets of data in real time, then results in sniffing extracted by using a program that was built using C language, after which the process of grouping data packets. The result obtained from the protocol that is widely used by the user is a standard protocol such as HTTP, DNS and DHCP, but this result was also obtained unknown protocol

5. Conclusion

Deep Packet inspection one of methods which can be used for identification of data packets based on application signatures. By using Single-Sign-On can be used in active hotspot user identification.

From the initial results obtained there are still many protocols that cannot be monitored (unknown), protocols which produced still standard such as DNS, HTTP, many applications on the internet that use SSL such as social media applications. But the results are still difficult to monitor data packets using SSL.

But the results of this study can be developed again in terms of process monitoring applications that use SSL/TLS

Reference

[1] A. Callado, A. Callado, C. K. Member, G. Szabó, B. Péter-gerö, and J. Kelner, "A Survey on Internet Traffic Identification . A Survey on Internet Traf fi c Identi fi cation," vol. 11, no. November 2015, pp. 37–52, 2009.

[2] D. Stiawan and A. Oklilas, Fali, "Intrusion Prevention in Heterogeneous System based on Behavior Approaches," *Int. Conf. Sci. Inf. Technol. (ICSITech)*, 2015.

[3] H. Nazief, T. Sabastian, A. Presekal, and G. Guarddin, "Development of University of Indonesia Next Generation Firewall Prototype and Access Control With Deep Packet Inspection," *ICACSIS*, pp. 47–52, 2014.

[4] T. T. T. Nguyen and G. Armitage, "A Survey of Techniques for Internet Traffic Classification Using Machine Learning," *Commun. Surv. Tutorials, IEEE*, vol. 10, no. 4, pp. 56–76, 2008.

[5] R. N. Jesudasan, P. Branch, and J. But, "Generic Attributes for Skype Identification Using Machine Learning," no. August, pp. 1–7, 2010.

[6] R. Alshammari and A. N. Zincir-Heywood, "An investigation on the identification of voip traffic: Case study on Gtalk and Skype," *Proc. 2010 Int. Conf. Netw. Serv. Manag. CNSM 2010*, pp. 310–313, 2010.

[7] H. Chen, F. You, X. Zhou, and C. Wang, "Algorithm Comparison of P2P Traffic Identification Based on Deep Packet Inspection," vol. 1, pp. 3–6, 2009.

[8] M. F. Sun and J. T. Chen, "Research of the traffic characteristics for the real time online traffic classification," *J. China Univ. Posts Telecommun.*, vol. 18, no. 3, pp. 92–98, 2011.

[9] Z. Cao, G. Xiong, Y. Zhao, Z. Li, and L. Guo, "A Survey on Encrypted Traffic Classification," *Appl. Tech. Inf. Secur.*, pp. 73–81, 2014.

[10] D. Smallwood and A. Vance, "Intrusion Analysis with Deep Packet Inspection Increasing Efficiency of Packet Based Investigations," pp. 342–347, 2011.

[11] M. Najam, U. Younis, and R. ur Rasool, "Speculative parallel pattern matching using stride-k DFA for deep packet inspection," *J. Netw. Comput. Appl.*, vol. 54, pp. 78–87, 2015.

[12] Thaksen j Parvat and P. Chandra, "A Novel Approach to Deep Packet Inspection for Intrusion Detection," *Int. Conf. Adv. Comput. Tecnol. Appl.*, vol. 45, pp. 506–513, 2015.

IOP Conf. Series: Materials Science and Engineering **190** (2017) 012012 doi:10.1088/1757-899X/190/1/012012

OPTIMIZING IT INFRASTRUCTURE BY VIRTUALIZATION APPROACH

Thomas Budiman[1], **Jarot S. Suroso**[2]

[1, 2] Graduate Program, Binus University, Jl. Kebon Jeruk Raya No. 27 Jakarta Barat 11530, Indonesia

thomas@rehobot.net[1], jsembodo@binus.edu[2]

Abstract. The goal of this paper is to get the best potential configuration which can be applied to a physical server without compromising service performance for the clients. Data were compiled by direct observation in the data center observed. Data was then analyzed using the hermeneutics approach to understand the condition by textual data gathered understanding. The results would be the best configuration for a physical server which contains several virtual machines logically separated by its functions. It can be concluded that indeed one physical server machine can be optimized using virtualization so that it may deliver the peak performance of the machine itself and the impact are throughout the organization.

1. Introduction

Virtualization is a way of using computer resources in such a way that no waste of computing resources during the idle time. Of course, it is also an active support to the Green Computing movement. The Green computing is the study and practices of using the computer resources efficiently [1].

Align with the movement, an organization may apply virtualization to either it's infrastructure or the software being used. Both type not only support the green computing movement, but also organization may gain some advantages by implementing virtualization. In the computing environment, there are two types of virtualization that commonly applied: hypervisor (kernel-based virtualization, ie. KVM) and paravirtualization/container based virtualization (ie. LXC or Linux Container). Both techniques has its own advantages and disadvantages [2].

The kernel based virtualization allows us to deploy multi operation system as a guest on top of the host machine, while in container based, the virtual machine will inherit host operating system, hence the guest operating system will be the same as the host machine.

This research has the goal instead of hosting solution to self-managed data center to fulfil the need of organization. The virtualization adoption to their data center, leads the organization to support the Green Computing movement. The purpose of research is to get a better understanding of how a physical machine potential is optimized by implementing the virtualization technique on top of it. It is true that relevant papers about virtualization and green computing has given the author a wide perspective of how this research are done.

2. Theory and Method

In today's dynamic age, businesses change rapidly and act to competition, hence requirements tend to change within projects. In such scenarios, costs of change can be overbearing [3].

Based on the survey results application development projects undertaken by the Standish Group in 2004 up to 2012, the average successful project (the project has been completed on-time and on-budget, with all features and functions as initially specified) ranges between 29 % up to 39%. As for the unsuccessful project (The project was cancelled at some point during the development cycle) ranges from 18% to 24%. For projects that challenged (The project is completed and operational but over-budget, over the time estimate, and offers fewer features and functions than originally specified) had a greater percentage ranging from 43% to 53% [4].

RESOLUTION

	2004	2006	2008	2010	2012
Successful	29%	35%	32%	37%	39%
Failed	18%	19%	24%	21%	18%
Challenged	53%	46%	44%	42%	43%

Project resolution results from CHAOS research for years 2004 to 2012.

Figure 1. Standish Group Chaos Research 2004 to 2012[4].

Scrum project management is one of the agile methodology that can be used as solutions in application development projects [4]. Scrum is a method or process comes from the concept of agile which is currently very popular used by companies in the field of industry or application software development around the world.

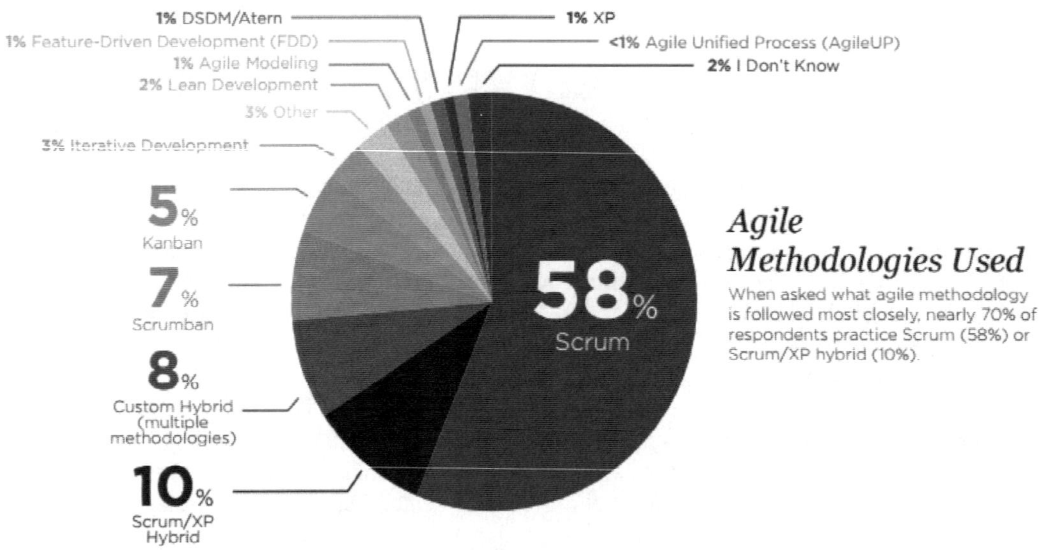

Figure 2. *Popular Agile Methodologies* [5].

Scrum project management is a project management methodology that has higher adaptability to the changes that occur in each of its elements than waterfall method. The advantages of agile project management and particularly the Scrum-based approach is its simplicity. Within an agile project, roles are clearly defined and do not cross boundaries. Features can be completely developed and tested in short iteration cycles. Because each team members bears major responsibility for their part of the project, ownership of the project is more broadly based. The methods of agile project management enforce extensive communication, which helps teams organize more effectively. This, ultimately, can lead to greater productivity for everyone involved [6]. SCRUM is best if the requirements frequently change [7].

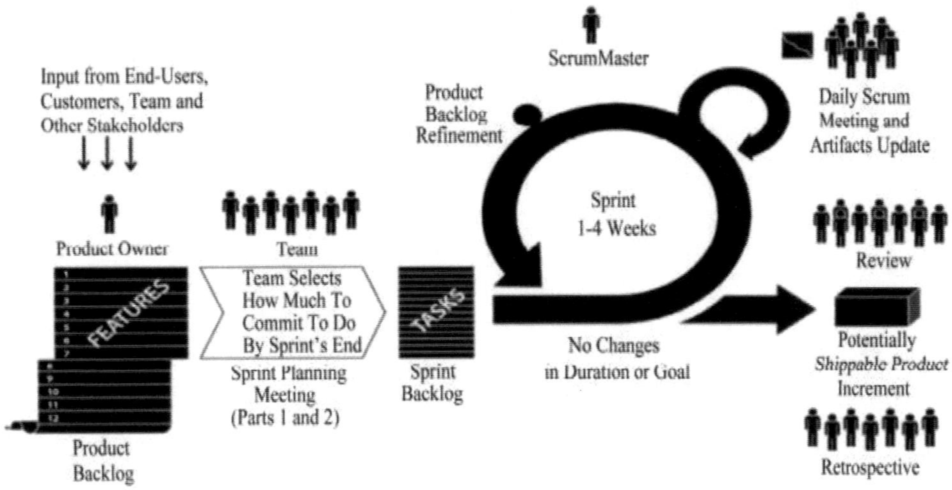

Figure 3. Scrum Framework [9].

The heart of Scrum is a Sprint, a time limit for a month or less, with an increment of "Done", function, potentially to be released and developed. Sprint typically have a duration that is consistent throughout the product development process. Sprint's new, begins immediately after the previous one ends Sprint. Sprint load and consist of the Sprint Planning, Daily Scrum, development, Sprint Review and Sprint Retrospective. Sprint's new, begins immediately after the previous one ends Sprint. Sprint load and consist of the Sprint Planning, Daily Scrum, development, Sprint Review and Sprint Retrospective [8].

There are several key roles in Scrum framework: 1) Stakeholders: the mission owners, possess the idea about why to build, what to build, and how processes should be; 2) Product owner: the product development owner, closely working with stakeholders, creates a prioritized wish list stories called a Product Backlog; 3) Scrum Master: the facilitator, closely working with product owner, makes sure the stories in the product backlog will successfully delivered as workable sub-products; 4) Scrum Team: the developers, closely working with Scrum Master, deliver workable sub-products according to the product backlog [9].

Scrum main characteristics may be summarized in [10]:

1. Software development is carried out through iterations, named Sprints, which last for 30 days. The result of each one of these is a feasible increase that is shown to the client.
2. Meetings all along the Project; it is noteworthy that the development team meets daily and briefly, only for 15 minutes, for coordination and integration.

The fundamental differences between traditional and agile concept is the traditional concepts, solutions and requirements (requirement) is very clearly defined early in the project so a large scope change was not expected. Traditional concept of work on the project on a regular basis and can be repeated by using a template that has been created. On the concept of agile solutions and the need (requirement) partially known and understood so that there is a possibility of additional features that were previously unknown to the project team. The concept of agile project teams preparing for big changes in the scope [11].

3. Method

When we choose to implement virtualization, we have only two options to choose, a full virtualization as known as Hypervisor, or a paravirtualization known as container based. Either one is very mature technology and widely used around the globe. In full virtualization, the guest machines might have a different operating system to use from the host machine, whereas no option in the paravirtualization. In paravirtualization the guest machines will inherits the host operating system. They will be configured as a block of user space (or container) to emulate the hardware limit. The following diagrams shows how these two are different from each other:

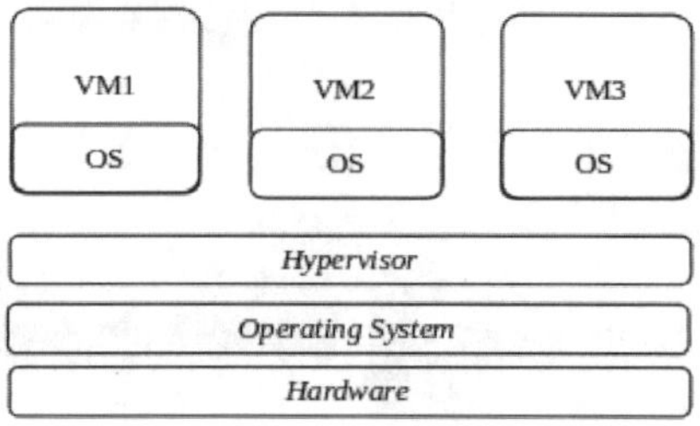

Figure 4. The Abstraction of a Hypervisor Virtualization

In hypervisor world, the virtual machine might have different operating system to the host machine, hence giving the flexibility to choose which OS to be deployed on the VM. It is possible because of the presence of the hypervisor abstraction layer, which mainly provides hardware emulation.

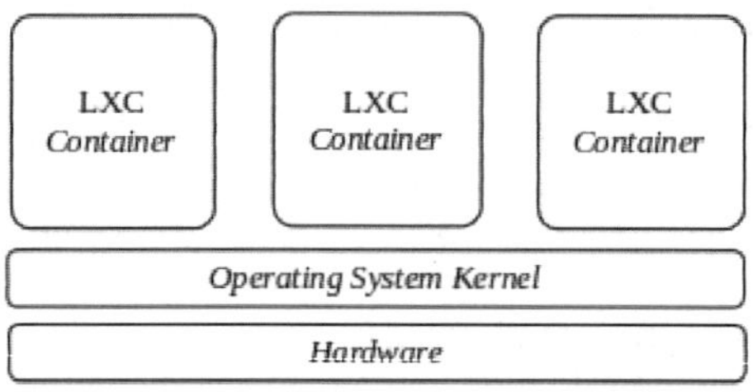

Figure 5. The Abstraction of a Paravirtualization

The paravirtualization a bit more simple than full virtualization, because they no longer employed the hypervisor abstraction layer. Because of this, the speed of hardware is nearly physical alike. The only drawback in the paravirtualization appeared that the guest OS will inherits what's available on the host, hence the guest OS will be the same as the host.

At some point the KVM might shows performance ahead of the paravirtualization, but in some cases the results shows differently where paravirtualization performance outperform KVM's.

The USE method are used in this research to help in setting up the virtual machines. It allows the authors to get step by step diagnose work for performance monitoring while converting and creating the virtual machines on the physical server [12]. This technique was introduced by Brendan Gregg, the performance engineer specialist or system performance engineer, a consultant and a trainer. He introduced the step of his method by the diagram in Figure 6.

All steps in the USE flowchart employed two built in linux command to measure the system health. Those two commands are: *vmstat* and *top*. VMSTAT were used to monitor the health of the storage health, while the TOP command helps to monitor the processor health and utilizations. Figure 4 and 5 below displays the results of *vmstat* and *top* command.

IOP Conf. Series: Materials Science and Engineering **190** (2017) 012012 doi:10.1088/1757-899X/190/1/012012

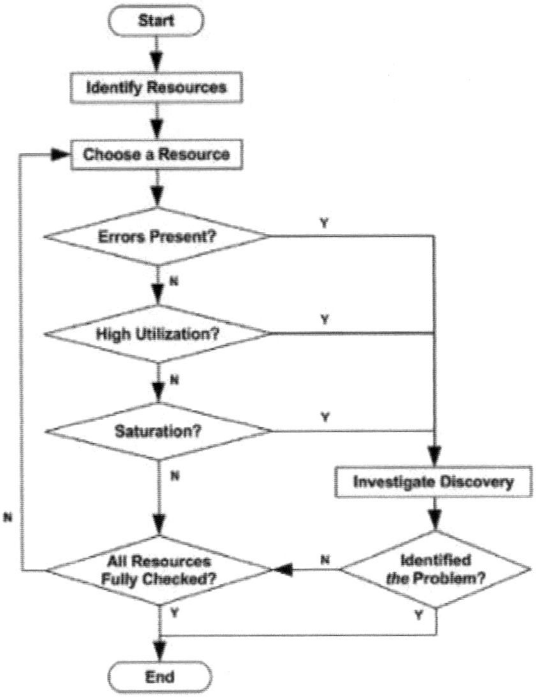

Figure 6. The USE Flowchart

```
+ X ... proxmox.rehobot.net
root@proxmox:~# vmstat
procs -----------memory---------- ---swap-- -----io---- -system-- ----cpu----
 r  b   swpd   free   buff  cache   si   so    bi    bo   in   cs us sy id wa
 1  0      0 4211732  78708 1304632    0    0    18    16   74   81  1  0 99  0
root@proxmox:~#
```

Figure 7. Output from *top* command

```
Tasks: 382 total,   1 running, 381 sleeping,   0 stopped,   0 zombie
%Cpu(s):  0.1 us,  0.1 sy,  0.0 ni, 99.9 id,  0.0 wa,  0.0 hi,  0.0 si,  0.0 st
KiB Mem:   8013876 total,  1963492 used,  6050384 free,    72588 buffers
KiB Swap:  7340028 total,        0 used,  7340028 free,  1280068 cached

  PID USER      PR  NI  VIRT  RES  SHR S  %CPU %MEM    TIME+  COMMAND
 3057 root      20   0  287m  61m 6192 S     0  0.8   0:00.59 pvedaemon worke
 3083 www-data  20   0  286m  61m 4420 S     0  0.8   0:01.53 pveproxy worker
    1 root      20   0 10612  880  744 S     0  0.0   0:01.44 init
    2 root      20   0     0    0    0 S     0  0.0   0:00.00 kthreadd
    3 root      rt   0     0    0    0 S     0  0.0   0:00.00 migration/0
```

Figure 8. Output from *vmstat* command

For more information about the results from both top and vmstat command, the Linux man page site serves full linux commands documentations: www.linux.die.net. To get feedback from user who engaged to one of the machine served by the virtual machine, a survey was set and deployed to the website. Further calculation of the questionnaire are using the LIKERT scale with the following scale:

Table 1. Questionaire Scale

Statement	**Scale**
Very Good	5
Good	4
Good Enough	3
Fair	2
Not Good	1

4. Result and Discussion

The first two tasks were converting two websites that were running in the hosting environment to the proxmox environment. The process smoothly done by copying all required files from hosting account to the local web server, and also the databases. The other virtual machines were created to suit the organization need. By the end of the research, 6 out of 7 virtual machines were employed in the physical server. The final results shown in the following table:

Table 2. VM Configuration Summary

VM #	HDD	RAM	Core
1	16 GB	2GB	2
2	8GB	4GB	1
3	150GB	4GB	2
4	8GB	4GB	2
5	10GB	1GB	4
6	10GB	1GB	1
7	4GB	1GB	1
TOTAL		**17GB**	**13**

The total configured memory and CPU cores were beyond the physically available. The physical available RAM were 8GB while CPU core were 12. Running both *vmstat* and *top* command every time a virtual machine were created to monitor the server health, produced results on Table 3.

Table 3. Server Health summary with 7 VMs

Σ Server	TOP (%CPU Utilization)		TOP (Load Average)			VMSTAT (Disk IO)	
	ID	WA	1	5	15	BI	BO
1	99.9	0	0.08	0.04	0.06	27	23
2	99.9	0	0.02	0.02	0.00	22	18
3	99.7	0	0.36	0.17	0.06	20	17
4	99.4	0	0.04	0.13	0.12	18	16
5	99.4	0	0.04	0.10	0.11	18	16
6	97.0	0.2	0.36	0.41	1.08	16	24
7	91.6	5.1	0.53	0.70	0.53	16	23

Saturation occurred when the virtual machine had over the limit of totally available cpu core. To avoid further errors to the server, the author had decided to stop the last virtual machine. From the survey results, it is considered no performance issues arise even when we run several virtual machines in one physical server.

5. Conclusion

The linux container applied in this research is truly helpful to the optimization of the physical server. One point should be noted that, even though it is true that guest machines could be configured beyond the physical resources limit on the host, it is better to keep the total configured virtual machine below the limit especially for the processor core. The fact that were found when this limit is ignored, had lead the host machine to suffer some of the saturations and IO waits that may leads to further errors if forced to run for a long period. To avoid this, the author(s) had decided to stop at the limit of 6 virtual machines configured and deployed.

References

[1] S. Aggarwal, M. Garg, and P. Kumar, "Green Computing is SMART COMPUTING – A Survey," *Int. J. Emerg. Technol. Adv. Eng.*, vol. 2, no. 2, pp. 297–303, 2012.
[2] W. Felter, A. Ferreira, R. Rajamony, and J. Rubio, "An Updated Performance Comparison of Virtual Machines and Linux Containers," Technology, vol. 25482, pp. 171–172, 2014.
[3] A. Jha, Agile in the IT World, KMPG, 2011.
[4] The Standish Group, "Chaos Manifesto," The Standish Group, 2012.
[5] Versionone.com, "The 10th Annual State of Agile Report," VersionOne.com, 2016.
[6] C. H. Frank, "Understanding agile project management methods using Scrum," International digital library, 2010.
[7] L. Rich C., "The Success Factors of Running Scrum: A Qualitative Perspective," Journal of Software Engineering and Applications, 2012.
[8] K. Schwaber and J. Sutherland, "Scrum Guide," 2014.
[9] L. Rich C., "The Success Factors of Running Scrum: A Qualitative Perspective," Journal of Software Engineering and Applications, 2012.
[10] W. G. Barrios and C. T. Zarrabeitia, "SCRUM: Application Experience in a Software Development PyME in the NEA," JCS&T, p. Vol. 12 No. 3, 2012.
[11] S. Aljaž, " Agile Project Management – a Future Approach to the Management of Projects?," Dynamic Relationships Management Journal, 2013.
[12] B. Gregg, Systems Performance - Enterprise and the Cloud. Prentice Hall, 2014.

IOP Conf. Series: Materials Science and Engineering **190** (2017) 012013 doi:10.1088/1757-899X/190/1/012013

Comparison of Various Similarity Measures for Average Image Hash in Mobile Phone Application

Sam Farisa Chaerul Haviana[1] **and Muhammad Taufik**[2]

[1, 2] Sultan Agung Islamic University, Semarang, Central Java, Indonesia

[1] samfch@unissula.ac.id, [2] mtaufik@unissula.ac.id

Abstract. One of the main issue in Content Based Image Retrieval (CIBR) is similarity measures for resulting image hashes. The main key challenge is to find the most benefits distance or similarity measures for calculating the similarity in term of speed and computing costs, specially under limited computing capabilities device like mobile phone. This study we utilize twelve most common and popular distance or similarity measures technique implemented in mobile phone application, to be compared and studied. The results show that all similarity measures implemented in this study was perform equally under mobile phone application. This gives more possibilities for method combinations to be implemented for image retrieval.

1. Introduction

One of the key issue in Content Based Image Retrieval (CIBR) is similarity measurement. After extracting image feature like image hash, next step is to calculate the distance or similarity of this hash with the others. Method like Average Hashing has consistent results in building the hashes. Those hashes can be represented as binary or hexadecimal. In its applications, those hashes could also be represented as string or character. This give possibilities to use more distance and similarity measurement method to find similarities between hashes. Numeric similarity measurement like Euclidean Distance, Manhattan Distance and Cosine Similarity or string similarity method like Damerau-Levenshtein are possible to be used as similarity measure in application. For binary representation or Boolean representation, method like Hamming distance is also suitable for this. Method like Q-gram [1], N-gram [2], Longest Common Subsequence (LCSS) [3] and Dynamic Time Warping (DTW) [4] were also give more possibilities for more method combination usage in application.

For certain purposes many distance or similarity measurement methods have been studied in terms of performance an accuracy. Like shown in [5], that studying some similarity measures for recommender system using Apache Mahout, Euclidean Distance Measure consistently performs well and produces better quality results compared to other similarity measures [5]. Another comparison has been studied in [6] that implement and evaluate 9 different similarity measures and their variant. The results, for small time series data DTW and LCSS were significantly more accurate than Euclidean [6]. Those studies was give the idea for this study to do more comparison of similarity measures, as similarity measures has grown into larger variations. Another purpose is to get more obvious implementation view of those similarity measures under limited computing capabilities device like mobile phone.

2. Implementation and Testing

For the purpose of this study, we design implementation and testing scenario to evaluate similarity measures. We compare each similarity measure in term of speed and memory cost. We rule out accuracy as accuracy will strongly related to the method to build hash that we utilize. Average hashing method has been chosen for this study as this method was one of the simplest method for building the image hashes. As it the one of the simplest method, the cost and speed of it will not take majority resources when application executed. By combining average hashing and similarity measures we try to achieve more obvious similarity measures performance in application.

For testing purpose we built an Android based application for perceptual image similarity. This application runs under testing device with specification of 2Gb of RAMs, Cortex-A7 Quad-core 1.6 GHz CPU, 13 MP primary back camera and 6.0.1 Marshmallow Android OS. This specification was common specification for mobile phone released the last few years. We use three category of images dataset. First category was online images we get from http://www.gettyimages.com and http://www.kapanlagi.com. We also tests the application using images found under DCIM folder as second category image dataset and images found under WhatsApp Media folder as the third category dataset. This categories was intended to test the application for various numbers and sizes of images. This three category were also considered as commonly used images in real usage of mobile phone. Each dataset category has different number of images. Online image category has 42 images with average size of image 145kb, image category under DCIM folder has 76 images with larger average size 2.656k (2.6Mb), taken with 13MP main device camera. The last category is WhatsApp Media images, has 721 images with average size of image is 132kb. We test by choosing a random query image then run the application for hashing and similarity measures function for each category. The results were recorded an analyzed. All similarity measures used in this study will be tested under the same scenario. Figure 1 show the flow of application for testing purpose.

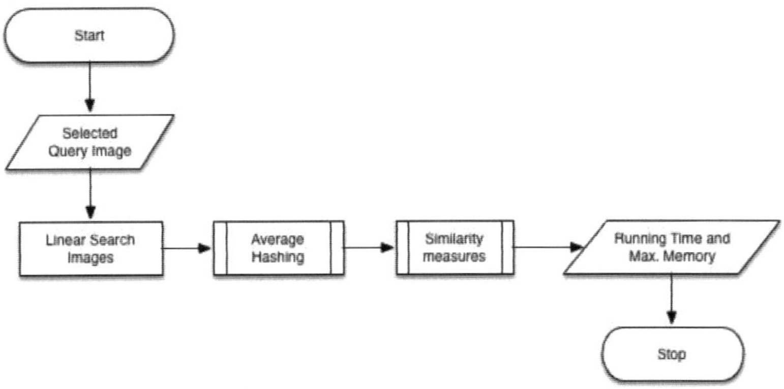

Figure 1. Application Flow for Testing

As shown in Figure 1, the process to find images was linear search the images under the specified path of image category folder. This means that we process images one by one for every image found. The process includes Average Hashing for building the hash of query image and every image found, then calculate the similarity of the hashes using specified similarity measures method. For every complete applications flows executed will output the running time and maximum allocated memory of single similarity measure. For every similarity measures method the application will be executed for five times then calculate the averages of running time and maximum allocated memory. The results will be analyzed for conclusion.

3. Results and Analysis

The first result we want to achieve is the comparison of running time of application for each similarity measures method. Figure 2 show the comparison of application running time in average for five times of testing for each similarity measures.

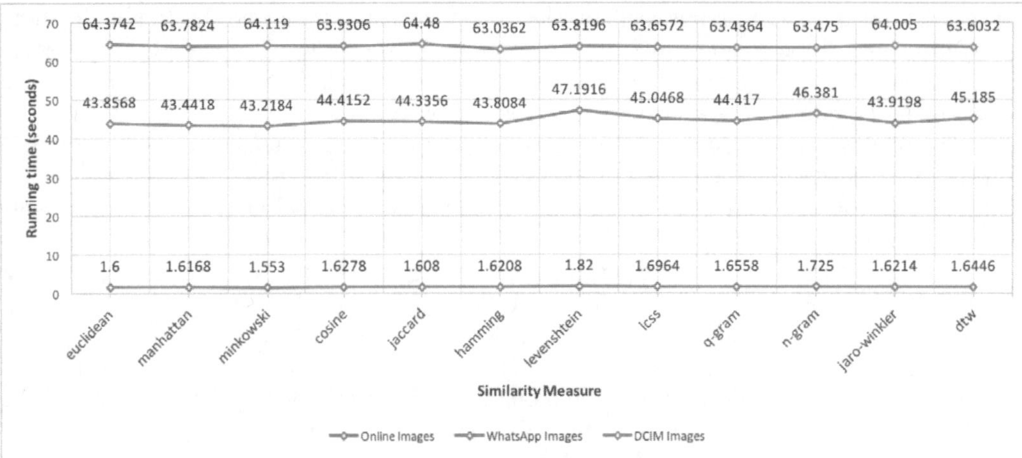

Figure 2. Running Time Comparison of Similarity Measures

From Figure 2 we got application running time for all similarity measures in each category were almost the same. This mean that in terms of speed, all similarity measures have similar performance under mobile application implementation. Although there is very slight differences running time in all similarity measures, but in general the running time is equal. The very significant difference is running time between categories. Category with smallest number and average size of images, Online image category has the fastest running time. In more various size and number of images like in WhatsApp images, the running time increased significantly. In largest size images category, DCIM images, running time was the highest. This show us that running time was increased in line with the size and number of images. The very dominant factor affecting running time is the Average hashing process itself. Since larger images need more time creating the hash. More number of images hash also affecting running time. From this results we can conclude that there is no fastest or slowest similarity measures method under mobile application implementation.

Our second test was to get maximum allocated memory for every similarity measures running under mobile application. Since Android has a very complicated memory management we are unable to get detailed results of allocated memory from each similarity measures implemented. What we have obtain was only average of maximum allocated memory for all similarity measures in each category. This number of allocated memory was affected by another android process like UI thread and other processes. So the number does not represent allocated memory only for each similarity measures process. Online image category has average of maximum allocated memory of 16.225 Mb. DCIM images was 111.51 Mb and WhatsApp image was 61.51 Mb. Since this results did not represent the allocated memory for similarity process only, we did not conclude any particular conclusion related to similarity measures memory cost.

4. Conclusion

Under mobile phone application, all similarity measures methods perform equally in terms of speed. Running time has no significant differences in all methods, although there is a very slight differences, but in general they are able to perform equally. Running time was very affected by the size and number of images processed, it increased in line with size and number of images. Memory usage for each running process of similarity measures under Android mobile phone was difficult to analyze, but from the tests we did not found any out of memory problem when running all methods.

IOP Conf. Series: Materials Science and Engineering **190** (2017) 012013 doi:10.1088/1757-899X/190/1/012013

References

[1] E. Ukkonen, "Approximate string-matching with q-grams and maximal matches," *Theor. Comput. Sci.*, vol. 92, no. 1, pp. 191–211, Jan. 1992.

[2] G. Kondrak, "N-Gram Similarity and Distance," *SPIRE 2005*, pp. 115–126, 2005.

[3] W. J. Hsu and M. W. Du, "Computing a longest common subsequence for a set of strings," *Computing*, vol. 24, no. March 1983, 1984.

[4] E. Keogh and C. A. Ratanamahatana, "Exact indexing of dynamic time warping Eamonn," *Knowl. Inf. Syst.*, vol. 7, pp. 358–386, 2005.

[5] S. Bagchi, "Performance and Quality Assessment of Similarity Measures in Collaborative Filtering Using Mahout," *Procedia - Procedia Comput. Sci.*, vol. 50, pp. 229–234, 2015.

[6] X. Wang, A. Mueen, H. Ding, G. Trajcevski, P. Scheuermann, and E. Keogh, "Experimental comparison of representation methods and distance measures for time series data," *Data Min. Knowl. Discov.*, vol. 26, no. 2, pp. 275–309, 2013.

Real Time Monitoring System of Pollution Waste on Musi River Using Support Vector Machine (SVM) Method

Muhammad Fachrurrozi, Saparudin, Erwin

Informatics Engineering, Computer Science Faculty, University of Sriwijaya.

mfachrz@unsri.ac.id, saparudin@unsri.ac.id, erwin@unsri.ac.id

Abstract. Real-time Monitoring and early detection system which measures the quality standard of waste in Musi River, Palembang, Indonesia is a system for determining air and water pollution level. This system was designed in order to create an integrated monitoring system and provide real time information that can be read. It is designed to measure acidity and water turbidity polluted by industrial waste, as well as to show and provide conditional data integrated in one system. This system consists of inputting and processing the data, and giving output based on processed data. Turbidity, substances, and pH sensor is used as a detector that produce analog electrical direct current voltage (DC). Early detection system works by determining the value of the ammonia threshold, acidity, and turbidity level of water in Musi River. The results is then presented based on the level group pollution by the Support Vector Machine classification method.

1. Introduction

Industrial development and construction resulted in an increasing number of industrial pollution on the environment. Industrial pollutions consist of waste in the form of water, gas and solid. In general, this waste is harmful because most of its components composed of addictive and its chemicals are hard to be degraded [1]. These substances have a bad impact on the environment and threaten the survival of living beings, according to Environmental Protection Agency. Several developed countries had already developed technologies to detect levels of waste in water, but because these are relatively expensive to adopt in Indonesia, it should be developed independently in the country in order to save expenses on imported goods.

Musi River flows in the center of Palembang, Indonesia, and is used for sort of needs, such as transportation, industry and household. Everyday people use its water for bathing, washing, and toilet facilities either directly from the river or trough Water Management Company of Palembang (PAM), as well as exploiting the products such as fish and other biota. On the other hand, some large companies and small companies are located around Musi River. Unfortunately, one of the routine is dumping waste into the river. Therefore, a monitoring and classification of water contamination system is needed around the Musi River. This system works in real time, so it can quickly minimize the negative impact of pollution. Water testing quality is required to determine whether the water is polluted or not, from the deviation limits of pollution. Some factors that affect water quality is the degree of acidity (pH), turbidity, and ammonia substance (NH_3).

Support Vector Machine (SVM) method applies hypothetical space on machine learning in the form of linear function in a high dimensional feature space. The learning technique based on optimization theory with learning bias implementation derived from statistical learning theory [2] [3]. This method is used to classify the level of water pollution based on the parameters of the factors affecting the level of water pollution.

2. Research Problems

The main problem of this research is the absence of public real-time monitoring of water pollution in the Musi river basin. Moreover, it is important to measure the result of the application of SVM method to classify the level of water pollution of the Musi river based on factors that affect water quality.

3. Research Objectives

This research aims to develop a real-time monitoring system of water pollution in Musi River and to know the accuracy of monitoring of water pollution level classification using Support Vector Machine (SVM).

4. Literature Review

4.1. Support Vector Machine (SVM)

The aim of learning process in SVM is to get a hypothesis in the form of best hyperplane which not only minimizes the empirical risk which is the average error on the training data, but also has a good Generalization [4]. Generalization is the ability of a hypothesis to classify data that is not contained in the training data correctly. SVM is a machine learning system that works on the principle of Structural Risk Minimization (SRM).

Figure 1 describes that SVM aims to find the best hyperplane that separates two classes in the input space. The best hyperplane separation between the two classes can be found by measuring the margin hyperplane and looking for the maximum points. The margin is the distance between the hyperplane to the nearest pattern of each class. The closest pattern is called a support vector. The solid line in Figure 1 shows the best hyperplane that is located right in the middle of both classes, while the red dots and yellow are in a black circle is a support vector. Attempts to locate hyperplane is the core of the learning process on SVM.

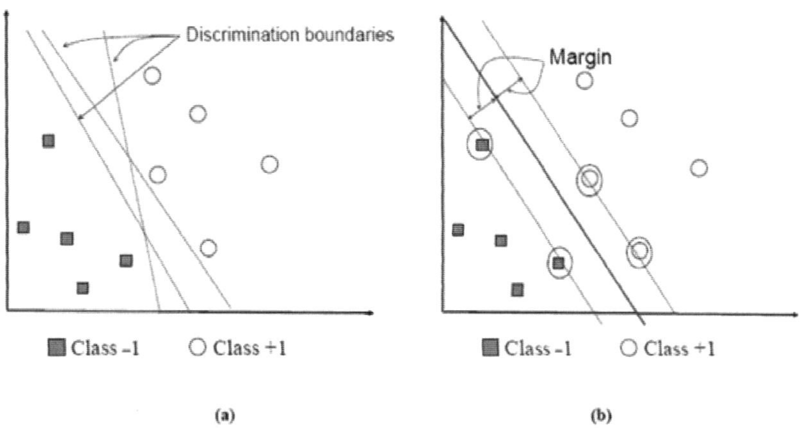

Figure 1. Class -1 and class +1 separated by Hyperplane. [5]

The provided data denoted into equation 1.

$$\vec{x_i} \in R^d \quad (1)$$

Whereas each label is denoted $Y i \in \{-1, +1\}$ for i = 1, 2, L, l, where l is the number of data. The both classes are assumed to -1 and +1 that can be completely separated by hyperplane with d dimension, which is defined in equation 2.

$$\vec{w}.\vec{x} + b = 0 \tag{2}$$

Pattern Xi which belongs to class -1 (negative samples) can be formulated as a pattern that equals to the equation 3.

$$\vec{w}.\vec{x_i} + b \leq -1 \tag{3}$$

Meanwhile x_i pattern that includes class +1 (positive samples) can be formulated as a pattern that satisfies the equation 4.

$$\vec{w}.\vec{x_i} + b \geq +1 \tag{4}$$

The maximum margin can be found by maximizing the value of the distance between the hyperplane and the closest point. It can be formulated as a Quadratic Programming (QP) problem, i.e. finding the minimum point of the equation (5), by observing the constraints in equation (6).

$$\min_{\vec{w}} \tau(w) = \frac{1}{2}\|\vec{w}\|^2 \tag{5}$$

$$y_i(\vec{x}_i.\vec{w} + b) - 1 \geq 0, \forall i \tag{6}$$

This problem can be solved with a variety of computational techniques, including Lagrange Multiplier.

$$L(\vec{w}, b, \alpha) = \frac{1}{2}\|\vec{w}\|^2 - \sum_{i=0}^{n} \alpha_i \left(y_i\left((\vec{x}_i.\vec{w} + b) - 1\right)\right) \tag{7}$$

$$(i = 1.2 \ldots, l) \tag{8}$$

α_i is Lagrange multipliers with value zero or positive ($\alpha i \geq 0$). The optimum value from equation 7, can be calculated with minimizing L against w and b, a maximize L against αi.
By assuming a characteristic that in the optimum gradient point L = 0, equation 7 can be modified as a maximizing problem that only contains αi which expressed in Equation 8.

Maximizing

$$\sum_{i=l}^{l} \alpha_i - \frac{1}{2}\sum_{i,j=l}^{l} \alpha_i\alpha_j y_i y_j \vec{x}_i.\vec{x}_j \tag{9}$$

With conditions (constraints),

$$\alpha_i \geq 0 \ (i = 1{,}2 \ldots, l) \ \sum_{i=l}^{l} \alpha_i y_i = 0 \tag{10}$$

The result of this calculation obtained some value of αi, most of them have a positive value. Data which correlated with the positive value of αi called support vector.

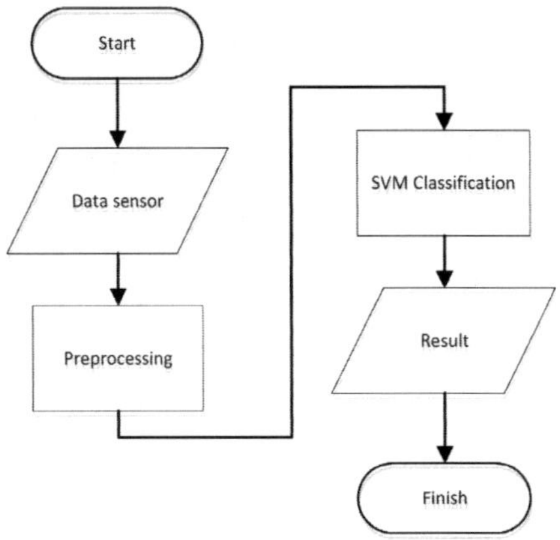

Figure 2. System Flowchart

5. Result and Discussion

Figure 2 shows a flowchart that illustrates the process of the study. Data is obtained from 3 sensors in real time, then send them to the system via GPRS, then the system display the classification result. The object research place was taken on 3 locations, they are Kalidoni, Sei Gerong, and Seberang Ulu 1 (SU 1).

Training Process

SVM method usually use training data. But in this research training process was skipped and replaced by the threshold based on Table 1.

Table 1. Parameter Threshold.

Parameter	Unit	Maximum Value
PH	Mg/l	6 - 9
NH3	Mg/l	0,5
Waste Water	Mg/l	20

Table 2. Measurement result: Seberang Ulu I

Date	**pH**	**Amonisk (%)**	**Keruh (NTU)**	**Location**	**State**
30-07-2016	4.82	3	8.48	Sebarang Ulu I	Bad
30-07-2016	4.94	2	9.02	Sebarang Ulu I	Bad
30-07-2016	5.94	2	8.44	Sebarang Ulu I	Bad
30-07-2016	5.06	2	7.09	Sebarang Ulu I	Bad
30-07-2016	4.82	1	5.68	Sebarang Ulu I	Bad
30-07-2016	4.94	2	5.98	Sebarang Ulu I	Bad
30-07-2016	4.94	2	5.92	Sebarang Ulu I	Bad
30-07-2016	5.06	2	7.27	Sebarang Ulu I	Bad
30-07-2016	5.06	2	8.19	Sebarang Ulu I	Bad
30-07-2016	4.82	2	8.95	Sebarang Ulu I	Bad
30-07-2016	5.06	2	9.58	Sebarang Ulu I	Bad
30-07-2016	5.06	3	9.73	Sebarang Ulu I	Bad
30-07-2016	5.18	3	9.79	Sebarang Ulu I	Bad

Date	pH	Amonisk (%)	Keruh (NTU)	Location	State
30-07-2016	4.94	2	8.81	Sebarang Ulu I	Bad
30-07-2016	4.82	3	8.75	Sebarang Ulu I	Bad
30-07-2016	5.06	3	8.46	Sebarang Ulu I	Bad
30-07-2016	5.06	3	9.19	Sebarang Ulu I	Bad
30-07-2016	4.94	2	8.65	Sebarang Ulu I	Bad
30-07-2016	4.94	2	8.63	Sebarang Ulu I	Bad
30-07-2016	4.94	2	7.91	Sebarang Ulu I	Bad
30-07-2016	4.94	3	7.41	Sebarang Ulu I	Bad
30-07-2016	4.94	3	7.65	Sebarang Ulu I	Bad
30-07-2016	4.94	3	7.63	Sebarang Ulu I	Bad
30-07-2016	5.06	2	7.25	Sebarang Ulu I	Bad
30-07-2016	5.06	2	6.65	Sebarang Ulu I	Bad
30-07-2016	5.06	3	6.54	Sebarang Ulu I	Bad
30-07-2016	4.94	2	5.34	Sebarang Ulu I	Bad
30-07-2016	5.06	2	4.73	Sebarang Ulu I	Bad
30-07-2016	5.06	3	5.70	Sebarang Ulu I	Bad
30-07-2016	4.94	2	6.95	Sebarang Ulu I	Bad
30-07-2016	4.94	3	4.94	Sebarang Ulu I	Bad
30-07-2016	4.94	2	8.89	Sebarang Ulu I	Bad
30-07-2016	4.94	1	8.82	Sebarang Ulu I	Bad
30-07-2016	4.82	1	8.53	Sebarang Ulu I	Bad
30-07-2016	4.70	1	8.22	Sebarang Ulu I	Bad
30-07-2016	4.46	2	8.13	Sebarang Ulu I	Bad
30-07-2016	4.22	2	8.03	Sebarang Ulu I	Bad
30-07-2016	4.10	2	8.48	Sebarang Ulu I	Bad
30-07-2016	4.10	2	4.01	Sebarang Ulu I	Bad
30-07-2016	4.58	4	4	Sebarang Ulu I	Bad
30-07-2016	4.58	1	4.87	Sebarang Ulu I	Bad
30-07-2016	5.06	2	4.03	Sebarang Ulu I	Bad
30-07-2016	5.30	3	3.97	Sebarang Ulu I	Bad
30-07-2016	5.42	2	3.94	Sebarang Ulu I	Bad
30-07-2016	5.30	3	3.94	Sebarang Ulu I	Bad
30-07-2016	5.30	2	3.96	Sebarang Ulu I	Bad
30-07-2016	5.42	2	3.931	Sebarang Ulu I	Bad
30-07-2016	5.42	2	3.946	Sebarang Ulu I	Bad
30-07-2016	5.42	2	3.931	Sebarang Ulu I	Bad
30-07-2016	5.42	2	4.011	Sebarang Ulu I	Bad
29-07-2016	9.26	8	7.57	Kalidoni	Bad
29-07-2016	9.14	8	7.549	Kalidoni	Bad
29-07-2016	9.14	6	6.492	Kalidoni	Bad
29-07-2016	9.4	10	6.072	Kalidoni	Bad
29-07-2016	9.02	8	5.561	Kalidoni	Bad
29-07-2016	9.14	8	5.078	Kalidoni	Bad
29-07-2016	9.14	9	6.086	Kalidoni	Bad
29-07-2016	9.14	8	6.219	Kalidoni	Bad
29-07-2016	9.14	9	6.73	Kalidoni	Bad
29-07-2016	9.02	8	7.18	Kalidoni	Bad
29-07-2016	9.02	8	6.25	Kalidoni	Bad
29-07-2016	9.02	7	5.91	Kalidoni	Bad

Date	pH	Amonisk (%)	Keruh (NTU)	Location	State
29-07-2016	9.02	9	5.82	Kalidoni	Bad
29-07-2016	9.02	10	5.97	Kalidoni	Bad
29-07-2016	9.02	6	6.20	Kalidoni	Bad
29-07-2016	9.02	6	6.41	Kalidoni	Bad
29-07-2016	8.90	6	6.37	Kalidoni	Good
29-07-2016	8.90	7	6.21	Kalidoni	Good
29-07-2016	8.78	7	6.77	Kalidoni	Good
29-07-2016	8.90	8	8.45	Kalidoni	Good
27-07-2016	8.78	6	8.49	Sei Gerong	Good
27-07-2016	8.66	7	8.52	Sei Gerong	Good
27-07-2016	8.30	5	8.16	Sei Gerong	Good
27-07-2016	8.30	6	8.00	Sei Gerong	Good
27-07-2016	7.94	5	7.73	Sei Gerong	Good
27-07-2016	7.94	4	7.69	Sei Gerong	Good
27-07-2016	7.70	4	7.07	Sei Gerong	Good
27-07-2016	7.58	4	7.24	Sei Gerong	Good
27-07-2016	5.90	5	5.92	Sei Gerong	Bad
27-07-2016	6.02	6	6.64	Sei Gerong	Good
27-07-2016	6.02	6	6.77	Sei Gerong	Good
27-07-2016	5.90	6	6.81	Sei Gerong	Good
27-07-2016	6.38	13	6.67	Sei Gerong	Good
27-07-2016	6.38	2	6.42	Sei Gerong	Good
27-07-2016	5.90	4	6.51	Sei Gerong	Bad
27-07-2016	5.90	4	6.47	Sei Gerong	Bad
27-07-2016	5.90	5	6.37	Sei Gerong	Bad
27-07-2016	5.90	6	11.81	Sei Gerong	Bad
27-07-2016	5.78	5	7.38	Sei Gerong	Bad
27-07-2016	5.78	5	7.40	Sei Gerong	Bad

Based on the threshold given in Table 1 and Table 2, from 150 samples, SU 1 is the highest level of pollutions as shown in Figure 3.

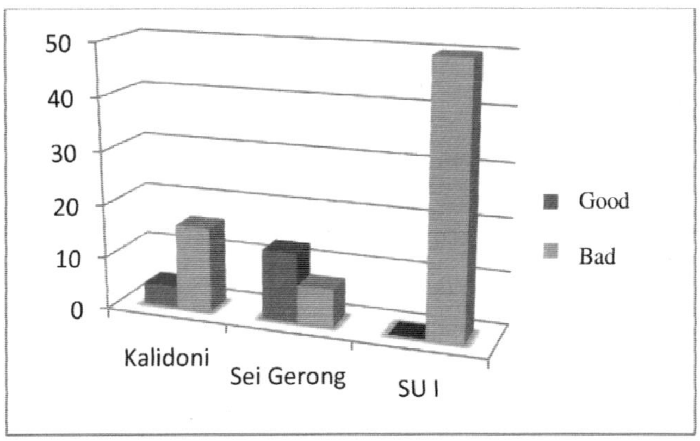

Figure 3. Measurement Results.

6. Conclusion

Based on the research experiments and analysis, it can be conducted the conclusions below:

1. The SVM classification shows good result, but lack justification value. This is because the assessment involves the threshold value only.
2. Seberang Ulu (SU) 1 has significant impact due to the waste pollution.

Acknowledgments

This research was supported by University of Sriwijaya. We thank our colleagues from Computer Science Faculty who provided insight and expertise that greatly assisted the research, although they may not agree with all of the interpretations/conclusions of this paper.

7. References

[1] P. J. Nkosi Nhlanhla, Muzenda Edison, Zvimba John, "The Waste tyre problem in South Africa: An analysis of the REDISA Plan," *Int. Conf. Chem. Environ. Eng.*, no. i, 2013.

[2] J. Shawe-Taylor and N. Cristianini, *Kernel Methods for Pattern Analysis*, vol. 47. 2004.

[3] M. A. Ahmadi and M. Pournik, "A Predictive Model of Chemical Flooding for Enhanced Oil Recovery Purposes: Application of Least Square Support Vector Machine," *Petroleum*, vol. 2, no. 2, pp. 177–182, 2015.

[4] M. Malvoni, M. G. De Giorgi, and P. M. Congedo, "Data on Support Vector Machines (SVM) model to forecast photovoltaic power," *Data Br.*, vol. 9, pp. 13–16, 2016.

[5] N. Cristianini and M. H. Hahn, *Introduction to Computational Genomics - A Case Studies Approach*. Cambridge University Press, 2006.

IOP Conf. Series: Materials Science and Engineering **190** (2017) 012015 doi:10.1088/1757-899X/190/1/012015

Fast Learning for Big Data Using Dynamic Function

T Alwajeeh[1], **A F Alharthi**[3], **R F Rahmat**[2], **R Budiarto**[3]

[1]Dept. of Computer Science & Engineering, College of CS&IT, Albaha University, Albaha P.O. Box 1988, Saudi Arabia
[2]Department of Information Technology, Faculty of Computer Science and Information Technology, University of Sumatera Utara, Medan, Indonesia
[3]Dept. of Computer Information System, College of CS&IT, Albaha University, Albaha P.O. Box 1988, Saudi Arabia

taa.2000@hotmail.com, afalharthi@bu.edu.sa, romi.fadillah@usu.ac.id, rahmat@bu.edu.sa

Abstract. This paper discusses an approach for fast learning in big data. The proposed approach combines momentum factor and training rate, where the momentum is a dynamic function of the training rate in order to avoid overshoot weight to speed up training time of the back propagation neural network engine. The two factors are adjusted dinamically to assure the fast convergence of the training process. Experiments on 2-bit XOR parity problem were conducted using Matlab and a sigmoid function. Experiments results show that the proposed approach signifcantly performs better compare to the standard back propagation neural network in terms of training time. Both, the maximum training time and the minimum training time are significantly faster than the standard algorithm at error threshold of 10^{-5}.

1. Introduction

Recently, we are entering the era of "big-data", and as the development of high-speed signal processing, fast and efficient learning and signal representation is becoming an emergent research topic. Extreme learning machine (ELM) [1] is one of the leading trends for fast learning. Unlike the other traditional learning algorithms, for example, Back Propagation-based neural networks, or support vector machine (SVM)], the parameters of hidden layers of ELM are randomly established and need not be tuned, thus the training of hidden nodes can be established before the inputs are acquired.

Feedforward neural networks have been widely used in various areas of machine learning. Hidden nodes in a neural network architecture work as universal approximation provided that all the parameters of the networks are adjustable. The most representative training method for Artificial Neural Networks is back propagation (BP) algorithm. BP calculates the gradient of a loss function with respect to all the weights in the network and updates the weights for minimizing the loss function. Nevertheless, the parameter tuning of BP-based neural networks is usually time consuming and cannot handle the overfitting problem.

2. Related Works

Research on fast learning started from back-propagation algorithm, introduced by Werbos [2], and later popularized by Rumelhart et al. [3], which calculates the error function based on the weights of every data in the neural network, and then updates the weight with a new value, based on the activation function, in order to minimize the value of error function. However, back-propagation is known for the slow computing time, also inefficient for maintaining data in big size [4].

Researches have been done to improve learning time in neural network. Chandra and Sharma [5] introduces parameterized multilayer perceptron to process big data, with trigonometric functions.They also proposed parameterized deep neural network to reduce time usage of the learning process, by applying periodic function to parameterize the neural network weights [6]. Wong and Xu proposed hierarchical fast learning artificial neural network, which process data using feature compression based on canonical covariance [7]. Hinton and Teh [8] proposed a learning algorithm to improve learning speed of deep belief nets. Another strategy to improve learning time also proposed by Pasha [9] and Rahmat [10] by using distributed adaptive nested neural network.

3. Proposed Design

This paper uses standard back propagation artificial neural network (BP-ANN) as illustrated inf Figure 1. The input layer has *i* neurons. The hidden layer consists of four neurons with two weights, while the output layer has one neurons with one weights. A sigmoid function is used as activation function.

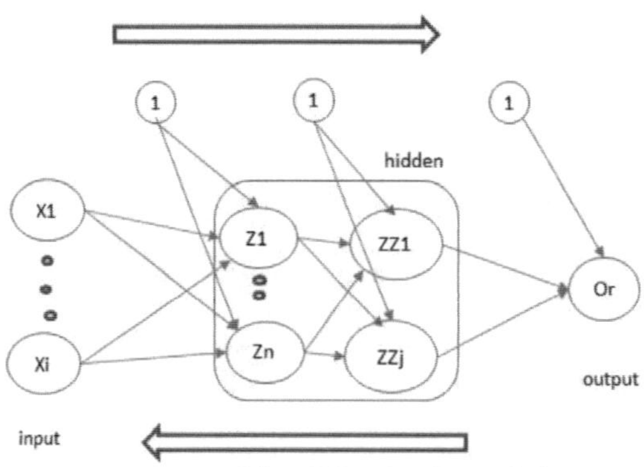

Figure 1. The BP-ANN model.

3.1. Momentum factor and traninig rate

Computing weights changes in neuron *k* of output layer and neuron *j* of hidden in BP-ANN involves two parameters; momentum factor (α) and training rate η. The larger value of η the faster the ANN converges. Usually, the value for η is randomly chosen by try and error from the value between 0 and 1. However, too big value of η may lead to oscillated training curve. The value of momentum factor α is also chosen randomly between 0 and 1. New weight adjustment Δw_{jk} at time t for each neuron *j* of hidden layer and neuron *k* of output layer is defined in equation (1).

$$\Delta w_{jk}(t+1) = w_{jk}(t) - \eta \frac{\partial E}{\partial w_{ik}(t)} + \alpha \Delta w_{jk}(t-1) \tag{1}$$

One important factor to speed up BP algorithm is the monotonicity of the error function during training for every epoch or iteration [11]. This paper uses an exponential function of error *E* for the dynamic training rate as defined in equation (2).

$$\eta_{DR}(E) = 1 + e^{(1+\sin(2E))} \tag{2}$$

Furthermore, in order to avoid overshoot weight in applying the training rate and momentum term as a dynamic function, a new approach is proposed by defining a relationship between the dynamic training rate and the dynamic momentum factor [12]. The momentum factor α_{DM} as an explicit function of training rate η_{DR} is defined in (3).

$$\alpha_{DM} = \sin(x + \varepsilon) + \sin\frac{1}{\eta_{DR}} \tag{3}$$

where $x = E \times f'(O_r)$, $f'(O_r)$ is the first derivation of activation function f which is defined as $f'(O_r) = O_r(1 - O_r)$ and ε is an absolute value, $0 < \varepsilon < 1$. We have investigated the best value of ε is 0.73. Substituting η_{DR} in equation (2) into equation (3) yield a new dynamic momentum factor as shown in equation (4).

$$\alpha_{DM} = [\,\sin(E(O_r\,(1 - O_r)) + \varepsilon) + \sin(\frac{1}{1 + e^{(1+\sin(2E))}})] \tag{4}$$

This approach maintains the weights are small as possible. Meanwhile, the momentum factor α and he training rate standard back propagation, SBP, manually train the training rate and momentum, where η and α are in the range of [0, 1].

3.2. The Proposed Training algorithm

The training algorithm of the proposed approach is given in Figure 2.

Step-1: Initialize randomly the weights
Step-2: Input: number of the neuron, hidden layer, the patterns, error threshold $E = 10^{-5}$
Step-3: While (*MSE>E*) do step 5-16
Step-4: For each training pair do step 6-15
Forward Propagation
Step-5: Compute input layer of hidden layer *Z*
Step-6: Compute input layer of hidden layer *ZZ*
Step-7: Compute input layer of output layer O_r and output value
Back Propagation
Step -8: Compute the error training
Step -9: Compute the error signal δ_r of the ANN
Step10: Compute the weight changes for each Δw_{jr} and bias Δw_{0r}
Step11: At zz_j compute the error signal δ_{-inj} and local gradient of error signal δ_j, using δ_r
Step12: Compute the weight correction for each Δv_{hj} and the bias Δv_{0j}
Step13: At z_h compute the error signal δ_{-inh} and local gradient of error signal δ_{η} using δ_r
Step14: At layer z_h compute the weight correction for each Δu_{ih}, and bias Δv_{0h}
Step15: Update the value for each output layer O_r, hidden layer zz_j and z_h
Step16: Compute the Mean Square Error: $MSE = 0.5 \times \frac{1}{P}\sum_{p=1}^{n}\sum_{k}^{i}(t_{kp} - o_{kp})$,
where P_i number of pattern for every epoch.

Figure 2. The training algorithm.

4. Experimental Results and Analysis

In this section we report the results obtained when experimenting our proposed method with 2-bits XOR Parity Problem. We use Matlab software running on Windows 8 machine with Intel Core i7 processor. The weights were generated randomly between [-0.4, 0.6] using equations (1-4). The obtained weights for *w1*, *w2*, *b1*, *b2* of hidden layer and *w3*, *b3* of output layer are:
w_1=[-0.3883 0.4175; -0.0872 0.2232]; w_2=[0.4575 -0.4463 ;0.1567 0.3496]; b_1= [0.2566; 0.2429]; b_2=[0.1101 0.1566] ; w_3=[0.4341 ;0.2878]; b_3= [- 0.3266].

The experiments were run 10 times for each value of error threshold and the average value is taken. The best results is shown in Table 1.

Table 1. The Best Result

Average Time (sec)	Average MSE	Average Epoch
0.4598	8.77E-06	269

Comparison to the standard BP training algorithm is depicted in Table 2. The value of α and η are varied between 0 and 1.

Table 2. The back progagation training algorithm results.

Time (sec)	MSE	Epoch	Value of η	Value of α
588.6040	1.0000e-05	551105	0.1	0.1
279.0730	1.0000e-05	269082	0.2	0.2
154.2040	1.0000e-05	150827	0.3	0.3
97.0920	1.0000e-05	91009	0.4	0.4
63.9200	9.9999e-06	59658	0.5	0.5
15.1220	9.9998e-06	11911	0.6	0.6
3590	0.0657	1501828	0.7	0.7
3590	0.0661	3250451	0.8	0.8
57.0780	9.9999e-06	52445	0.9	0.4
33.9220	1.0000e-05	23760	1	0.5
27.6790	9.9998e-06	28389	0.8	0.5
56.6350	9.9998e-06	50429	0.8	0.4
32.0330	1.0000e-05	23765	1	0.5

The best time for the standard BP is 15.1220 seconds, which means that the proposed training algorithm is 30 times faster than the standard BP training algorithm.

Furthermore, the proposed training algorithm converges fast to the global minimum adn provides the best results at ε = 0.75 as shown in Figure 3.

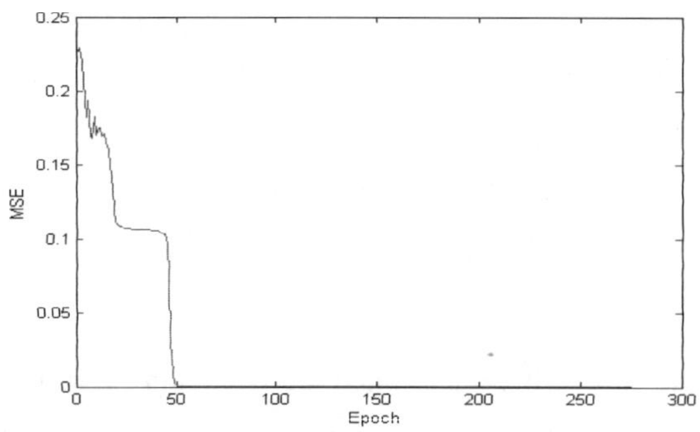

Figure 3. The convergence of the proposed training algorithm.

The proposed algorithm performs much better compare to the standard BP due the fact that the weights are automatically adjusted for every epoch in every layers through the dynamic dynamic momentum factor α_{DM} as defined in equation (4) as well as training rate η_{DR} as defined in equation (2). The use of implicit momentum function in the training rate in equation (3) makes the proposed algorithm able to control the growth of the weights. The dynamic momentum factor and training rate affect the weight for each hidden layer and output layer and eliminate the saturation training in the proposed algorithm. In addition, it uses an initial weight from interval [-0.4, 0.6] that narrows the search space compare to the standard Back Propagation.

5. Conclusion

An approach for fast learning on big data is proposed. The approach introduced a training algorithm to speed up the training time of Back Propagation neural network. The proposed algorithm considers a incorporates a dynamic function for auto-adjust two parameters: the momentum factor and the training rate. The proposed dynamic function eliminates the saturation of training time in Back Propagation artifical neural networks. In future, the proposed training algorithm can be applied for analyzing big data which needs extreme learning algorithm.

.

Acknowledgment

The author would like to thank Albaha University for the research funding under research contract number 1437/02.

References

[1] Deng C W, Huang G B, Xu J 2015 *Science China Information Sciences* **58** 020301(16), doi: 10.1007/s11432-014-5269-3.

[2] Werbos P 1974 *Beyond regression: new tools for prediction and analysis in the behavioral sciences*, (Cambridge MA: Harvard University Press).

[3] Rumelhart D E, Hinton G E and Williams R J 1986 *Nature* 323, pp. 533-536.

[4] Huang G B, Deng C W, Xu J and Tang J X 2015 *Science China Information Sciences* **58** 020301 pp 1-16.

[5] Chandra B and Sharma R K 2014 *Proc. of IEEE International Conference on Big Data (Anchorage, Alaska, USA)* pp 17-22.

[6] Chandra B and Sharma R K 2016 *Neurocomputing* 171 pp. 1205-1215.

[7] Wong L P and Xu J 2006 *Proc. of Int. Joint Conf. on Neural Networks (Montreal, Canada)* pp

[8] Hinton G E and Teh Y W 2006 *Neural Computation* 18 pp 1527-1554.

[9] Pasha M F, Rahmat R F, Budiarto R, and Syukur M 2010 *Int. Journal of Innovative Computing, Information and Control* 6 pp 1005-1022.

[10] Rahmat R F, Pasha M F, Syukur M, Budiarto R 2015 *Int. Arab Journal of Information Technology* **12** 6 pp 532-539.

[11] Li J, Lian L and Jiang Y 2010 *3rd Int. Conf. on Information and Computing (Wuxi, China)*, 4 pp 292–295.

[12] Shao H and Zheng G 2011 *Neurocomputing* **74** 5 pp 749–752.

IOP Conf. Series: Materials Science and Engineering **190** (2017) 012016 doi:10.1088/1757-899X/190/1/012016

Designing Business System Model using System Modeling Approach to the Small and Medium Enterprises (SME) of Furniture in Indonesia

Irwan Sukendar[1], Wiwiek Fatmawati[2], Imam Much Ibnu Subroto[3], Rizki Arigama[4]

[1, 2]Industrial Engineering Department, Universitas Islam Sultan Agung Semarang
[3, 4]Electrical Engineering Department, Universitas Islam Sultan Agung Semarang

Email : irwan@unissula.ac.id

Abstract. This paper studies the design of business system model with System Modeling Approach on small and medium enterprises (SMEs) of furniture. Methods used consists of five phases: phase of identification of business processes actual on SMEs of Furniture, phase of identification of deficiencies and improvement of business processes, phase of design algorithm and flowchart business processes, phase of analysis of the elements of the system, and phase of the design of data flow diagram (DFD) , The results of the analysis of the elements of the system are: Products and quantities ordered product consumers and DP paid by consumers identified as elements of system inputs 1,2 and 3. The result of the calculation, payment slips and mail order (SO) are identified as elements of system output 1, 2 and 3. Acceptance of orders, stocks checking of raw materials at the warehouse, calculating raw material requirements, adequacy of raw materials, the price of the contract, and the due date, as well as the submission of the results of calculations to consumers were identified as elements of system components 1, 2, 3, and 4. Admin taking orders, Admin check stocks of raw materials at the warehouse, Admin making calculation, and Admin convey the results of calculations to consumers were identified as an element of interaction system 1, 2, 3, and 4. Consumers were identified as element of environmental systems. Moreover, the boundary between SMEs and consumers were identified as elements of the system boundary.

1. Introduction

According to the Law on SMEs / 2008, micro-enterprises are the efforts that have a net worth of more than Rp 50 million to Rp 500 million, excluding land and building and the Start annual net sales of Rp 300 million to Rp 2.5 billion. Moreover, small business has a net worth of more than Rp 50 million to Rp 500 million, excluding land and building and the Start annual net sales of Rp 300 million to Rp 2.5 billion. The medium-sized businesses have a net worth of more than Rp 500 million up to Rp 10 billion, excluding land and building and the Start annual net sales of Rp 2.5 billion up to Rp 50 billion.

Some of the existing business processes in SMEs in general are: receipt of orders from consumers, preparation of raw materials, production processes, delivery of goods, and report generation. Some characteristics that can be identified on the business process of SMEs are that they still use manual way in the overall business process. SMEs usually do not maximally use computer facilities and the use of information system for supporting business processes. As the result, the weaknesses and shortcomings include: manual data collection on stocks of raw materials written in the book affect data to be recorded in computerized database to be duplication of data, the amount of raw materials and number of products in the warehouse are not up-to-date and real time, production delay occurs, the data less accurate information and not integrated.

Weaknesses and shortcomings mentioned above hypothetically can be overcome with the use of computer and information system implementation of Enterprise Resource Planning (ERP). ERP is a major innovation in the field of management information systems and organizations that enhance the organization's ability to create a more integrated, consistent, and control information [1]. ERP is defined as a software solution that integrates information and business processes to allow the sharing of information between departments within an organization [2].

Research on ERP is growing rapidly as ERP fairly rapid growth worldwide. Research on ERP so far can be classified into several themes: the theme of the pre-ERP implementation, ERP software design theme, theme, ERP implementation and post-implementation theme. The case study or research objects ERP, among others: the company in general, large companies, SMEs, and educational organizations.

On the theme of the pre-implementation of ERP, the researchers discuss about business processes and Business Process Re-engineering (BPR). Two research on this theme can be found in [3] and [4] which were using a common approach. On the theme of ERP software design, some of the researches include: Gelogo and Kim [5] that designed the architecture of mobile ERP. On the theme of ERP implementation, the researchers discuss: procedures, phases and methodology, Critical Success Factors (CSF), user participation, factors and issues, management and governance, as well as cost analysis in ERP implementation. CSF is defined as factors that significantly improve project implementation [6] .

Studies that discusses the theme of the pre-implementation, namely: [6], [2], [7], [8], [9], [1], [10], [11], [12] and [13]. Refs. [8] and [13] discusses procedures, phases and implementation methodology of the study ERP. Both studies took a case study on the company in general. On the other hand, Refs. [6], [1], [14], and [13] discuss Critical Success Factors (CSF) implementation ERP. The first two research case studies were on large-scale enterprises, while the next two studies were case studies on company in general. Meanwhile, Ref. [9] discusses the role of user participation in the implementation of this ERP. [7] and [10] discusses the factors and issues that affect the implementation of ERP. The study is the first case study on the company in general, while the second study takes a case study on the Small and Medium Enterprise (SME).

The management and governance in the implementation of the first ERP have been elaborated by [1], [11] and [12]. The research is a case study on a large-scale enterprises, while two further research took a case on companies in general. The cost of implementing this ERP was analysed in [2]. Last theme is the theme of post-ERP implementation. Research on this theme was carried by [15] as case study on the company in general.

ERP in the international research so far can be classified in several themes, namely: pre-implementation, software design, implementation, and post-implementation. In the pre-implementation phase, the researchers discuss about business processes and business process re-engineering (BPR). BPR is a concept of process improvement in a dramatic approach. Keywords BPR is fundamental, radical, dramatic and processes. BPR must make fundamental changes to improve productivity and quality [4]. Several studies on the scope of the BPR mostly take general. Meanwhile, this paper conducts research within the scope of the BPR took the approach of system modelling. The modelling system is a simple representation of a real system by eliminating things that are not important, but retains the character of the system [16]. In order to improve and intensify research in the areas of ERP, this paper models to the study design of business systems with the System Modeling Approach on SMES of Furniture. Using System Modeling approaches in the design of BPR, this paper reports performance on ERP research.

IOP Conf. Series: Materials Science and Engineering **190** (2017) 012016 doi:10.1088/1757-899X/190/1/012016

2. Methods

This study is divided into five stages. Five phases of business processes identification can be explained as follows:

2.1. Phase of Identification of the actual business process IKM Furniture

At this stage, the actual business process IKM Furniture identified

2.2. *Phase of Identification of deficiencies and improvement of business processes (Business Process* Reengineering)

At this stage, deficiencies in the actual business processes are identified. Furthermore, the proposed improvement is provided.

2.3. Phase of Design Algorithm and process flowchart binis

At this stage, to design algorithms and flowcharts on the basis of improved business processes.

2.4. Phase of Analysis of the business system elements

The design method of business system model in this study using system modeling approach. Based on this approach, business processes will be analyzed for every element of the system. Therefore, at this stage, the business processes have been improved and designed in the form of algorithms and flowcharts, will further analyze the elements of his system. The elements of the system are:

2.4.1. input

Input is everything that comes from the environment to the system

2.4.2. output

Output is anything that comes out of the system towards the environment

2.4.3. Component

Components are all things that exist in the system and interact processing inputs into outputs

2.4.4. Interaction between components

Interaction is the relationship between components in the system in order to process the input into output

2.4.5. Environment.

Environment is everything that is outside the system. There environments provide input to the system, and there is also an environment that accepts the output of the system

2.4.6. Boundary

Boundary is the boundary between the system and the environment

2.5. Stage Design Data Flow Diagrams (DFD)

In order to prepare for the manufacture of computer program, then the next stage, the business system model will be designed in the form of Data Flow Diagrams (DFD)

3. Results

3.1. Phase of Identification of the actual business process IKM Furniture

The actual business process IKM Furniture can be outlined in the following:

First time customer will order to admin. Furthermore, the admin will check stock levels of raw materials in the warehouse and calculate the price. After that the admin will return to the office by indicating the amount of costs to be paid by consumers. If the consumer does not agree, the transaction will be stopped. Meanwhile, if the consumer agrees, the admin will make mail order as many as 3 copies of admin, owner, and production parts. After the production received a letter from the admin orders, it will soon be the production process in accordance with the order. Once in the form of finished products, the production department will make a note number 2 copies of finished products for admins and owners. Furthermore, the admin will send the products to the customer order.

The actual business process can be described as in Figure 1.

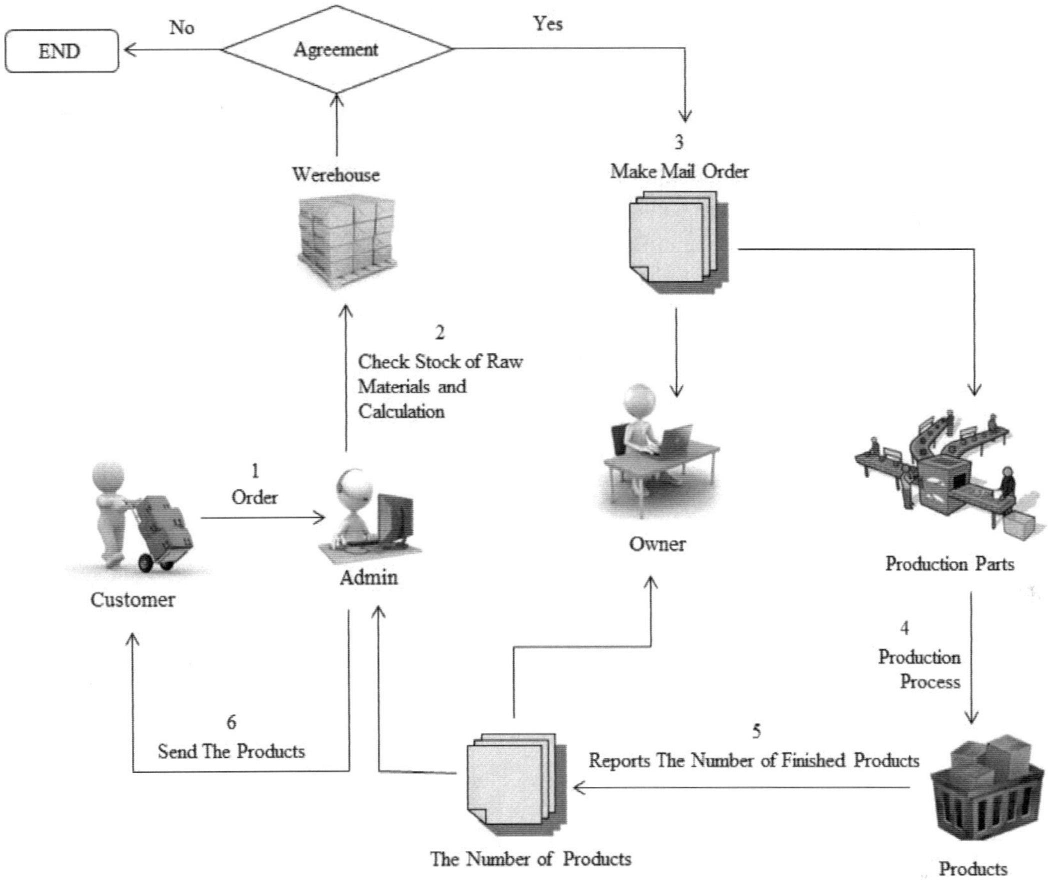

Figure 1. The actual business process in SME of Furniture

3.2. Phase of Identification of Lack and improvement of business processes
Shipping information occurs in the admin or customer orders are as follows:
Consumers come to the company to make a reservation. After that the admin has to check the stocks of raw materials in the warehouse and price calculation. This process takes as long as 20 minutes. If the consumer does not agree then the process will stop, whereas if the consumer agrees admin will convey the information to the owner. This process is performed for 5 minutes. Then the admin will make mail order as many as three sets that are for admin, owner, and production parts. The process of the making of the order carried out for 15 minutes. Thus, if the consumer agrees, then the time required is 40 minutes.

From the explanations above, there are frequent duplications of mail order and reports of finished products, as well as the workflow for ordering customer is too long and convoluted. The other drawback is the owner is in place only at certain times, while ordering customer can not be predicted. Customer orders resulting in frequent delays due to waiting for the owner to be in place and also admin poorly informed about the amount of raw material available.

If duplication of mail order and report the finished product can be minimized, and the process is too long and convoluted can be shortened, there will be an increase in work time. Moreover, it can reduce the use of paper that is too much and could reduce the error rate.

3.3. Phase of the design of algorithms and business process flowchart
3.3.1. The design of algorithms of business processes

Algorithm of Business processes designed on SME Furniture is as follows:
a. Customer order products to admin.
b. Admin will check stock levels of raw materials in the system.
c. If the stock in the warehouse fewer than the number of orders, then the admin will negotiate with the customer.
d. Once that happens the price agreement between the admin and customer. If the customer refuses, then the transaction is stopped. Meanwhile, if the customer agrees, then the admin will enter the data of customer orders menu and menu order supplier.

3.3.2. Phase of Designing business process flowchart

Flowchart Business processes are designed on SME Furniture is shown in Figure 2.

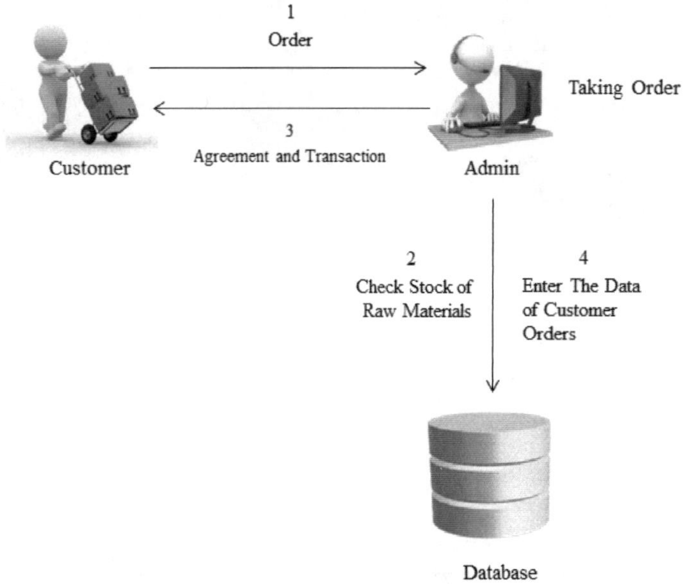

Figure 2. Flowchart design business processes

3.4. Phase of analysis of elements of a business system

With the approach to modeling the system, the next step is the analysis of the elements of a business system. The following table lists the elements of the business system:

Table 1. Elements of the business system of SMEs Furniture

No	Item	elements of the system
1	Model of consumer products ordered	input 1
2	The number of products ordered Consumers	input 2
3	DP of consumers	input 3
4	The results of calculations	output 1
5	payment slip	output 2
6	SO (mail order)	output 3
7	receipt of order	component 1
8	Checking the stock of raw materials in the warehouse	component 2
9	Calculating raw material requirements, adequacy of raw materials, the price of the contract, due date	component 3
10	Presentation of the calculation to the consumer	component 4
11	Admin accepting orders	interaction 1

12	Admin check stocks of raw materials at the warehouse	interaction 2
13	Admin makes calculations	interaction 3
14	Admin convey the results of calculations to consumers	interactions 4
15	Consumer	environment 1
16	The boundary between the SME and Consumer	boundary

3.5. Phase of Design of Data Flow Diagrams (DFD)

The last stage of this research is the design of Data Flow Diagrams (DFD). DFD business systems at IKM Furniture are designed as in Figure 3.

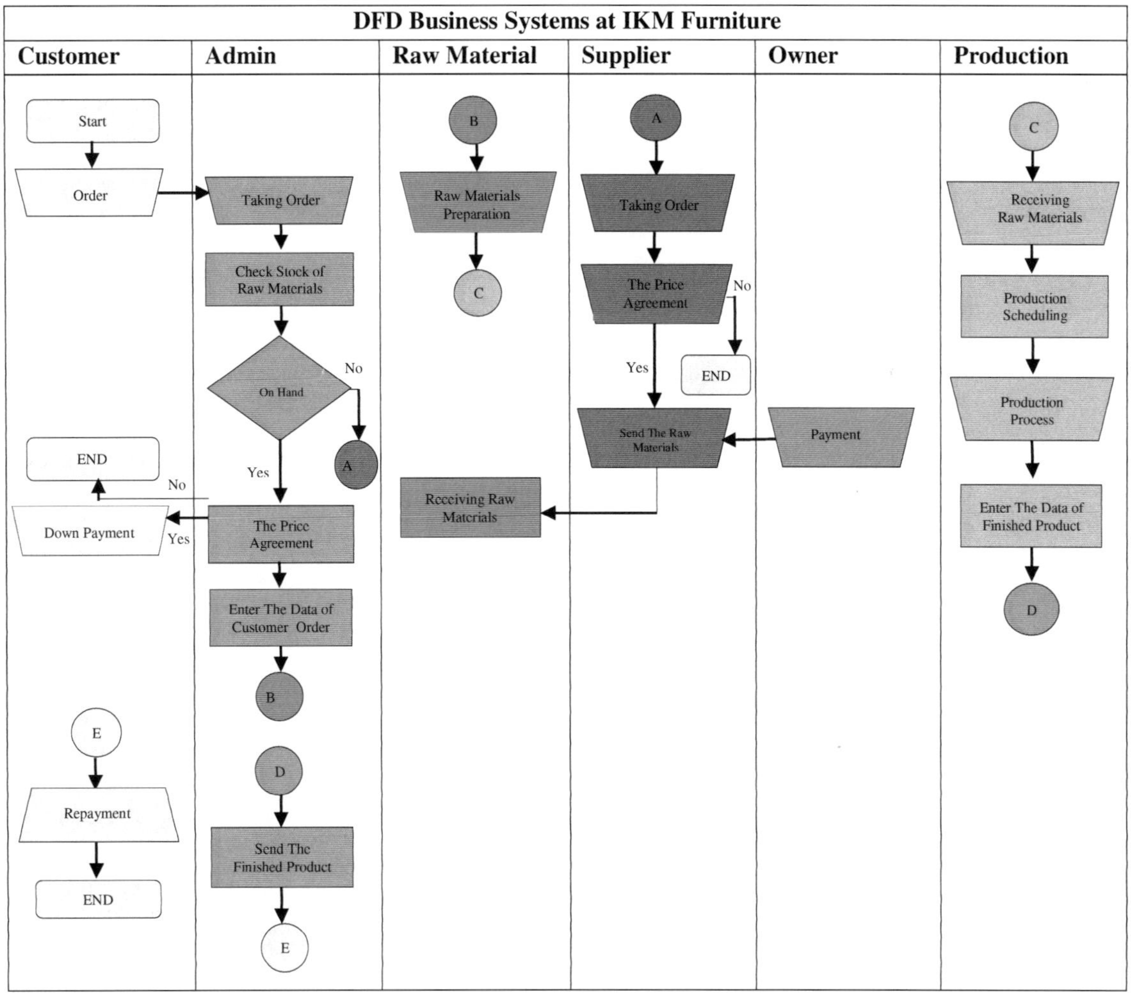

Figure 3. DFD business system

4. Conclusion

There are some shortcomings in the initial system, such as: the time required to provide the results to the admin consumers about 40 minutes too long. Furthermore, the final decision to order the production of raw materials in the hands of leaders who are more often not be in the SME. The system built in this paper, is planned to reduce the waiting time for consumers only be about 5 minutes.

in addition, the new system is built on this study, make recommendations to the deputy leader or manager to be able to take a decision to purchase raw materials production, or allow the leader to make decisions online wherever he is.

In the business system of SME of Furniture, Products and the number of products ordered customers and DP paid by consumers identified as elements of system inputs 1,2 and 3. The result of the calculation, payment slips and mail order (SO) are identified as elements of system output of 1.2, and 3. Acceptance of orders, stocks checking of raw materials at the warehouse, calculating raw material requirements, adequacy of raw materials, the price of the contract, and the due date, as well as the submission of the results of calculations to consumers identified as elements of system components 1, 2, 3, and 4. Admin accept orders, Admin check stocks of raw materials at the warehouse, making calculation, and Admin convey the calculation results to consumers identified as an element of interaction system 1, 2, 3, and 4. Consumers identified as an element of environmental systems. While the boundary between SMEs and consumers identified as elements of the system boundary.

References

1 Zaitar, Y., and Ouzarf, M.: 'ERP Projects: Key success factors and risk of failure a proposed model of governance of enterprise resource planning', International Journal of Computer Applications, 2012, 46, (8), pp. 34-39

2 Supramaniam, M., Abdullah, A., and Ponnan, R.: 'Cost Analysis on ERP System Implementation amongst Malaysian SMEs', International Journal of Trade, Economics and Finance, 2014, 5, (1), pp. 72

3 Tsai, W.-H., Chen, S.-P., Hwang, E.T., and Hsu, J.-L.: 'A study of the impact of business process on the ERP system effectiveness', International Journal of Business and Management, 2010, 5, (9), pp. 26

4 Yahya, B.N.: 'BUSINESS PROCESS REENGINEERING: CONCEPTS CAUSES AND EFFECT', Jurnal Teknik Industri, 2004, 4, (2), pp. pp. 102-110

5 Gelogo, Y.E., and Kim, H.: 'Mobile integrated enterprise resource planning system architecture', vol, 2014, 7, pp. 379-388

6 Ijaz, A., Malik, R., Lodhi, R.N., Habiba, U., and Irfan, S.M.: 'A qualitative study of the critical success factors of ERP system-A case study approach', in Editor (Ed.)^(Eds.): 'Book A qualitative study of the critical success factors of ERP system-A case study approach' (2014, edn.), pp. 2556-2566

7 Jayawickrama, U., and Yapa, S.: 'Factors affecting ERP implementations: client and consultant perspectives', Journal of Enterprise Resource Planning Studies, 2013, 2013, pp. 1

8 Manoj, P.: 'ERP Implementation', International Journal of Computer Science & Applications (TIJCSA), 2013, 1, (11)

9 Matende, S., and Ogao, P.: 'Enterprise resource planning (ERP) system implementation: a case for user participation', Procedia Technology, 2013, 9, pp. 518-526

10 Dixit, A.K., and Prakash, O.: 'A study of issues affecting ERP implementation in SMEs', Researchers World, 2011, 2, (2), pp. 77

11 Jidong, Z., and Liyan, W.: 'ERP Implementation: A Corporate Governance Perspective', International Journal of Public Information Systems, 2010, 6, (1)

12 Mabert, V.A., Soni, A., and Venkataramanan, M.: 'Enterprise resource planning: Managing the implementation process', European journal of operational research, 2003, 146, (2), pp. 302-314

13 Umble, E.J., Haft, R.R., and Umble, M.M.: 'Enterprise resource planning: Implementation procedures and critical success factors', European journal of operational research, 2003, 146, (2), pp. 241-257

14 Al-Mashari, M., Al-Mudimigh, A., and Zairi, M.: 'Enterprise resource planning: A taxonomy of critical factors', European journal of operational research, 2003, 146, (2), pp. 352-364
15 Daoud, H., and Triki, M.: 'Accounting information systems in an ERP environment and Tunisian firm performance', 2013
16 Daellenbach, H.G.: 'Systems and Decision Making: A Management Science', 1994

IOP Conf. Series: Materials Science and Engineering **190** (2017) 012017 doi:10.1088/1757-899X/190/1/012017

Joule-Thief Circuit Performance for Electricity Energy Saving of Emergency Lamps

Eka Nuryanto Budisusila[1], Bustanul Arifin[2]

[1, 2] Electrical Engineering Department of Industrial Engineering Faculty
of Sultan Agung Islamic University (UNISSULA)

[1] ekanbs@unissula.ac.id, [2] bustanul@unissula.ac.id

Abstract. The alternative energy such as battery as power source is required as energy source failures. The other need is outdoor lighting. The electrical power source is expected to be a power saving, optimum and has long life operating. The Joule-Thief circuit is one of solution method for energy saving by using raised electromagnetic force on cored coil when there is back-current. This circuit has a transistor operated as a switch to cut voltage and current flowing along the coils. The present of current causing magnetic induction and generates energy. Experimental prototype was designed by using battery 1.5V to activate Light Emitting Diode or LED as load. The LED was connected in parallel or serial circuit configuration. The result show that the joule-thief circuit able to supply LED circuits up to 40 LEDs.

1. Background

Along with the improvement of technology, electrical devices that initially require great energy in operation gradually began to be replaced with electricity-saving devices, and other devices that is more environmentally friendly. The effort to save electricity is likely to be continued to develop. The availability of electrical energy given is unbalanced with the needs and demands of tremendous power supplies [1][2][3]. From problems above, an idea to save electrical energy in a way to use it optimally was occurred. One of solution is by exploiting the potential of basic electric energy generation through the development of electrical components that generate electromagnetic fields in the form of windings or coils. The basic working principle of the proposed circuit is by applying electromotive force (EMF) or electro mechanics force (EMF) in optimal way through the concept of magnetic induction. One of application of the concept above is a circuit called as joule-thief. This circuit is also known as simple blocking transistor circuit that is working as a thief of energy that can generate electrical energy when the energy sources are very small [4]. A joule-thief circuit prototype is proposed testing by considering many practical terms without any in-depth analysis in the laboratory to know the performance of the circuit [5][6].

2. System Design

In this research, system is designed as the basic of Joule-thief circuit including power supply, solenoid in toroidal form, transistor and output loads.

Each block or subsystem in the system is varied according to the characteristics of each component. The source is variable voltage input. The transistor circuit is constructed from different types of resistance and the current base settings. Solenoid in toroidal form is made in various amounts of primary and secondary windings. The load is Light Emitting Diode (LED) in series and parallel configuration as variable load [3][4][5].

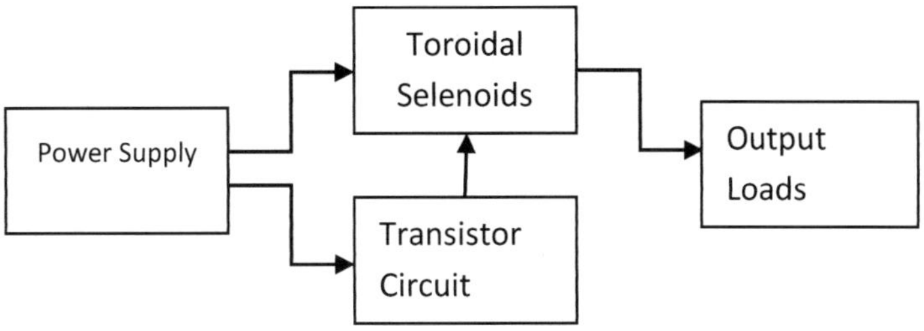

Figure 1. System Design Block

3. Research Method

As the block on the system, circuit and components are already constructed, then measurement is carried out by measuring voltage and current using multimeter. Waveform of signals and electromagnetic waves are observed by using an oscilloscope. The intensity LED lights is measured by the digital luxmeter.

4. Data And Analysis

Joule-thief circuit testing is made from several part of components in circuits. Many variation in the prototype is applied by changing the number of windings and the type of transistor. In Figure 2, a DC power supply is 1.5 volt b. Oscilloscope GW-Instek GOS-6200 is used to display observed signals. The oscilloscope can measure the frequency up to 200 MHz. Measurement of the output signal is done in two ways i.e. first measurement is the pulse output with no load and a second measurement is pulse output with LED loads

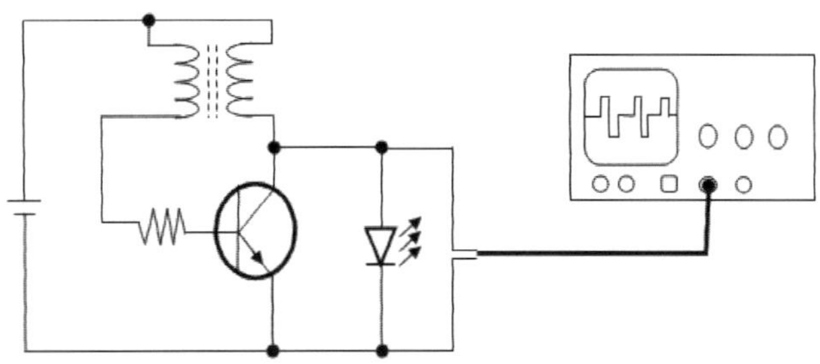

Figure 2. Joule-thief circuit test

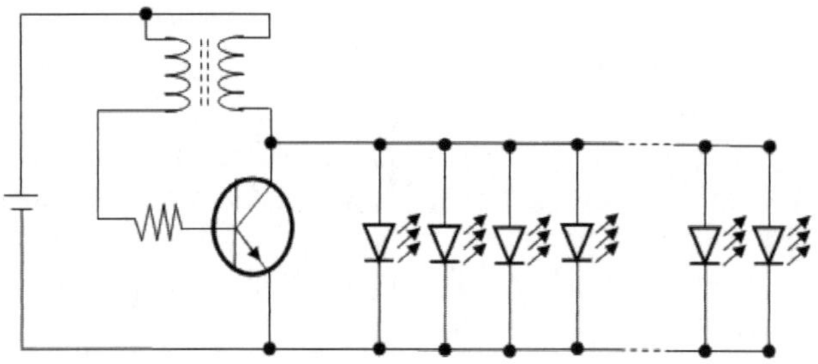

Figure 3. Parallel LEDs Load Circuit

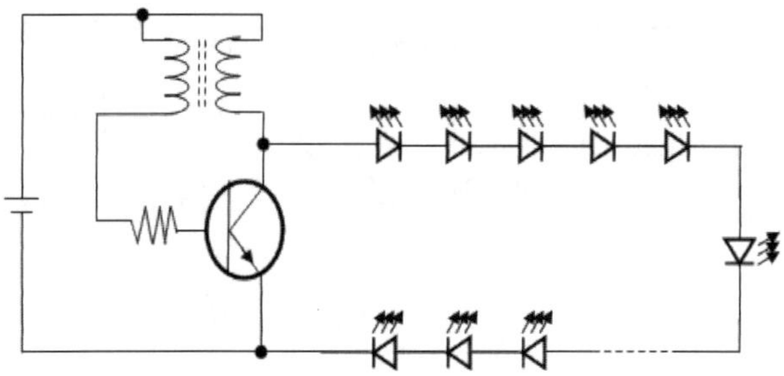

Figure 4. Serial LEDs Load Circuit

Testing is conducted by creating multiple variations of component such as the changes in the number of windings on ferrite toroid by keeping diameter remains the same. Circuit-1 use the equal coil between the primary and secondary windings. The primary winding are 22 turns and the secondary windings are 22 turns. Circuit-2 use ratio winding of primary and secondary coil in 30: 50. Winding ratio for the circuit-3 is 15: 100 and 20: 40 for circuit-4.

Measurement Analysis :

At no load condition the measurements results for circuit shows that the peak to peak voltage (Vpp) is 60 V, with a 353.75 kHz frequency at 20 V/div. On load condition as the LED is plug in, voltage recorded is 6 V, by adjusting LED load then the an average voltage is 3 volts. On the second circuit, no-load measurement Vpp measured up to 140 V (7 x 20V), with a frequency of 50 kHz. On load condition, the measurement of voltage is in about 6 volts. For circuit-3, the measurement results show the voltage 70 V with 70.15 kHz. On load condition the voltage drops up to 4 volts. At no-load measurements for circuit-4, the produced voltage is30 volts with of 331.04 kHz frequency and on load condition the voltage is 3.5 voltages.

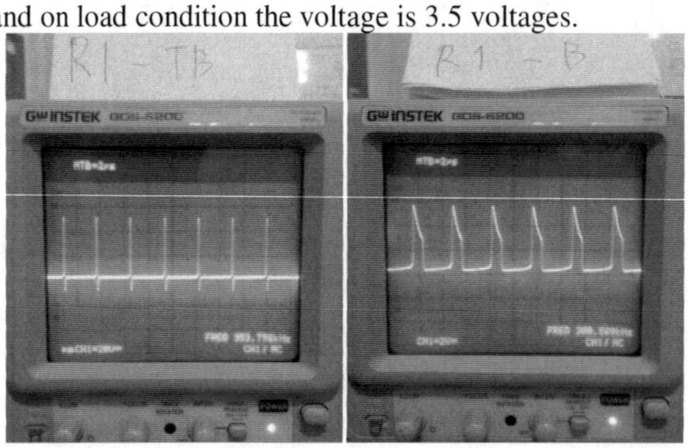

Figure 5. Measurement Result of Circuit 1 without and within load

IOP Conf. Series: Materials Science and Engineering **190** (2017) 012017 doi:10.1088/1757-899X/190/1/012017

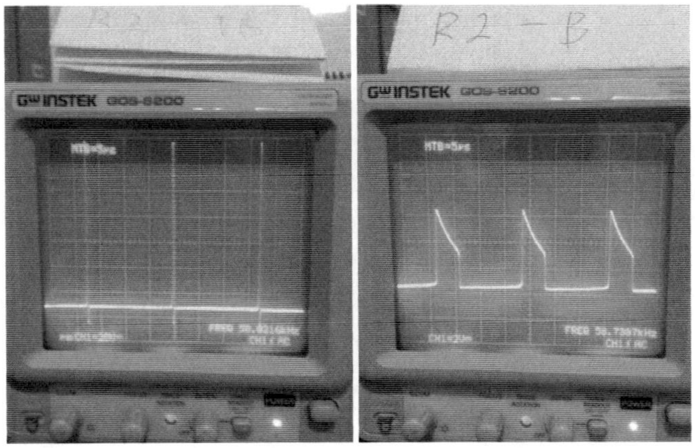

Figure 6. Measurement Result of Circuit 2 without and within load

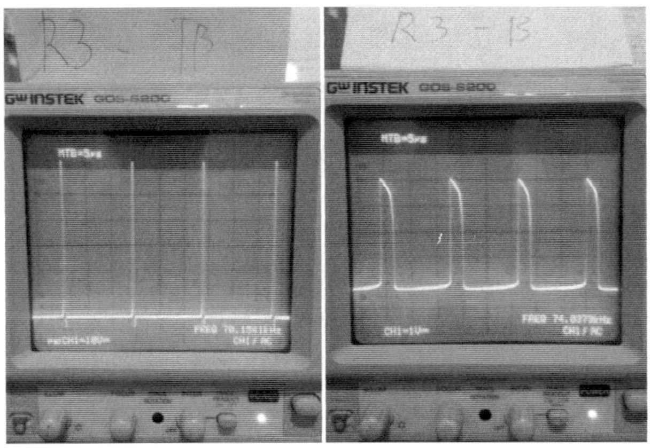

Figure 7. Measurement Result of Circuit 3 without and within load

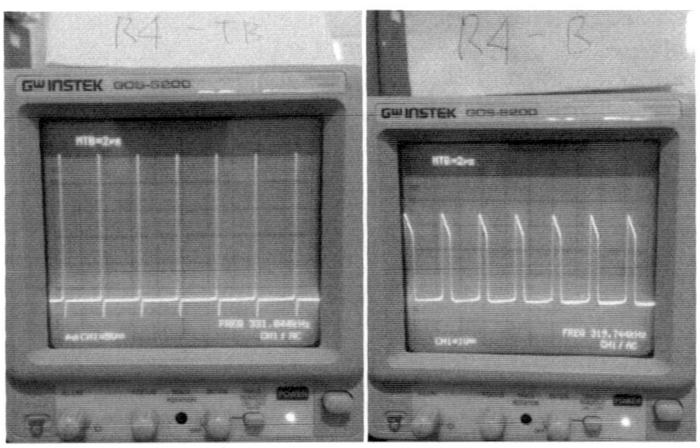

Figure 8. Measurement Result of Circuit 4 without and within load

The test condition is continued by arranging all LED in series or parallel. There are 33 LEDs installed in the circuit. Circuit-1 provides the dimmest lighting, while the other circuit tends to light brighter.

When 26 LEDs are installed in series circuit, all LED circuits can light all over, except for circuit 1. The circuit-1 is only capable to ignite 10 LEDs.

(a) (b) (c)

Figure 9. (a) All Circuit LED Flame in Parallel Load
(b) Maximum LED Flame in serial Load on Circuit 1
(c) LED Flame in serial Load on Circuit 2, 3 and 4

Based on experimental results, number of coils wound around a ferrite toroidal determines the magnitude voltage and current generated on the output side. More turns gives more voltage and current at out put side in consequently more LED can be ignited. At series circuit, higher voltage with constant current is indicated by the number of LEDs which are ignited with constant light intensity. In parallel configuration, the large amount of current produced by joule thief circuit is indicated by the number of LED with bright light or high light intensity.

The comparison of the number of primary and secondary windings also affect the performance of the joule thief circuit. The more windings on the primary side than the secondary side provide strong currents in the output circuit. Instead, a more number of windings in the secondary will generate higher voltage at the output side.

Although the joule thief circuit generates a high voltage, this proposed circuit does not make the LED lights as the load to be damaged or burnt out due to excessive voltage because the given load automatically adjust the voltage corresponding to average obtained voltage. In addition, the output voltage in the form of pulses with a certain frequency gives the average voltage rise time (ON) and the down time (OFF) and it is multiplied to the peak voltage (Vpp).

5. Conclusion and Sugestion

5.1. Conclusion

1. The Joule thief circuit can provide reinforcement of current and voltage from 1.5 V DC battery source to supply the load LEDs by utilizing the reverse flow current that occurs at the time of toroidal winding current source or voltage through cut-breaking conducted by the NPN transistor.
2. More coils wound on toroidal ferrite give the higher voltage and current.
3. Voltage performance test is successfully conducted by using series LED circuit and current performance test is effectively conducted by using parallel LED circuit.

IOP Conf. Series: Materials Science and Engineering **190** (2017) 012017 doi:10.1088/1757-899X/190/1/012017

4. The output voltage of joule thief circuit can be adjusted to the LED load and it does not cause damage to the LEDs.

5.2. Suggestion

1. To improve the performance of the Joule-Thief circuit for larger power requirements, the use of a ferrite toroid with a larger diameter or more winding can be proposed together with the use of a switching transistor with the larger currents type such as 2N3055.
2. Further testing to determine the ability of joule-thief circuit in energy savings and battery life for supplying circuit and load is interesting because the exhausted battery or battery with voltage less than 1 volt still able to ignite the LED load as transistor switching is kept to switch.

References

[1] Hongxing Li, Tsutomu Wakimoto Keita Murase Kenji Tomiyoshi, 2012, 0.25 μm 1.2 MHz BOOST-PFM-CONTINUOUS triple mode LED driver integrated in analog front end IC for portable application, Analog Integrated Circuits and Signal Processing International Journal, September 2012, Volume 72, Issue 3, pp 595–603

[2] Zhidong Liu, Hoi Lee, 2015, Design of high-performance integrated dimmable LED driver for high-brightness solid-state lighting applications, Analog Integrated Circuits and Signal Processing International Journal, March 2015, Volume 82, Issue 3, pp 519–532

[3] Carson A. Pope, 2013, Testing the Efficiency of a Joule Thief LED Light Circuit, California State Science Fair 2013 Project Summary

[4] Haidar, Sajjad, 2015, MOSFET-based Joule Thief steps up voltage, February 23, 2015, EDN Network

[5] Harjunowibowo, Dewanto, 2016, Simple Blocking Oscillator for Waste Battery's Voltage Enhancement, International Journal of Information and Electronics Engineering, Vol. 6, No. 2, March 2016

[6] Harjunowibowo, Dewanto, 2015, Simple Blocking Oscillator Performance Analysis for Battery Voltage Enhancement, Journal of Mobile Multimedia, Vol. 11, No.3&4 (2015) 321-329

[7] Setiyani, Riyana Indah, Prototipe Sistem Otomatis pada Lampu Berbasis Sel Surya yang Terintegrasi Blocking Oscillator, Prosiding Seminar Nasional Fisika dan Pendidikan Fisika (SNFPF) Ke-6 2015 Volume 6 Nomor 1 2015 ISSN : 2302-7827

[8] Francis Weston Sears, 1986, Fisika untuk Universitas 2 : Listrik dan Magnet, saduran : Nabris Chatib, Penerbit Bina Cipta, Jakarta

[9] Malvino, 2000, Prinsip-prinsip Elektronika, Jilid 1, Translation : M. Barnawi, Penerbit Erlangga, Jakarta

[10] Umar Yahdi, 1997, Pengantar Fisika Listrik dan Magnet, Penerbit Gunadharma, Jakarta

[11] Yohanes Surya, 2012, Listrik dan Magnet, Penerbit Grasindo, Jakarta

[12] Zuhal, 2007, Prinsip Dasar Elektroteknik, Penerbit PT Gramedia Pustaka Tama, Jakarta

IOP Conf. Series: Materials Science and Engineering **190** (2017) 012018 doi:10.1088/1757-899X/190/1/012018

Development of Bio-impedance Analyzer (BIA) for Body Fat Calculation

Munawar A Riyadi*, A Nugraha, M B Santoso, D Septaditya and T Prakoso

Department of Electrical Engineering, Diponegoro University, Kampus Tembalang, Semarang 50275, Indonesia

*corresponding author, e-mail address: munawar@elektro.undip.ac.id

Abstract. Common weight scales cannot assess body composition or determine fat mass and fat-fress mass that make up the body weight. This research propose bio-impedance analysis (BIA) tool capable to body composition assessment. This tool uses four electrodes, two of which are used for 50 kHz sine wave current flow to the body and the rest are used to measure the voltage produced by the body for impedance analysis. Parameters such as height, weight, age, and gender are provided individually. These parameters together with impedance measurements are then in the process to produce a body fat percentage. The experimental result shows impressive repeatability for successive measurements (stdev $\leq$ 0.25% fat mass). Moreover, result on the hand to hand node scheme reveals average absolute difference of total subjects between two analyzer tools of 0.48% (fat mass) with maximum absolute discrepancy of 1.22% (fat mass). On the other hand, the relative error normalized to Omron's HBF-306 as comparison tool reveals less than 2% relative error. As a result, the system performance offers good evaluation tool for fat mass in the body.

1. Introduction

Household bathroom scales is commonly used for measuring body weight in everyday life. However, these scales can not distinguish between fat mass and non-fat mass (e.g. water, bones, flesh etc). Recently, the need to find out the composition of body mass, especially fat, is gaining interest due to potential health consequences behind the fat composition. Excessive fat in human body is thought to increase the likelihood of developing the disease, such as type-2 diabetes, heart problems and cancer. On the other hand, body fat deficiency could also lead to things that are not beneficial for the body [1]. The complete information of body composition would make it easier to break the pattern to maintain the health. Therefore a measurement instrument for body composition assessment that is able to distinguish fat mass and non-fat mass is deemed important and demanded [2].

Body composition assessment is divided into several categories based on the models and methods. Based on the model, there are atomic models, molecular models, cellular models, tissue-system models, and multi-component models [3]. Based on the method, it consists of underwater weighing, dilution method, dual energy x-ray absorpiometry, computed tomography, magnetic resonance imaging, neutron activation analysis, bioelectric impedance analysis, water displacement plethysmography, skinfold thickness, near-infrared measurement, waist circumference, waist-hip ratio, sagittal abdominal diameter and body mass index [1,4]. Bioelectric Impedance Analysis is one method of body composition widely used to determine body fat. This is because the BIA is easy to use, fast, noninvasive, portable and can be operated alone. It has been used in many hospitals and institutions to measure a wide variety of age ranges and conditions of the human body. In addition, BIA is also more accurate than the caliper tests, easier to use than densitometry and safer than dual energy x-ray absorpiometry (DXA) [5].

2. Method

BIA measures the composition of the human body by using the difference in the electrical conductivity of tissue. By modelling the human body into two compartments, the human body consists of fat mass (FM) and non-fat or fat-free mass (FFM) [1,6]. Fat free mass is divided into intracellular water, extracellular water, mineral bone and visceral protein as shown in Figure 1 [7]. There are differences between man and woman in terms of body composition. Generally, woman has higher percentage of fat in her body than man, while man's body usually contains higher content of minerals manifested in larger and denser bones than woman. These differences take great influence in the impedance of the body.

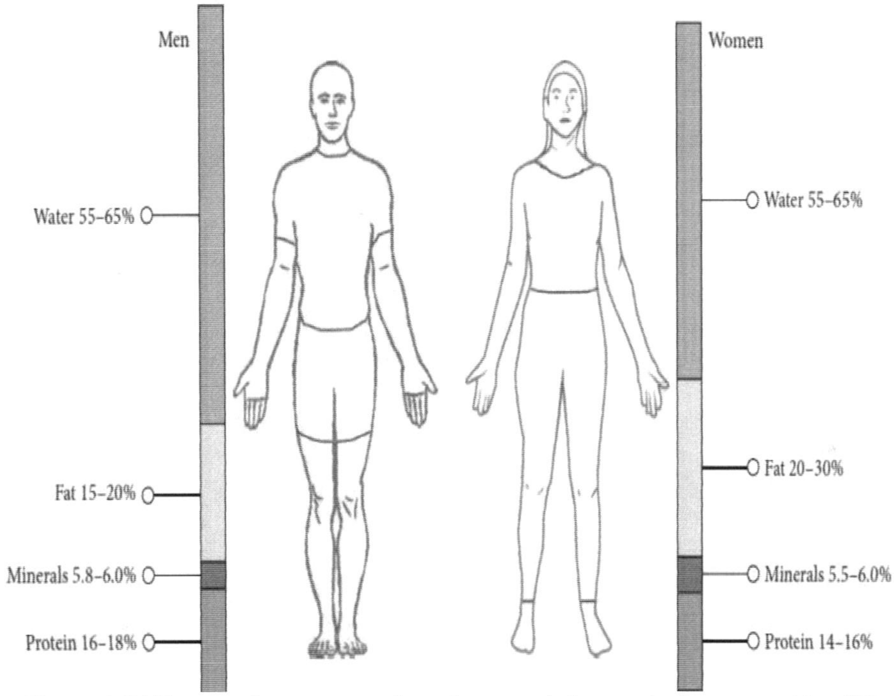

Figure 1 Differences between men's and women's human body composition [7]

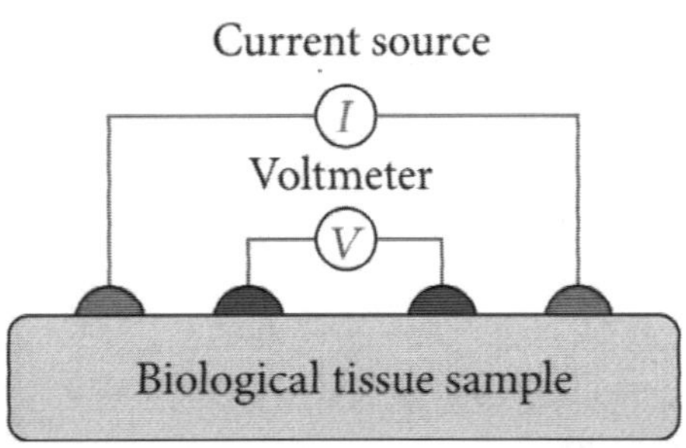

Figure 2 Concept of body impedance determination

The impedance of the body is calculated by introducing small current into the body and measuring the potential difference between two nodes on the body. Figure 2 shows the simplified concept of the impedance determination. The magnitude of current should be regulated and limited to less than 0.25 mA to avoid unwanted impact on the health of the user. It is noteworthy that the amount and variety of cells directly determine the resistance along the path taken by the current. Direct current only flow in the fluids between cells, while the alternating current can also flow through the cells, thus the cell characteristics are taken into account with this type of current [7,8]. Therefore, the alternating current (f=50 kHz) [1] was applied to be able to calculate the impedance effect in the cells.

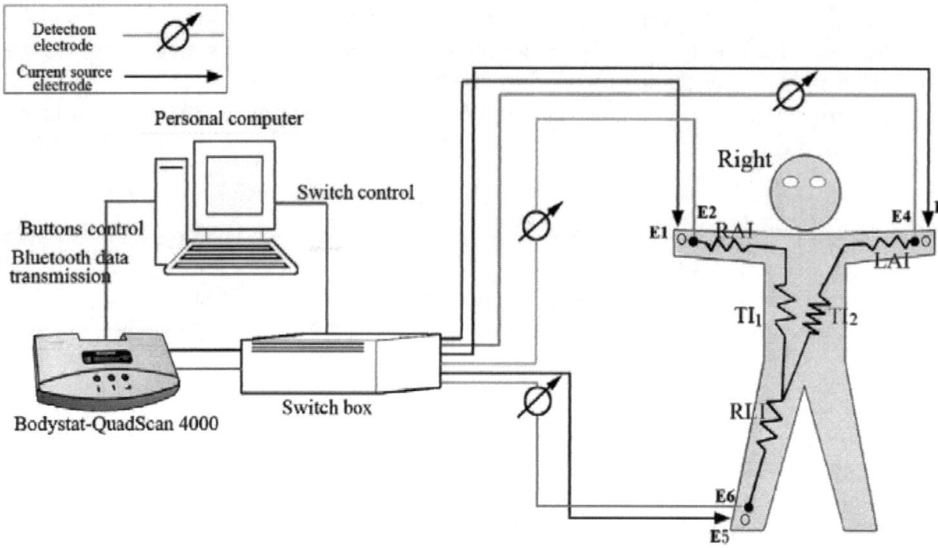

Figure 3. The measurement system of hand-to-foot mode and cross mode [4]

The simple model of body impedance network is shown in Figure 3. The model shows that different parts of the body provide different impedances. The notation of RAI is for right arm impedance, LAI for left arm impedance, TI1 and TI2 are for trunk impedances, and RLI is for right leg impedance. These impedances are influenced by many factors, including the mass of fat, and each part may provide various impedance values. Therefore, current flowing along different pairs of node into the body passes different path and eventually produces different impedance as well. For example, current flows to the body from node E1 and out from E5 passes RAI, TI1 and RLI, while current from E3 to E5 takes the path of LAI, TI2 and RLI. Consequently, the selection of current nodes and voltage detection point determines the outcome of the impedance measurement.

To determine the body fat percentage, the value of FM and FFM needs to be calculated first. FFM value is obtained by using the results of impedance measurement by the BIA correlated with empirical data for adjusting factors by considering several body parameters [9]. The value of FM is then calculated simply by using the following equation:

$$FM = w - FFM \tag{1}$$

where w is the body weight (in kg).

One published formula to calculate FFM was proposed by E. Myloot et al [6], based on the result of their experiments. In their experiments, they calculated the FFM with the comparison of BIA issued by Omron. The formula is as following:

$$FFM = 0{,}360 \left(\frac{h^2}{Z}\right) + 0{,}162h + 0{,}289w - 0{,}134a + 4{,}83g - 6{,}83 \tag{2}$$

where h, Z, a and g are body height, impedance, age and gender (g=1 for man, 0 for woman), respectively. By determining the impedance, FFM can be calculated, and later the fat mass is evaluated using equation (1). Usually, the fat mass is shown in percentage over body weight.

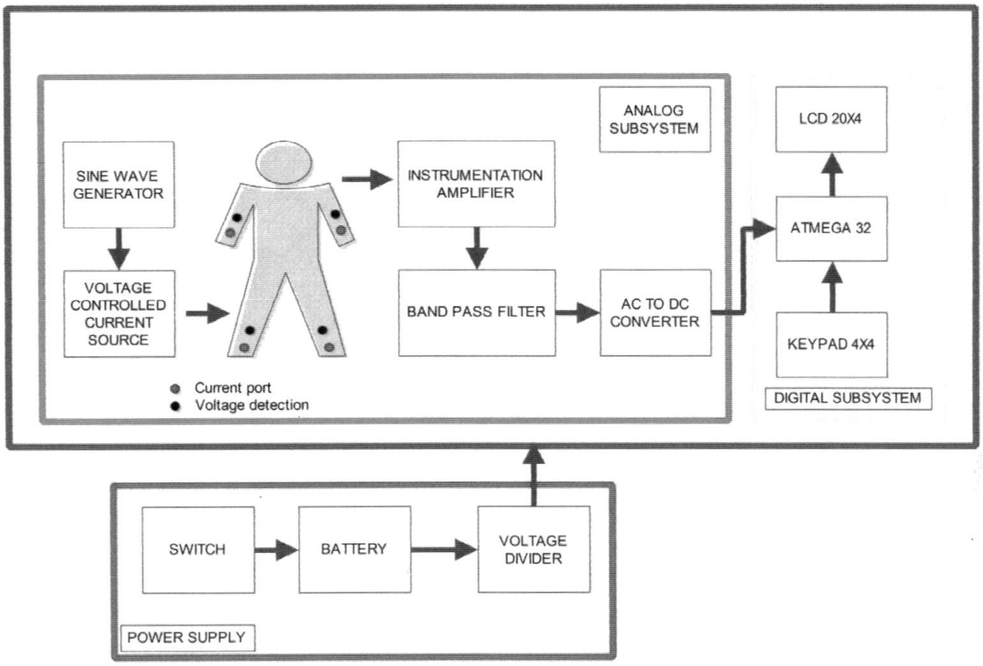

Figure 4. Block diagram of proposed bio-impedance analyzer

In order to determine the body impedance, a Bio-Impedance Analyzer was designed. The system consists of three parts, namely power supply, analog subsystem and digital subsystem blocks. Power supply consists of voltage divider, switches and batteries which provide voltage supply to the analog and digital blocks. The analog subsystem consists of sine wave generator (f=50 kHz) along with voltage-controlled current source, instrumentation amplifiers, band pass filter and AC to DC converter. This block aims to transmit currents into the human body and measure the voltage which later used to calculate the body impedance. The current source is designed to produce constant current of 0.25 mA regardless of the load impedance, in order to avoid harm to the body. The digital subsystem is ATmega32 microcontroller with 4x4 keypad and LCD functioned to control the entire measurement system. The impedance as well as body fat percentage are also calculated in this subsystem. Figure 4 shows the overall picture of the measuring system with a body fat percentage measurement method of whole body bioelectrical impedance analysis (BIA).

3. Result and discussion

The designed bioimpedance analyzer was tested on ten subjects, with equal numbers between man and woman. The subjects were of broad age range between 20-64 years old with various heights and weights as well. Their fat mass percentages were obtained using formula from equation (2), and were compared with the result from similar measurement using Omron's HBF-306. Table 1 shows the statistical analysis of repeatability of 10 repeated measurements for each subject. In addition, Table 2 shows the experimental result (P) in comparison with HBF-306 (Q), from the average of repeated measurements for each subject which consists of 10 times of repetition. For the purpose of fair comparison, the data taken were for hand to hand measurement, i.e. current ports as well as nodes are right and left arms. Although the analyzer is capable to do the impedance analysis for different pair of ports (i.e. hand to foot, foot to foot, or cross sectional analysis), the data is focused on hand to hand analysis due to the limited capability of the comparison tool, as Omron's HBF-306 can only measure hand to hand body impedance.

Table 1. Statistical analysis of fat mass percentage from repeated measurements

Subject No.	Fat Mass Percentage(%)			
	Average	Min	Max	Std Dev
1	31.07	31	31.2	0.08
2	38.19	38.1	38.3	0.10
3	20.16	20.1	20.2	0.05
4	49.43	49.3	49.8	0.16
5	44.52	44.4	44.6	0.09
6	14.35	14.1	14.5	0.12
7	15.74	15.7	15.9	0.07
8	21.27	21.1	21.5	0.13
9	37.24	37.1	37.9	0.25
10	28.08	28	28.2	0.08

Table 2. Result of *hand to hand* measurement compared to Omron's HBF-306

Subject No.	*h* (cm)	*w* (kg)	*g* (M/F)	*a* (yr)	Fat Mass Percentage		|P-Q| (%)	|(P-Q)/Q| (%)
					P[#] (%)	Q[$] (%)		
1	157	56	F	23	31.07	30.65	0.42	1.37
2	149	58	F	49	38.18	38.86	0.68	1.75
3	145	40	F	20	20.16	20.19	0.03	0.15
4	147	66	F	64	49.43	48.36	1.07	2.21
5	162	74	F	51	44.52	44.51	0.01	0.02
6	175	56	M	22	14.35	14.18	0.17	1.20
7	165	54	M	23	15.75	15.09	0.66	4.37
8	161	62	M	24	21.27	22.49	1.22	5.42
9	168	80	M	42	37.24	36.73	0.51	1.39
10	158	64	M	31	28.08	28.06	0.02	0.07
						Average error	0.48	1.03

[#] *P is the result from the designed bioimpedance analyzer*
[$] *Q is the result from Omron HBF-306*

The result shows that the analyzer was able to produce output of fat mass percentage for diverse range of input. The fat mass percentage produced by our bio impedance analyzer were in the range of 14.1 (min) to 49.8 (max) on the overall individual measurement, as indicated from Table 1. The statistical analysis also indicates that the repeatability is impressive with the maximum standard deviation of 0.25% (fat mass).

In addition, the comparison with Omron's HBF-306 shows that the average absolute difference of total subjects between two analyzer tools is 0.48% (fat mass) with maximum absolute discrepancy of 1.22% (fat mass). On the other hand, the relative error normalized to HBF-306 reveals less than 2% relative error. This means that for subject with 25% fat mass, we can confidently say that the average error only takes about ± 0.5% from the measured value. In other word, the subject's fat mass is 25±0.5%.

Analysis using hand-to-hand probes can only reveal the overall fat mass, but cannot indicate which part of body that possess more fat. The more specific analysis can be obtained by combining different measurement nodes with more enhanced calculation, as in [9]. Our analyzer is actually capable of doing the measurement from different node schemes. Other testing result shows that the average relative error rate from measurement of body fat percentage is 3.28% with right hand to right foot node, 3.15% with right hand to left foot node, 7.13% with foot to foot node, 4.73% with left hand to right foot node, and 3.89% with left hand to left foot method, all compared to HBF 306. However, HBF 306 is only for the hand-to-hand probes. Therefore, due to different node schemes between our analyzer and the comparison tools, the fat mass value obtained from all schemes other than hand-to-hand probes cannot be validated.

The experiment data shows promising data in comparison with commercial BIA. The small error found, either in the absolute or normalized error, is within acceptable value. Moreover, the usage of relatively cheap components in this system provides good alternative for evaluation of fat mass in the body.

4. Conclusion

We have designed a body impedance analyzer using single frequency, alternating current source (f=50 kHz, 0.25 mA max) with ATMega32 as primary microcontroller for calculation of body fat percentage. It is designed with the capability of fat mass analysis using several schemes of measurement, i.e. hand to hand, foot to foot, hand to foot as well as the cross sectional nodes. The experimental result indicates that the analyzer provides excellent performance for the scheme of hand to hand node, in comparison with Omron's HBF 306 as verification tool. The average discrepancy between our analyzer and the comparison tool is less than 0.5% (fat mass), or less than 2% of normalized error, within acceptable value. Other node schemes have not been validated so far due to limited capability of the comparison tool. Overall, this system shows promising performance that offers good evaluation tool for fat mass in the body.

References

[1] Kyle UG, Bosaeus I, De Lorenzo AD, Deurenberg P, Elia M, Gómez JM, Heitmann BL, Kent-Smith L, Melchior JC, Pirlich M, Scharfetter H. Bioelectrical impedance analysis—part I: review of principles and methods. *Clinical nutrition*. 2004; **23**(5):1226-43.

[2] National Institutes of Health Technology Assessment Conference Statement. Bioelectrical Impedance Analysis in Body Composition Measurement. 1996

[3] Wang ZM, Heshka S, Pierson RN, Heymsfield SB. Systematic organization of body-composition methodology: an overview with emphasis on component-based methods. *The American journal of clinical nutrition*. 1995;**61**(3):457-65

[4] Huang Y, Chen Y, Chuang C, Chiang L, Lu H, Lin H, Chen K, Hsiao A, Hsieh K. Cross-mode bioelectrical impedance analysis in a standing position for estimating fat-free mass validated against dual-energy x-ray absorptiometry. *Nutr. Res.*, 2015 **35**(11): 982–989

[5] Völgyi, E., Tylavsky, F. A., Lyytikäinen, A., Suominen, H., Alén, M., Cheng, S. Assessing body composition with DXA and bioimpedance: effects of obesity, physical activity, and age. *Obesity*. 2008; **16**(3), 700-705.

[6] Mylott E, Kutschera EM, Widenhorn R. Bioelectrical Impedance Analysis as a Laboratory Activity : At the Interface of Physics and the Body. *Physics (College. Park. Md).*, 2014 **82**: 521–528

[7] Bera TK. Bioelectrical impedance methods for noninvasive health monitoring: a review. *Journal of medical engineering*. 2014 **2014**: 381251.

[8] Ryu B, Jeong H, Ryu H. Bioelectrical Impedance Analysis by Multiple Frequencies for Health Care Refrigerator. *PIERS ONLINE,* 2010 **6**(7):640–645.

[9] Shafer KJ, Siders WA, Johnson LK, Lukaski HC. Validity of segmental multiple-frequency bioelectrical impedance analysis to estimate body composition of adults across a range of body mass indexes. *Nutrition*. 2009 Jan 31;25(1):25-32.

Investigation of Passive Filter for LED Lamp

Edi Sarwono[1,2], Mochammad Facta[2], Susatyo Handoko[2]
[1]Vocational High School 1 Semarang, Indonesia
[2]Electrical Engineering Department, Diponegoro University, Semarang, Indonesia

edisarwonono@gmail.com; facta@elektro.undip.ac.id; susatyo@elektro.undip.ac.id

Abstract. Light Emitting Diode lamp or LED lamp is one of the energy saving lamps nowadays widely used by consumers. However, LED lamp has contained harmonics caused by the rectifier circuit inside the lamp. Harmonics cause a quality problem in power system. As the harmonics present in current or voltage, the waveforms are distorted. Harmonics can lead to overheating in magnetic core of electrical equipments. In this paper, several tests are carried out to investigate the harmonic content of voltage and currents, and also the level of light intensity of the two brands of LED lamps. Measurements in this study are conducted by using HIOKI Power Quality Analyzer 3197. The test results show that the total harmonic distortion or THD of voltage on various brands of LED lamps did not exceed 5% as in compliance to the limit of IEEE standard 519-1992. The largest harmonic voltage is 2.9%, while maximum harmonic current for tested brands of LED lamp is 170.6%. The use of *low pass filter* in the form of LC filter was proposed. Based on experimental results, the application of LC filter at input side of LED lamp has successfully reduced THD current in the range of 85%-88%.

Keyword: LED lamp, harmonic distortion, power quality, THD current, THD voltage, lighting intensity

1. Introduction

Light Emitting Diode lamp or LED lamp is one of the energy saving lamps used currently widely by consumers such as households, offices and industry [1]. However, LED lamp contains harmonics caused by the rectifier circuit attached inside the lamp [2]. Harmonics can cause interference on the electrical system, overheating in magnetic core of electrical equipments and it can be detrimental in terms of both producers and consumers [3]. To reduce harmonics, many references suggest the use of filter [4,5,6]. In this work passive filter has been proposed to reduce harmonics in the two selected brands of LED lamps. In this work passive filter is proposed to reduce harmonics in the two selected brands of LED lamps. These two brands represent the LED lamps commercialized in Indonesia. Passive filter is chosen to be applied because this circuit required no additional power supply to work [7]. Small dimension of passive filter can be obtained by selected the appropriate components.

The work in this paper starts with investigation of harmonics in the two selected brands of LED lamps. The early investigation is required to get harmonics spectrum and to find out the most dominant harmonics order. The most dominant order becomes a target to be eliminated by choosing a suitable component for passive filter. According to several references passive filter can be constructed by combining two or more components of inductor (L) and capacitor (C) [8]. In this work, LC filter is selected to be applied as additional circuit to eliminate the most dominant harmonics of selected LED lamps.

IOP Conf. Series: Materials Science and Engineering **190** (2017) 012019 doi:10.1088/1757-899X/190/1/012019

2. Led Lamp Specification

In this paper, two brands of LED lamps with three different rating power are investigated. Their specifications are:

- Brand-A (3, 4, and 7 Watt)
- Brand-B (3, 5, and 9 Watt)

Both LED lamps required alternating current (AC) 220 volt and 50 Hz to turn on [9].

3. Experiments and Results

To find the dominant parameters of LED lamps such as voltage, current, power factor, real power factor, percentage of total harmonics distortion for voltage and current, the power quality analyzer (PQA) HIOKI 3197 is used in this work. The lighting intensity of the LED lamps in the value of lux is measured by using lux meter. The illustration of experimental circuit in general is shown in Figure 1. The values of electrical parameters as the dominant parameters of LED lamps is read and recorded directly from PQA to personal computer by using USB.

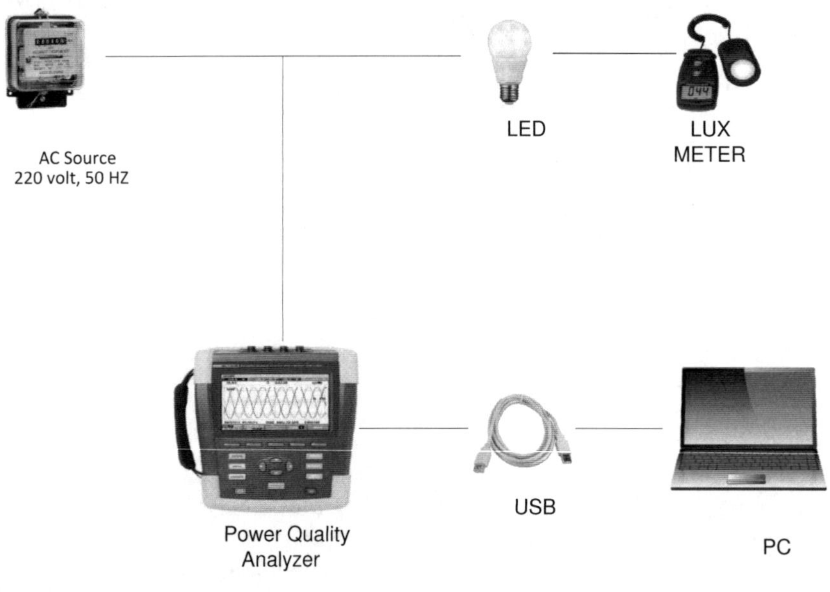

Figure 1. Harmonics & Lumen measurement LED lamps

Table 1 and 2 show the result of measurement for every brand of LED lamps. Every lamp is measured separately to get the value of voltage and current harmonics. It is shown that the largest harmonics of voltage is 2.9% and the largest of harmonics of current is 170.6%. In this experiment the highest light intensity 1007 lux is given by 9 watt of LED lamps brand-B.

Table 1. Percentage Of Total Harmonic Distortion Of Current (THD_I) LED Lamp Brand-A

No	Rating Power	THD_V (%)	THD_I (%)	Lumen
1	3W	2.7	154	288
2	4W	2.8	148.6	365
3	7W	2.8	128.9	787

Table 2. Percentage Of Total Harmonic Distortion Of Current (THD_I) LED Lamp Brand-B

No	Power	THD_V (%)	THD_I (%)	Lumen
1	3W	2.9	163.7	398
2	5W	2.9	170.6	514
3	9W	2.8	158.7	1007

Table 3. Harmonics order of LED Lamp Brand-A and Brand-B

Order	U [V]	I [A]
1	223.0	14.0m
2	0.8	0.3m
3	4.8	11.8m
4	0.2	0.2m
5	0.9	8.8m
6	0.1	0.1m
7	1.5	6.6m
8	0.1	0.2m
9	2.4	6.4m
10	0.1	0.2m
11	1.4	6.5m
12	0.1	0.2m
13	0.5	5.9m
14	0.1	0.2m
15	0.9	4.9m
16	0.1	0.3m
17	0.2	4.2m

Harmonic order Brand-A

Order	U [V]	I [A]
1	220.6	22.2m
2	0.3	0.2m
3	5.2	19.2m
4	0.1	0.2m
5	0.8	14.5m
6	0.1	0.2m
7	1.8	10.5m
8	0.1	0.1m
9	2.4	9.6m
10	0.1	0.2m
11	1.2	10.3m
12	0.1	0.3m
13	0.5	10.3m
14	0.1	0.3m
15	0.9	9.3m
16	0.0	0.3m
17	0.2	8.2m

Harmonic order Brand-B

Table 3 shows the order of current harmonics for LED lamps brands A and B consecutively as they are displyaed in PQA. The highest current harmonics for brand-A is 11.8 mA and it is present in the 3rd order of harmonics. For brand-B, the highest order also present in the 3rd order with magnitude of current 19.2 mA.

The shapes of waveform of voltage and current for brand-A and brand-B LED lamps are shown in Figure 2. It is clearly seen that current waveform for both lamps is close to impulse rather than sinusoid. These shapes imply the content of harmonics mostly dominant in current due to voltage waveform for both lamps remain sinusoid.

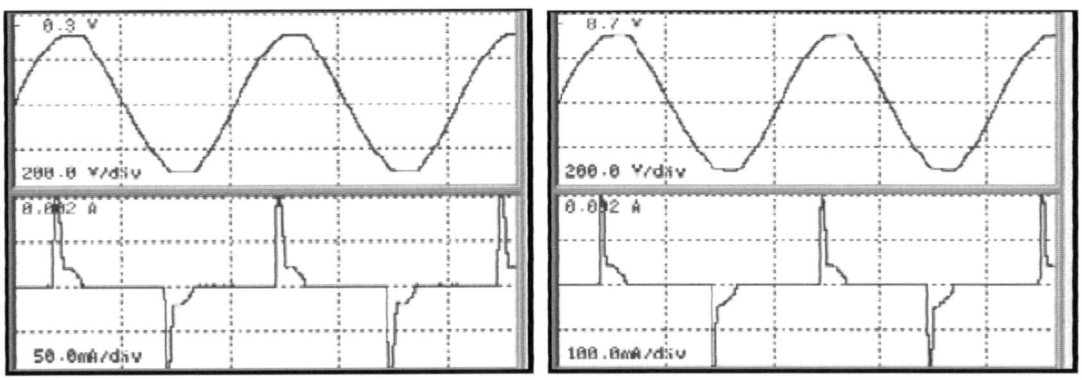

(a) Waveform LED lamp brand-A (b) Waveform LED lamp brand-B

Figure 2. Waveform voltage and current

According to the results above, it is found that both of LED lamp of brand-A and Brand-B have dominant high harmonics distortion for current, so that the work is continued to eliminate the highest harmonics in the current. It is also discovered that the highest magnitude of current was presented in the fundamental and 3rd order. To reduce harmonics significantly the 3rd order harmonics should be eliminated by applying the suitable filter for 3rd order.

4. Design of LC Filter

Passive filter is most commonly used as filtering techniques for mitigation of harmonics, because it offers low impedance path to divert harmonic current caused by non-linear load [5].

In this work, the passive filter is simply composed of one inductor in series with one capacitor to form LC filter because this topology require no power supply to eliminate harmonics. No additional power supply gives advantage to the small dimension of filter. This LC filter only passes low frequency and reduce high frequency so that this circuit is also none as low pass filter [8].

Based on finding in the previous chapter, the 3rd harmonics has the highest magnitude of current, then the filter was designed to eliminate the 3rd order. By using the resonance principle in [5,8], the value of inductor (L) in Henry and capacitor (C) in Farad is determined by equation (1).

$$C = \frac{1}{(2\pi f)^2 h_0{}^2 L} \quad (1)$$

Where, h_o is order of harmonic to which filter is designed to tune. In this paper, the cut-off frequency of harmonic filter is fixed at 3rd order or 150 Hz because fundamental frequency is 50 Hz.

Based on equation (1), the value of inductor is 115 mH and the value of capacitor is 9.8μF. the inductor and capacitor is connected in series as it is shown in Figure 3.

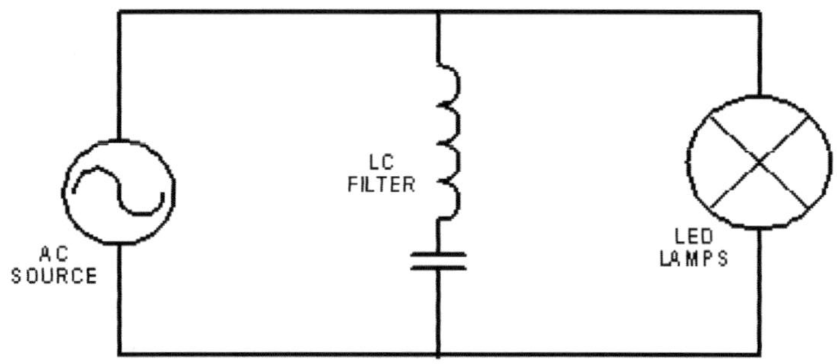

Figure 3. LED lamps with LC filter and AC source

The next experiment is carried out by using circuit in Figure 3 for every lamps of brand-A and brand-B. The measurement result show that after LC filter attached to LED lamps than the voltage waveform remain sinusoid and the current waveform becomes more sinusoidal as shown in Figure 4.

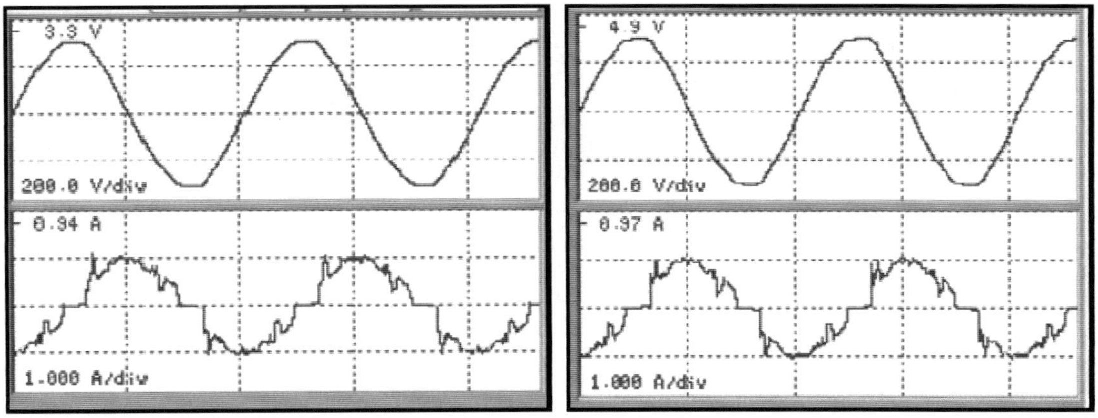

(a) Waveform LED lamp brand-A (b) Waveform LED lamp brand-B

Figure 4. Waveform voltage and current after filtering

Table 4. Percentage Of THD_I LED Lamp Brand-A after filtering

No	Power	THD_V (%)	THD_I (%)	Lumen
1	3W	2.8	19.1	293
2	4W	2.8	18	367
3	7W	2.7	19.2	808

Table 5. Percentage Of THD_I LED Lamp Brand-B after filtering

No	Power	THD_V (%)	THD_I (%)	Lumen
1	3W	2.8	19	403
2	5W	2.8	19.1	529
3	9W	2.9	19	1024

The use of LC filter in LED lamps for both brands-a and brand-B has significantly reduced the value of THDi as it is shown in table 4 and table 5. For example, THD_i decrease from 154 % to 19.1 % for brand-A 3 watt and from 170.6% to 19.1% for brand-B 5 watt. Based on the results in same tables, the lighting intensity for every lamp was found to increase as the current shape become more sinusoid.

5. Conclusion

The harmonics issued in the power system due to the presence of nonlinear loads like LED lamps has been presented in this paper. Harmonics arefound more dominant in current than voltage because THD_i is found greater than THD_v.

The use of LC filter to reduce the dominant order in harmonics is successfully decreased the value of THD_i for every sample of LED lamps without affecting the shape of voltage waveform. The voltage remains sinusoid with lowest THD_v as it was required by standard.

References

[1] Kshitu S, Arti N, Neelam S, and Raghuvir S 2010 *International Journal of Engineering Science and Technology*, **2** 8 pp 3955-3963.

[2] Chaudhary D D, Nayse S P, and Waghmare L M 2011 *Journal of Wireless & Mobile Networks (IJWMN)* **3** 10 pp 140-149.

[3] Syed D, Masood U, Farwa U and Khurram M 2015 *Journal of Computer Engineering and*

IOP Conf. Series: Materials Science and Engineering **190** (2017) 012019 doi:10.1088/1757-899X/190/1/012019

Applications **4** 3 pp 153-163.

[4] Jiao J, Ma H, Qiao Y, Du Y, Kong W, and Wu Z-C 2014 *Advance Journal of Food Science and Technology* **6** 3 pp 368-373.

[5] Kaloxylosa A, Groumasb A, Sarrisb V, Katsikasb L, Magdalinosb P, Antoniouc E, Politopoulouc Z, Wolfertd S, Brewstere C, Eigenmannf R and Tero C-M, *Journal of Computers and Electronics in Agriculture* 100 pp 168–179.

[6] Malik R F, Hafiz M, Nopransyah A, Zalbina MR and Septian L M, 2015 *Proc. of Int. Conf. on Electrical Engineering, Computer Science and* Informatics (*Palembang, Indonesia*).

[7] Lindsey S and Raghavendra C S, 2002 *IEEE Aerospace Conference* (*Montana, USA*) **3** 3 p 1125.

[8] Perera T A and Collins J 2010 *Auckland: Auckland University of Technology. Thesis*.

[9] Mohanan V, Budiarto R, Zainon W M N W 2012 *Proc. 18th Asia-Pacific Conference on Communications (Jeju island, Korea)* pp 53-58.

[10] Mohanan V, Budiarto R 2013 *Proc. 19th Asia-Pacific Conference on Communications (Bali, Indonesia)* pp 196-201.

IOP Conf. Series: Materials Science and Engineering **190** (2017) 012020 doi:10.1088/1757-899X/190/1/012020

Evaluation of skew-planar antenna for UAV communication at 2.4 GHz band

T Prakoso, A A Yaqin, Ibrahim, I Santoso, A Triwiyatno, M A Riyadi

Department of Electrical Engineering, Faculty of Engineering, Universitas Diponegoro, Semarang 50275, Indonesia

teguhprakoso@elektro.undip.ac.id

Abstract. Skew-planar antenna is suitable for unmanned aerial vehicles (UAVs) since it has circular polarization (avoiding polarization-mismatched), omnidirectional radiation pattern (low pointing loss) thus comply with maneuver angles of UAV, and also small size. Investigation with electromagnetic simulations shows that tilt angle and coaxial length are the most sensitive parameters. Modified design with 55° tilt angle (original design: 45°) and 100 mm coaxial length (original: 60 mm) potentially provides wider bandwidth (400 MHz vs. 290 MHz), better impedance-matched ($|S_{11}|$: -47.1 dB vs. -13.6 dB), and better gain (1.98 dBi vs 1.61 dBi). The antenna is predicted capable to support service distance of 3 km in 2.4 GHz band if used with communication module with 18 dBm transmit power and -100 dBm receiver sensitivity. Therefore, the modified antenna is suitable to be applied in UAV.

1. Introduction

Communication system is one of the most important systems on an unmanned aerial vehicle (UAV). The performance of this system determines the maximum flight range and the adaptability in complex environment [1]. UAVs have several challenges to keep reliable communication through long distance, wide area, and flight maneuver and height angles, in all flight condition. UAVs fly on route by adjusting their pitch, yaw, and roll angles during their missions. It means that relative angle between UAVs and their ground station controllers (GSCs) are changing during the mission. If both UAVs and GSCs communicate to each other using linearly-polarized antennas, the UAVs' maneuvers are expected to produce polarization-mismatched. This effect reduce signal power captured by receive antenna and may result in communication interruption that in some situation are fatal for a mission. To avoid this problem, circularly-polarized antennas are installed at UAVs and GSCs. The wide angle maneuvers, variability of height and distance also suggest the use antennas with wide angle of radiation pattern, i.e. omnidirectional antennas. One of antenna type that owns circular polarization and omnidirectional radiation pattern is skew-planar antenna.

Skew-planar antenna is modification of cloverleaf antenna [2]. The original cloverleaf antenna basically is loop antenna which is divided into four segments (leaves) to ensure uniformity of current phase along the loop. By employing this method omnidirectional radiation pattern and horizontal polarization as required by FM broadcasting system can be produced. Planar version of cloverleaf antenna (dual polarization) is applied in Canadian Hydrogen Intensity Experiment (CHIME) [3]. Another modification to cloverleaf antenna is by skewing the leaves upward (45°) to produce omnidirectional, circularly-polarized antenna namely skew-planar antenna [4].

With capability to provide circular polarization, omnidirectional radiation pattern, enough beamwidth and gain, as well as small size, skew-planar antenna is popular in UAV community. However, an investigation on the effect of antenna parameters to antenna performance is still needed. This study can lead to design optimization and provide trade-off antenna parameters and performance measures (reflection coefficient (S_{11}), gain, radiation pattern, bandwidth, axial ratio). Section II of this paper presents the effect of tilt angles and coaxial length to the antenna performance ($|S_{11}|$, gain). Second part of this paper (Section III) discusses service distance prediction of this antenna employed in communication between UAV and ground controller.

2. Antenna design and parametric study

2.1. Antenna design

The design of skew-planar antenna is shown in figures 1 and 2. This antenna consists of four elements (leaves) that are tilted at θ degrees and connected to inner conductor and outer conductor of coaxial cable. *Lb* is the length of straight part of the leaf. *La* is the length of curved segment of the leaf. Combination of *La* and *Lb* produces "φ." *Lc* is the length of coaxial cable. *S* is the gap between connected antenna element inner conductor and outer conductor of coaxial cable, *D* is diameter of copper wire. *Lt* is length total of element that equal 1 λ at 2.4 GHz which is

$$Lt = 2Lb + La = \lambda \quad (1)$$

For the following parametric study and antenna simulations, coaxial cable RG-316/U is used. The material of the antenna element (leaf) is copper.

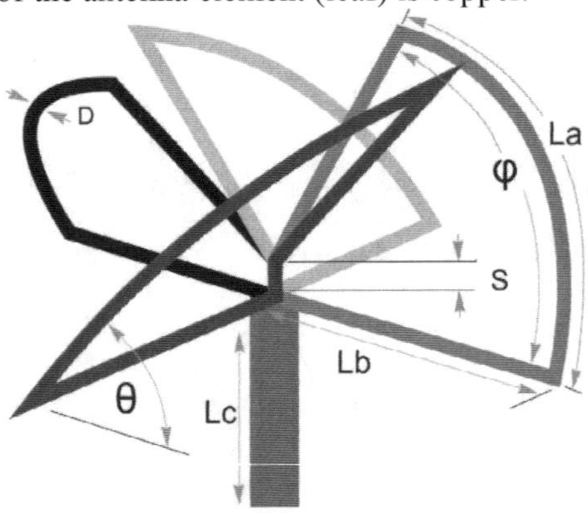

Figure 1. Design of skew-planar antenna.

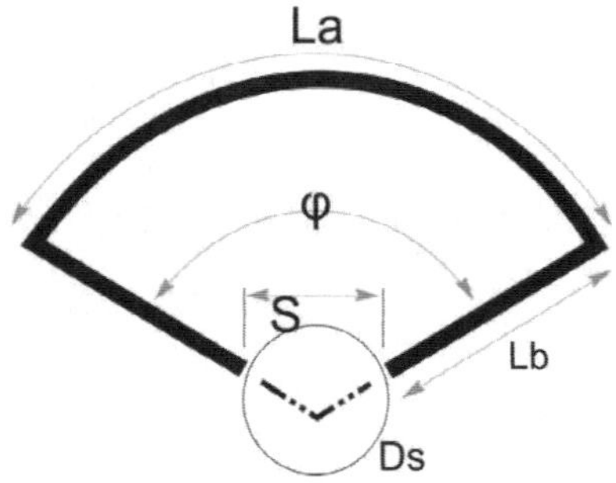

Figure 2. Detail of a leaf.

2.2. Parametric study

Some parameters have been varied and their effects on antenna performance metrics have been analyzed. The parameters are antenna element angle (φ), antenna element tilt angle (θ), diameter of copper wire (*D*), coaxial cable length (*Lc*), and leaf length (*Lt*). The antenna performance metrics are reflection coefficient ($|S_{11}|$) and gain. In this study, CST Microwave Studio was used.

It is found that skew-planar antenna is very sensitive to leaf tilt angle (θ) and coaxial cable length (*Lc*), quite sensitive to wire diameter (*D*), and less sensitive to total length of a leaf (*Lt*) and leaf angle (φ). With this consideration, results of parametric sweep on θ and *Lc* are presented here.

2.2.1. Tilt angle (θ). Generally this antenna is designed with θ = 45° to produce orthogonal electric fields which the same in magnitude; combined with phase difference of 90° circular polarization radiation can be resulted. Figure 3 shows that tilt angle affects reflection coefficient greatly. Varying θ from 35° to 65° shifts the resonant frequency (f_r) in the range of 114 MHz; f_r shifts to low frequency until θ = 50° (label 5) and then goes to higher frequency. It also affects $|S_{11}|$ at resonant frequencies and their respective bandwidth ($|S_{11}| \leq$ -10 dB). The figure shows that θ = 45° does not provide widest bandwidth and the best matched at resonant frequency. In term of $|S_{11}|$, θ = 55° demonstrates the best matched and bandwidth. Directivity as function of θ is depicted in figure 4 and 5. In the azimuth-cut, this antenna omnidirectional pattern with directivity variation around 0.15 dB. The highest value of directivity owns by θ = 30° and the lowest by θ = 65°, and θ = 55° has value in the middle. The elevation-cut shows similar values except for angle -150° to -100° and 100° to 150°. Considering these facts, tilt angle of 55° (label "6") potentially produce better performance than commonly-known tilt angle (θ = 45°): (a) operational frequency from 2.31 – 2.67 GHz, (b) bandwidth of 360 MHz, (c) maximum directivity of 2.17 dBi, and (d) beamwidth of 113°.

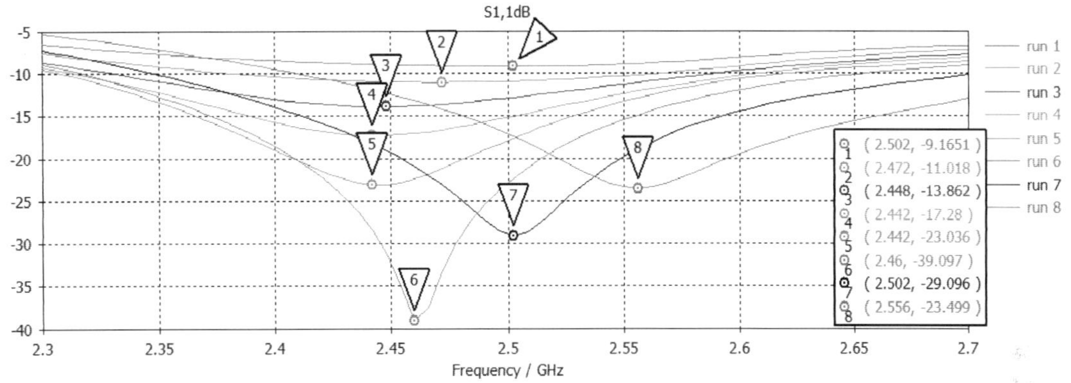

Figure 3. Effect of element's tilt angle variation on reflection coefficient ($|S_{11}|$) values (label "1": 30°, "2": 35°, "3": 40°, "4": 45°, "5": 50°, "6": 55°, "7": 60°, and "8": 65°).

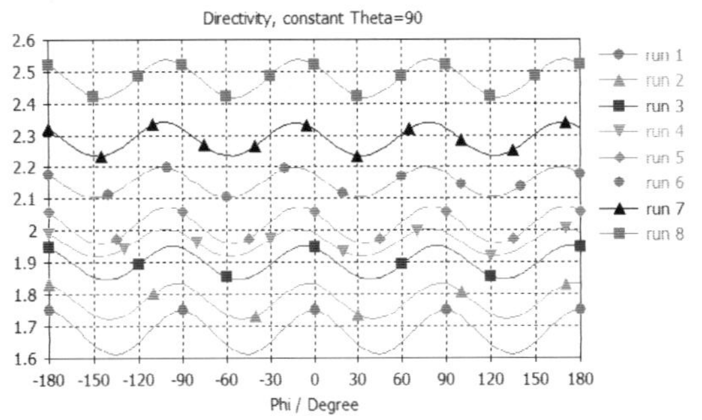

Figure 4. Directivity (dBi, vertical axis) at 2.4 GHz vs. tilt angle at azimuth-cut (labels are the same with Figure 3).

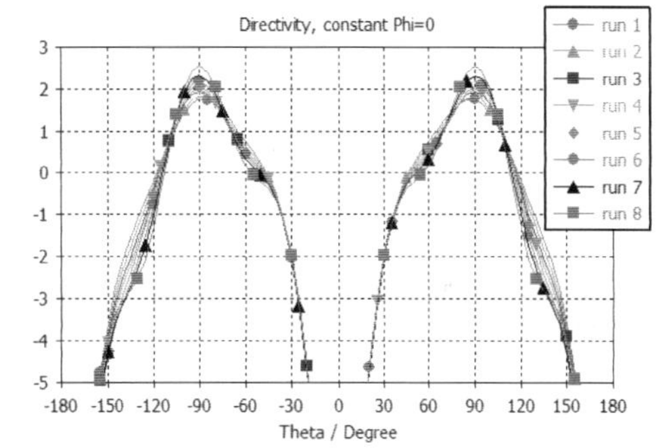

Figure 5. Directivity (dBi, vertical axis) at 2.4 GHz vs. tilt angle at elevation-cut (labels are the same with Figure 3).

2.2.2. Coaxial cable length (Lc). The length variation from 30 to 150 mm shifts the resonant frequency around 100 MHz, see figure 6. Directivity values in azimuth-cut as presented in figure 7 shows that generally the radiation pattern is omnidirectional with variation around 0.2 dB except for length of 150 mm. Combining figure 6 and 7, the best coaxial length setting is 100 mm. Figure shows that the maximum directivity of this length is 2.17 dBi and beamwidth of 113°. This figure also demonstrates that the antenna has doughnut-shaped radiation pattern with nulls at zenith and nadir.

2.3. Proposed design

From the parametric study, combination parameters that produce the best performance in terms of reflection coefficient and radiation are selected, see table 1. As comparison, original design of skew-planar antenna is also presented in the table. Figure 9 compares realized gain, right-hand polarization, in azimuth-cut. It shows that the proposed design ("modified") and original design ("original") has omnidirectional radiation pattern. The modified design has higher realized gain than the original one.

Figure 10 depicts comparison of realized gain, right-hand polarization, in elevation-cut. The proposed design has stronger radiation at main lobe. Figure 9 and 10 indicate that 3D radiation patterns of both antennas are doughnut-shape.

Reflection coefficient comparison of proposed and original skew-planar is presented in figure 11. The proposed design shows significantly better bandwidth and impedance-match degree compared to the original design. The operational frequency for modified one is 2.313 – 2.712 GHz (bandwidth: 400 MHz) and original one is 2.276 – 2.561 GHz (bandwidth: 290 MHz) hence both antennas are suitable to be used at 2.4 GHz band. The resonant frequency for the proposed design is 2.47 GHz (|S11| = -47.1 dB) and the original one is 2.41 GHz (|S11| = -13.6 dB).

Axial ratio comparisons of both antennas are depicted in figure 12 (azimuth-cut) and figure 13 (elevation-cut). In horizontal-cut, axial ratio of original antenna is around 2 dB and axial ratio of modified one is around 4 dB. It means that the original one is closer to perfect circular polarization than the modified one. Finally, all of these comparisons are summarized in table 2.

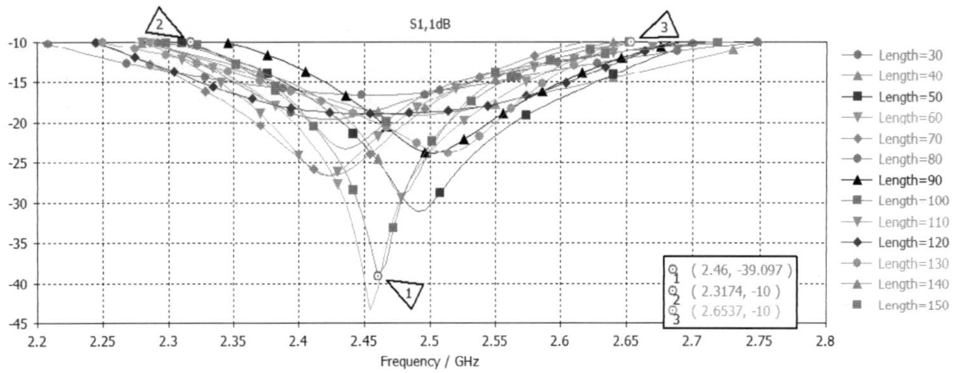

Figure 6. Effect of coaxial length variation on reflection coefficient ($|S_{11}|$) values. The length unit is mm, and $|S_{11}|$ is in dB.

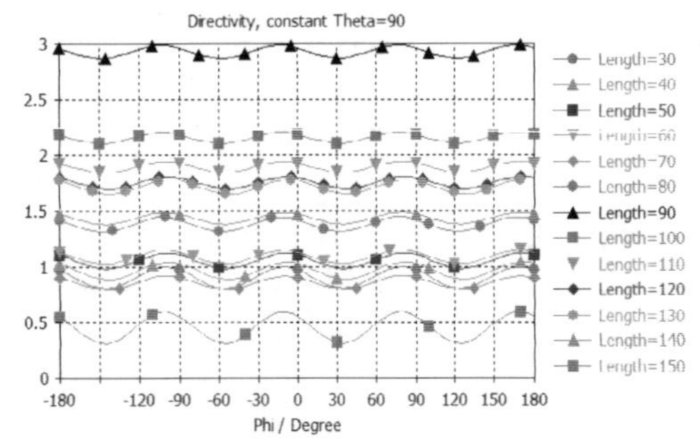

Figure 7. Directivity (dBi, vertical axis) at 2.4 GHz vs. tilt angle at azimuth-cut. The unit of length is mm.

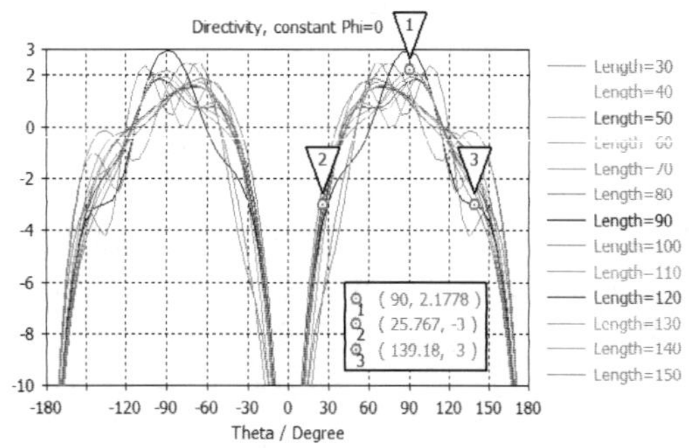

Figure 8. Directivity (dBi, vertical axis) at 2.4 GHz vs. tilt angle at elevation-cut. The unit of length is mm.

3. Prediction of antenna performance in UAV link

Based on the simulation results of the proposed design, service distance of UAV can be predicted. The calculation assumes small-scale fading margin of 6 dB, pointing error loss of 3 dB, gain variation penalty of 2.1 dB, connector and cable loss of 0.19 dB, and the antenna is used in pair with cloverleaf antenna in the link. The predicted service distance employing various communication modules working at 2.4 GHz band is presented in table 3. For low altitude fixed-wing UAV, this antenna can be used with communication module with 16 – 18 dBm transmit power (P_{tx}) and -98 to -100 dBm receiver

IOP Conf. Series: Materials Science and Engineering **190** (2017) 012020 doi:10.1088/1757-899X/190/1/012020

sensitivity(S_{rx}) to get more than 2 km service distance. Module with only 100 dB difference between P_{tx} and S_{rx} only capable to provide around 1 km distance. For short distance UAV such as quadcopter, this antenna is enough; however, further distance mission needs the utilization of high gain circular antenna. Skew-planar antenna in pair with 10 dBi axial mode helical antenna can support 10 km distance. Far distance mission uses ground controller antenna which can reach 40 dBi gain [5].

4. Conclusion

The most sensitive parameter of skew-planar antenna is leaf tilt angle and then coaxial length. Modifying the tilt angle to 55° and optimizing other parameters potentially produce wider bandwidth and higher gain, in expense of axial ratio, compared to original design. The proposed antenna is predicted to support service distance of 3 km if employed with communication module which has 18 dBm transmit power and -100 dBm receiver sensitivity. Therefore, this antenna is suitable for fixed-wing UAV.

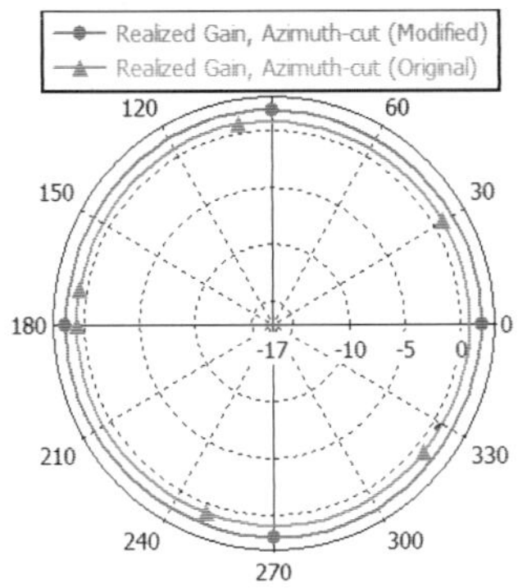

Figure 9. Comparison of realized-gain values, right-hand polarization, azimuth-cut at 2.4 GHz.

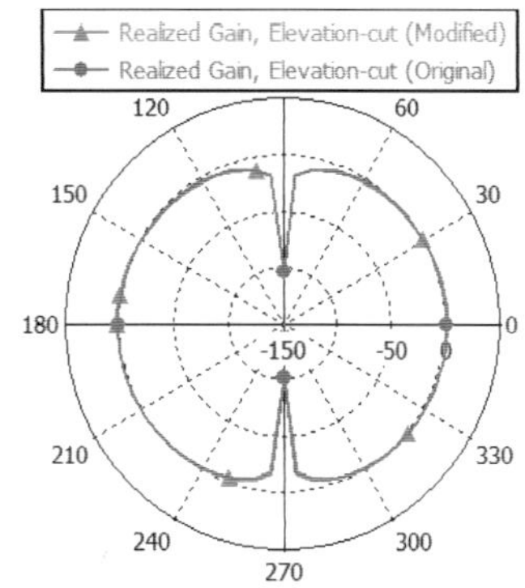

Figure 10. Comparison of realized-gain values, right-hand polarization, elevation-cut at 2.4 GHz.

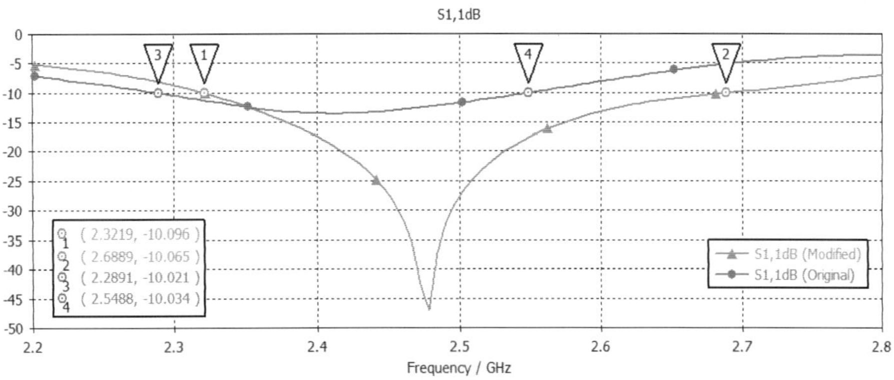

Figure 11. Comparison of reflection coefficient (|S11| in dB, vertical axis) values of modified and original designs.

Table 1. Dimension comparison of modified and original designs.

Parameter	Modified Design	Original Design
Leaf length (*Lt*)	125 mm	125 mm
Antenna element angle (φ)	103.4°	108.3°
Antenna element tilt angle (θ)	55°	45°
Length coaxial cable (*Lc*)	100 mm	60 mm
Diameter of copper wire (*D*)	1 mm	0.8 mm

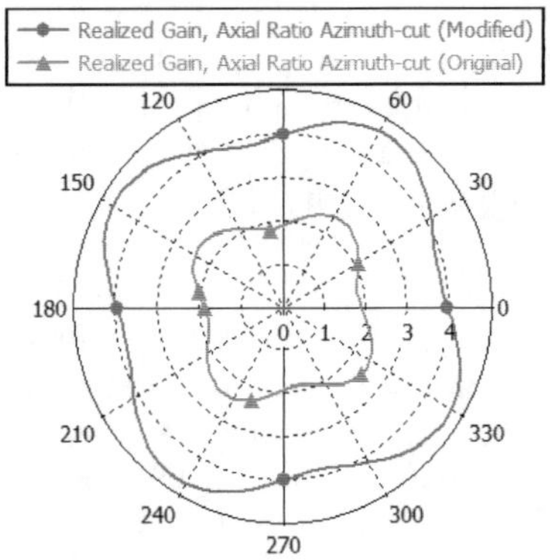

Figure 12. Comparison of realized-gain values, right-hand polarization, azimuth-cut at 2.4 GHz.

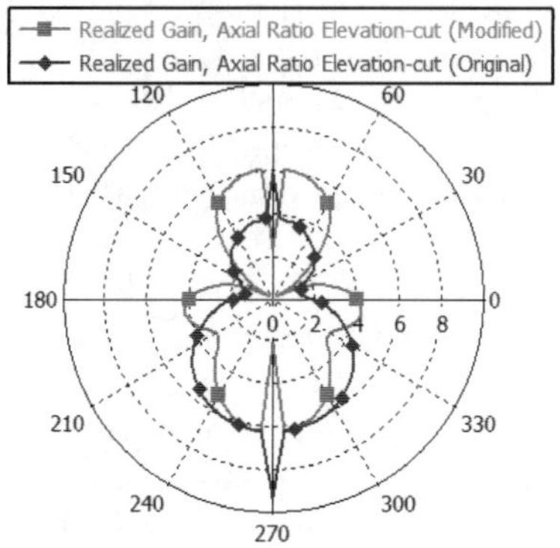

Figure 13. Comparison of realized-gain values, right-hand polarization, elevation-cut at 2.4 GHz.

Table 2. Performance metric comparison of modified and original designs.

Performance metric	Modified	Original
Resonant frequency	2.47 GHz	2.41 GHz
$\|S_{11}\|$ at resonant frequency	-47.1 dB	-13.6 dB
VSWR at resonant frequency	1.0	1.5
Operational frequency	2.313 – 2.712 GHz	2.276 – 2.561 GHz
Bandwidth (VSWR≤2)	400 MHz	290 MHz
Beamwidth (≥-3dB)	110.9°	116.7 °
Gain (main lobe)	1.98 dBi	1.61 dBi
Radiation pattern	Omnidirectional	Omnidirectional
Polarization	Right hand circular	Right hand circular

Table 3. Predicted service distance.

Communication transceiver module	Elevation Angle (°)							
	unit	0	15	30	45	60	75	90
Xbee zigbee s2c	km	1.26	1.07	0.74	0.55	0.36	0.12	0.00
Xbee Pro zigbee s2c	km	3.56	3.03	2.10	1.56	1.02	0.33	0.00
Xbeee Pro Digimesh 2.4	km	3.17	2.70	1.87	1.39	0.91	0.30	0.00
Xbee wifi	km	1.12	0.96	0.66	0.49	0.32	0.10	0.00
DJT Fry Sky telemetry	km	2.45	2.09	1.45	1.08	0.70	0.23	0.00

References

[1] Barnard J 2007 *IET Seminar on Communicating with UAV's* p 75
[2] Smith PH 1947 *Proceedings of the IRE* **35** p 1556
[3] Deng M and Campbell-Wilson D 2014 *16th International Symposium on Antenna Technology and Applied Electromagnetics (ANTEM)* (New Jersey: IEEE) p 1.
[4] ARRL 1964 *The A. R. R. L. Antenna Book* (Newington: American Radio Relay League) p 319
[5] Alonso JMI and Perez MS 2015 *IEEE Antennas and Wireless Propagation Letters* **14** 771

A Comparisson of Synchronous And Nonsynchronous Boost Converter

Mohamad Isnaeni Romadhon, Trias Andromeda, Mochammad Facta, Agung Warsito
Departemen Teknik Elektro, Universitas Diponegoro Semarang

Jl. Prof. Sudharto, SH, Kampus UNDIP Tembalang, Semarang 50275, Indonesia
mohamadisnaeni116@gmail.com, trias1972@gmail.com, facta@elektro.undip.ac.id

Abstract. Modern electronic systems require resources with high efficiency. The efficiency of direct current to dicrect current converters as a power source can be increased by replacing a diode with MOSFET. The use of MOSFET is expected to reduce power loss as the internal resistance of MOSFET is lower than a diode. To implement the propossed idea, a boost type direct current chopper and TL494 as PWM generator circuit were applied in this work. MOSFET is used in synchronization mode to replace diode at conventional topology of chopper.. The proposed circuit and conventional topology were made and their performance were observed. The efficiency of both circuit were compared and analyzed. The result of the experiments showed that the efficiency of converter within MOSFET at synchronization mode is proportional with the increment of duty cycle, while at conventional topology the efficiency remain stable at any duty cycle. Synchronous boost converter is more efficient than nonsynchronous boost converter at duty cycle over than 40%.

1. Introduction

Modern electronic systems require resources with high efficiency [1]. The efficiency of DC to DC converter as power source of electronic equipment can be increased by replacing diode with MOSFET [2]. DC to DC converter within MOSFET synchronization mode called synchronous DC to DC converter. Both of MOSFET on synchronous converter must have similar frequency and does not work simultaneously. MOSFET is expected to reduce power loss at DC to DC converter because MOSFET has low internal resistance ($R_{ds(on)}$) during conduction mode [3]. The experiment will calculate and compare conduction loss at TL494 synchronous and nonsynchronous boost converter then analyze the efficiency [4, 5].

2. Method

2.1. Nonsynchronous Boost Converter

IOP Conf. Series: Materials Science and Engineering **190** (2017) 012021 doi:10.1088/1757-899X/190/1/012021

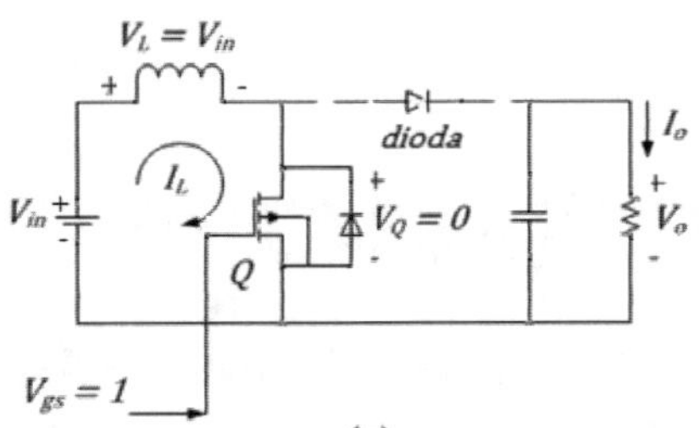

(a)

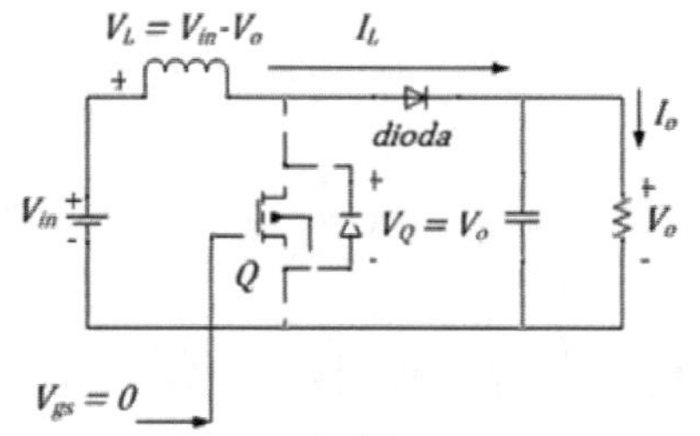

(b)

Figure 1a. Nonsynchronous boost converter schematic equivalent circuit when Q active but diode inactive.

Figure 1b. Nonsynchronous boost converter schematicequivalent circuitwhen Q inactive but diode active.

The MOSFET (*Q*) is active when the nonsynchronous converter run at *D* duration. The source only supplies the inductor (*L*), so the inductor voltage (V_L) equals the input voltage (V_{in}) as shown in Figure 1a. The inductor current (I_L) linearly ramps up $\left(dI_L/dt\right)$ at *D* duration and increasing the energy in inductor *L* (Figure 2). The $\left(dI_L/dt\right)$ shifting current may call asΔI_L.

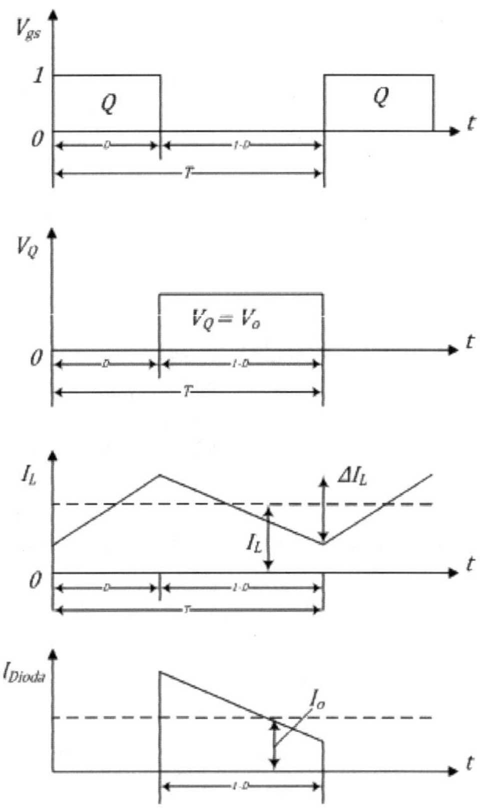

Figure 2. Ideal waveform of nonsynchronous boost converter.

Q is inactive. When the nonsynchronous boost converter run at *1-D* duration. I_L flows through the diode as shown in Figure 1b. The inductively stored energy transferred to the output stage. I_L linearly ramps down $\left(-dI_L/dt\right)$ at *1-D* duration. The $\left(-dI_L/dt\right)$ shifting current may call as $-\Delta I_L$ (Figure 2).

In a boost converter, I_L equals the input current (I_{in}). The average of I_L can be calculated from the output current (I_{out}) by equating the input and the output power [4]:

$$Vin \times Iin = Vout \times Iout \quad (1)$$

Equation 1 replaced by Equation 2:

$$Iin = \frac{Vout}{Vin} \times Iout = \frac{Iout}{(1-D)} \quad (2)$$

2.2. Synchronous Boost Converter

The low-side MOSFET (*Q1*) is active when synchronous boost converter run at *D* duration. The source only supplies the inductor (*L*). The inductor voltage (V_L) equals the input voltage (V_{in}) as shown in Figure 3a. The inductor current (I_L) linearly ramps up,and increasing the energy in *L* (Figure 4). Then the synchronous boost converter run at dead time (D_t) duration, in this case *Q1* is inactive. I_L flows through the high-side MOSFET body diode (*D2*), because in this period the high-side MOSFET (*Q2*) is inactive. The inductively stored energy transferred to the output stage through *D2* (Figure 3b).

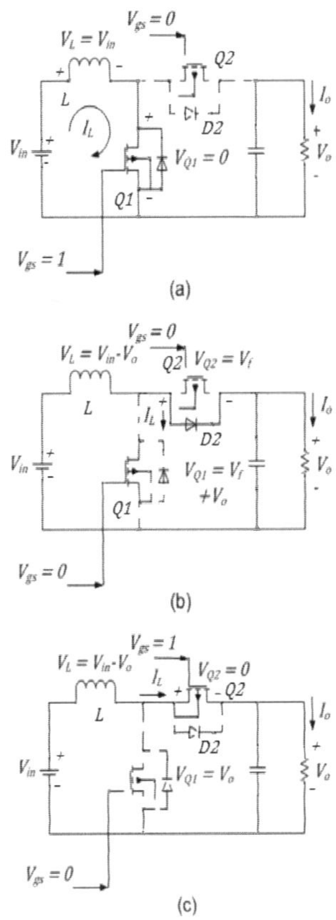

Figure 3. Synchronous boost converter schematic
(a) Q1 is active and Q2 is inactive
(b) Q1 is inactive and D2 is active
(c) Q1 is inactive and Q2 is active

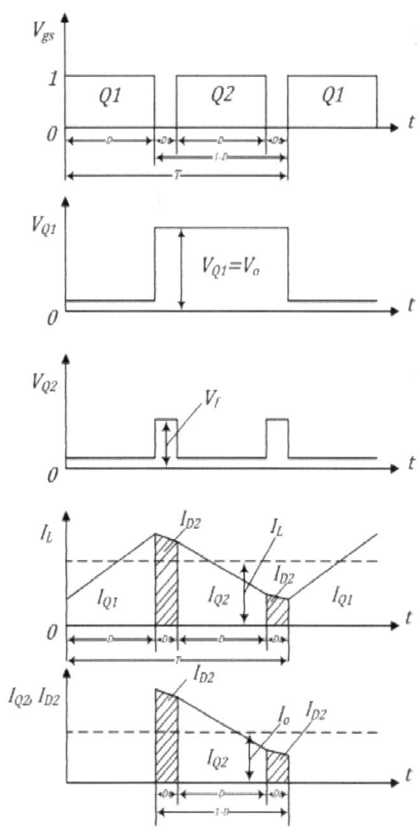

Figure 4 Ideal waveform of synchronous boost converter

The synchronous boost converter runs at the second D duration, in this condition $Q2$ is active and $Q1$ is inactive (Figure 3c). I_L flows through $Q2$. The inductively stored energy transferred to the output stage, so I_L linearly ramps down decreasing the energy in L (Figure 4).

3. Result

The power dissipation at MOSFET conduction process is major loss in DC converter system [5]. The power dissipation at MOSFET conduction process comes about drain to source resistance at MOSFET saturation mode. In this case the output current is directly proportional to the power dissipation. Basically,$I^2R_{ds(on)}$ at synchronous mode is smaller than IV_{fD} at nonsynchronous. This point ensues the low of MOSFET$R_{ds(on)}$.
Efficiency at the general system can be represented by Equation 3:

$$\eta = \frac{P_{out}}{P_{in}} \tag{3}$$

$P_{in}$is the input power, which consist of the output power (P_{out}), the power rectifier losses (P_{Rec}) and the power loss exclude the rectifier loss (P_{loss}) [5-7]:

$$P_{in} = P_{out} + P_{loss} + P_{\mathrm{Re}\,c} \tag{4}$$

P_{loss}can be ignore, so Equation 4 replaced by Equation 5 [5-7]:

$$\eta = \frac{P_{out}}{P_{out} + P_{\mathrm{Re}\,c}} \tag{5}$$

P_{Rec} at the nonsynchronous boost converter or P_{RecD} is accumulated losses from the diode and MOSFET when both of them conduct. P_{RecD} can be calculated by Equation 6 [5]:

$$P_{\mathrm{Re}\,cD} = P_{conQ} + P_{swQ} + P_{conD} + P_{gateQ} + P_{\mathrm{r}\,recQ,\,D} \tag{6}$$

At the low frequency DC converter whose lower than 300kHz, P_{gateQ} dan$P_{rrecQ,D}$ can be eliminated [5]. Equation 5 is replaced by Equation 7 [5-7]:

$$\eta = \frac{P_{out}}{P_{out} + P_{conQ} + P_{swQ} + P_{conD}} \tag{7}$$

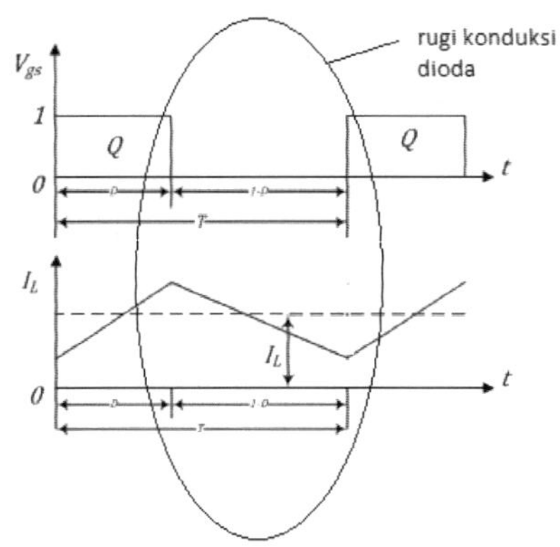

Figure 5 Ideal diode power dissipation area of nonsynchronous boost converter

P_{conQ} is power dissipation at Q while$Q$ is active. Q is active when DC converter is working at D duration, so P_{conQ} can be calculated by I_L, D, and $R_{ds(on)Q}$ parameters [2]:

$$P_{conQ} = R_{ds(on)Q} \times \frac{I_O}{(1-D)} \times D \tag{8}$$

where:
P_{conQ} = Q power dissipation
$R_{ds(on)Q}$ = Q drain to source resistance
I_o = output current
D = duty cycle

P_{conD} is power dissipation at the diode while Q is inactive. Q is inactive when DC converter is working at $1\text{-}D$duration, as shown at Figure 5. P_{conD} can be calculated by I_L, $1\text{-}D$, and the diode forward voltage drop (V_{fD}) parameters [2]:

$$P_{conD} = V_{fd} \times I_O \tag{9}$$

P_{Rec}at the synchronous boost converteror P_{RecSM} is accumulated losses from all loss at $Q1$dan$Q2$, soP_{RecSM} can be calculated by Equation 10 [5]:

$$P_{RecSM} = P_{conQ1} + P_{swQ1} + P_{conQ2} + P_{conD2} + P_{gateQ1,Q2} + P_{rrecQ1,D2} \tag{10}$$

Equation 10 replaced by Equation 11 [5]:

$$\eta = \frac{P_{out}}{P_{out} + P_{conQ1} + P_{swQ1} + P_{conQ2} + P_{conD2}} \tag{11}$$

$P_{conQ1}$is$Q1$loss while $Q1$is active. $Q1$is active when the DC converter is working at D duration. P_{conQ1}can be calculated byI_L, D and $R_{ds(on)Q1}$ parameters [2]:

$$P_{conQ1} = R_{ds(on)Q1} \times \frac{I_O}{(1-D)} \times D \tag{12}$$

where:
P_{conQ1} = Q1 power dissipation
$R_{ds(on)Q1}$ = Q1 drain to source resistance
I_o = output current
D = duty cycle

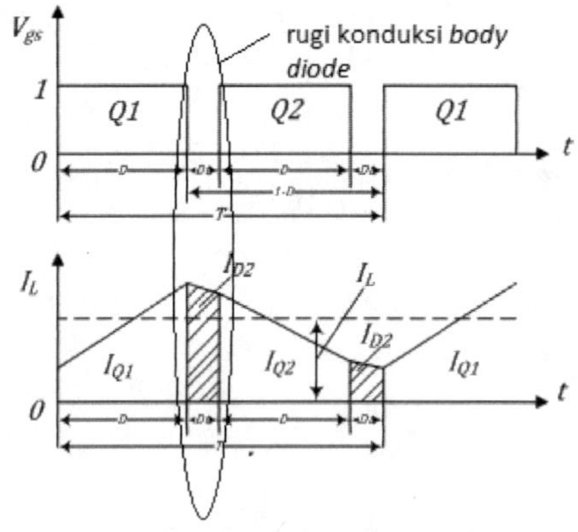

Figure 6 Ideal body diode power dissipation area of synchronous boost converter

$P_{conD2}$is$D2$power dissipation while $Q1$and $Q2$ is inactive. In this case the DC converter is working at D_t duration (Figure 6). The inductively stored current flows through diode, soP_{conD2}can be calculated by I_L, D_t and the forward voltage drop $D2$ (V_{fD2}) parameters [2,5]:

$$P_{conD2} = 2 \times D_t \times V_{fD2} \times \frac{I_o}{(1-D)} \quad (13)$$

P_{conQ2} is Q2 power dissipation while Q1 is inactive and Q2 is active. In this case, the DC converter is working at D duration (Figure 7). P_{conQ2}can be calculated byI_L, D, and$R_{ds(on)Q2}$ parameters [2]:

$$P_{conQ2} = R_{ds(on)Q2} \times \frac{I_o}{(1-D)} \times D \quad (14)$$

where:
P_{conQ2} = Q2 power dissipation
$R_{ds(on)Q2}$ = Q2 drain to source resistance
I_o = output current
D = duty cycle

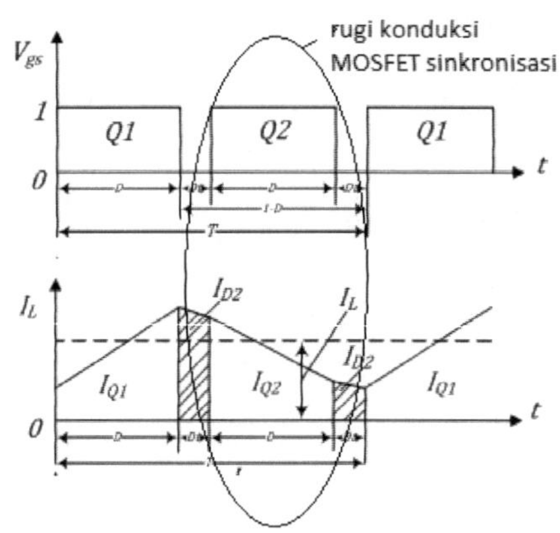

Figure 7 Ideal MOSFET $Q2$ power dissipation area of synchronous boost converter

The main block for experimental circuits for nonsynchronous and synchronous boost converters are shown in Figure 8.

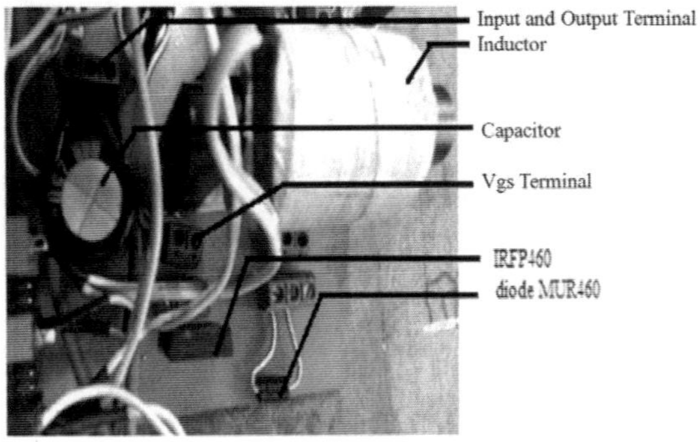

(a)

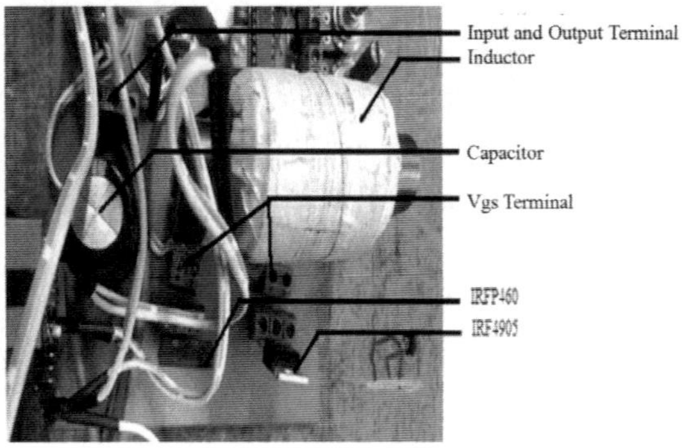

(b)

Figure 8 (a) nonsynchronous and (b)synchronous boost converter

The conduction loss comparison result of the synchronous and nonsynchronous boost convereter showed at Table 1:

Table1 Conduction loss comparison result of the synchronous and nonsynchronous boost converter.

Duty cycle (%)	Synchronous boost converter (W)	Nonsynchronous boost converter (W)
6,8	1,411719	0,531607
10	1,44159	0,50175
15	1,479826	0,575158
20	1,312618	0,595362
25	1,271958	0,63057
30	1,134797	0,658106
35	0,957773	0,704584
40	0,856632	0,804452
43,5	0,729494	0,943483

Table 1 present the result of the synchronous and nonsynchronous boost converter power dissipation comparison. The power dissipation of nonsynchronous boost converter increases proportionally with duty cycle. The power dissipation of synchronous boost converter is inversely proportional with duty cycle.

The output characteristic of TL494 PWM generator at push pull mode where deadtime grown at small duty cycle and dead time alight at high duty cycle will affect to conduction loss of the synchronous boost converter.

IOP Conf. Series: Materials Science and Engineering **190** (2017) 012021 doi:10.1088/1757-899X/190/1/012021

When the duty cycle is enlarged, the output current flowing through the body diode goes up. The power dissipation at body diode will be rise. The efficiency of synchronous and nonsynchronous boost converter can be calculated with Equation 11 and 7, as shown at Figure 9:

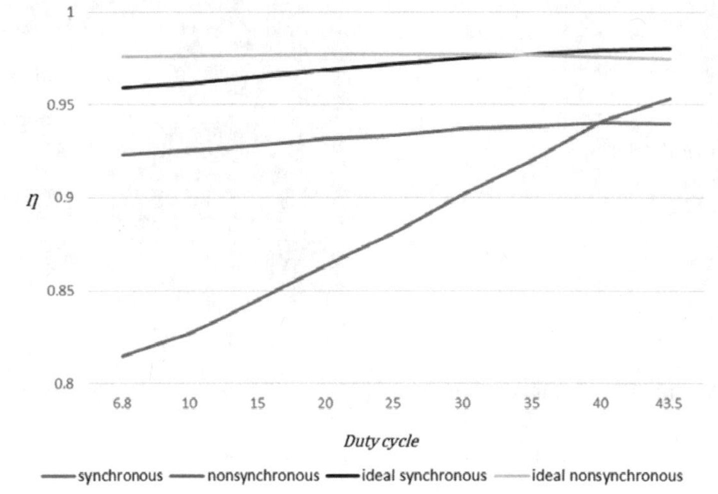

Figure 9 Efficiency comparison of synchronous and nonsynchronous boost converter

4. Conclusion

At the TL494 synchronous boost converter, the efficiency linearly ramps up when duty cycle increase. The efficiency is steady on various duty cycles. The efficiency of synchronous boost converter is greater than nonsynchronous when duty cycle is more than 40%. Minimizing duty cycle has made dead time increases, so conduction loss of body diode at the synchronous converter topology is less. Selected appropriate component at synchronous converter topology can maximize the efficiency.

5. References

[1] Rashid Muhammad H, 2001, *Power Electronics Handbook* (Florida: University of Florida)

[2] Jaunay Serge and Brown Jess, *DC to DC Design Guide* (Vishay Siliconix AN607)

[3] Hart Daniel W, 2001, *Power Electronics – DC-DC Converter,* (Indiana: Valparaiso University)

[4] Mohan Ned, 2012, *Power Electronic – A First Course* (Minneapolis: University of Minnesota)

[5] Zhang M.T, Jovanovic M.M , Lee F.C.Y , 1998, "Design Considerations and Performance Evaluations of Synchronous Rectification in Flyback Converters", *IEEE Transactions on Power Electronics,* vol. 13 No. 3, pp. 538 - 546

[6] Dittmer Greg, 2008, *Synchronous Boost Converter Provide High Voltage Without The Heat Linear Technology Magazine*

Geometrical Shape Investigation For Electrodes in Silent Discharge Chamber

Mochammad Facta[1], Ngurah Ayu Ketut Umiati[2] , Agung Warsito[1]

[1] Departemen Teknik Elektro, Universitas Diponegoro Semarang, Indonesia
[2] Departemen Fisika, Universitas Diponegoro Semarang
Jl. Prof. Sudharto, SH, Kampus UNDIP Tembalang, Semarang 50275, Indonesia

facta@elektro.undip.ac.id; mochfacta@gmail.com;

Abstract. Silent discharge is the most prominent method to carry out plasma reaction because. discharge is easily initiated by injecting alternating current in high voltage to the pair of separated electrodes. The electron emission from surface of dielectric placed on instantaneous cathode is stimulated by ion induced electron emission. In this method, spark is avoided by placing insulation material to either one or both of the electrodes. In practical, it is very difficult to determine the exact limit of the voltage that initiate the discharge by mathematical analysis because it depends on many factors, namely dimensions, type and geometrical shapes of electrode, thickness of insulation, and type of electric field inside the discharge gap. To get lower initial voltage for discharge, it is important to find the best geometrical shape of electrode in relation to skin effect that trigger electron emission. This work investigates the behaviour of charges, current, electric field and voltage surrounding electrodes with various geometrical shape.

1. Introduction

Silent discharge method covered larger area of excitation than pulse discharge method, and this meant that silent discharge method generated is able to generate more gas discharges inform of micro discharge than pulse. Basically the silent discharge chamber is constructed by two pair of electrode, air gap in between and a dielectric layer that covers at least one or both electrode. This paper presents an investigation for the construction of the chamber which is support the micro discharge with minimal discharge deformation. It can be shown through computer simulation that geometrical shape of electrode and chamber construction give influence to micro discharge initiation and prevent the deformation of plasma such as ozone.

2. Geometrical Construction for Electrodes

In design of the geometrical shape of the electrode for silent discharge application, it must be noted that alternating current (AC) high frequency will be injected inside the chamber and this event will create skin effect. Skin effect is physical phenomena which show the tendency of AC to distribute itself within a conductor so that the current density near the surface of the conductor is greater than that at its core. The electric current tends to flow at the skin of the conductor at average depth which is known as skin depth. The skin depth is defined as the depth below the surface of the conductor at which current density decays. The skin depth is calculated by empirical formula in Equation (1) [1-4].

$$\delta = \sqrt{\frac{2\rho}{\omega\mu}} \tag{1}$$

Where ρ is resistivity of conductor, ω is angular frequency of current that is equal to $2\pi \times$ frequency, and μ is absolute magnetic permeability of conductor that is equal to $\mu_o \cdot \mu_r$. The Value of permeability of free space (μ_o) is $4\pi \times 10\text{-}7$ and relative permeability of the material conductor is symbolized as μ_r.

Due to skin effect, the current flows in an extremely thin region near the surface of the electrode. When the electrode is injected by an external voltage source, an electric field (*E*) presents in the conductive material. The relationship between current density (*J*) and *E* can be written as [5-8]:

$$E = \frac{1}{\sigma} J \tag{2}$$

and

$$\nabla^2 E = -\nabla \frac{\partial B}{dt} \tag{3}$$

while

$$B = \mu H \tag{4}$$

By replacing E and B in equation (3) with equation (2) and (4) then new equation is obtain in equation (5).

$$\frac{1}{\sigma} \nabla^2 J = -\mu \nabla \frac{\partial H}{dt} \tag{5}$$

Equation (5) is possible to be modified in both side and it become equation

$$-\nabla^2 J = -\sigma\mu \frac{\partial}{dt} \left\{ J + \frac{\partial(\varepsilon E)}{dt} \right\} \tag{6}$$

where ε and σ are the permittivity and conductivity if the conductor, respectively. By considering the first and second derivative of *E* and *J*, the equation (6) can be written as [1,4,8,9]

$$\frac{1}{\sigma\mu\omega} \left(\frac{\partial^2 J}{dt} \right) + (j\sigma - \omega) E = 0 \tag{7}$$

To have an idea on what would be the best geometrical shape of electrode in relation to skin effect; it is useful to investigate the behaviour of *J* and E. This can be carried by executing Equation (7) using the Partial Differential Equation (PDE) toolbox provided in Matlab [10]. The Simulation is conducted for two cases of geometrical configuration: (1) copper plane and (2) copper mesh electrodes. The conductivity and permittivity of copper are 57×10^{-6} S/m, and $8.8 \times 10^{□12}$ F/m, respectively.

Figure 1 shows the distribution of *J* for a copper plane electrode, when the electrode is injected by electric field 10kV/m at 30 kHz ac supply. As can be seen the current concentrates at the edge of the electrode, while low current density are distributed at the centre of the plate. The low current density means that only few amount of charges or electrons per area provided for ionization which leads to less microdischarge to be generated. Since skin effect at high frequency cannot be avoided, the electrode need to be modified in order to increase the number of microdischarge per unit area.

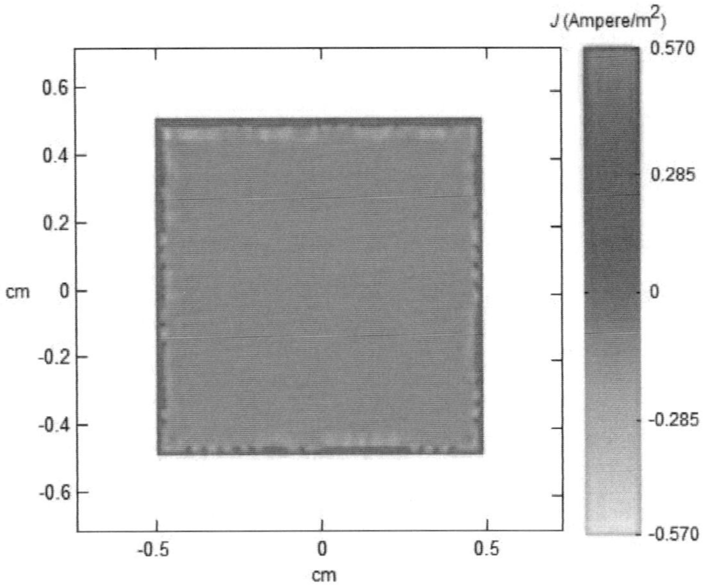

Figure 1. *J* in a copper plane electrode (1 x 1 cm) at 30 kHz 10 kV/

The copper electrode is modified as a mesh. The mesh configuration allows for numerous current paths to be established in a wider area. The distributions of *J* for the mesh electrode injected by electric field 10kV/m at 30kHz simulated by PDE toolbox in Matlab is shown in Figure 2. It can be observed higher *J* present not only in the most outer edge of the mesh, but instead it spreads in all region, i.e. in every small rectangular path. The larger J provides better ionization and generate more microdicharges intensively. This mesh configuration decreases the free path of electrons of injected gas and increase the number of collision among ions and electrons of injected gas to create more chemical reaction.

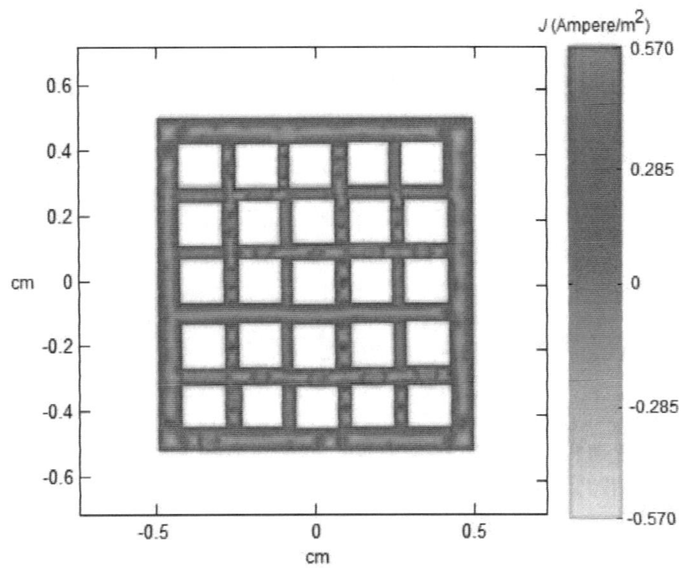

Figure 2. *J* in a copper mesh electrode (1 x 1 cm) at 30 kHz 10 kV/m

3. Experimental Result

To verify the performance of the chamber under various combinations of electrode and their geometrical shapes, a simple 50 Hz ozone generator is set up. Chambers with plane and mesh electrodes are prepared. Three types of electrode materials are used i.e. (1) copper plate, (2) aluminium plate, and (3) aluminium mesh. The dimension of the chamber is 230 × 130 × 20 mm. All the electrodes and dielectric was installed chamber with 2 mm air discharge gap. A mica as an insulator is attached to one of the electrodes.

Oxygen gas is used in this experiment. When there are more oxygen gas is formatted into ozone gas inside the experimental chamber, then the micro discharge with particular geometrical shape is successful initiated. The experimental set up is shown in Figure 3.

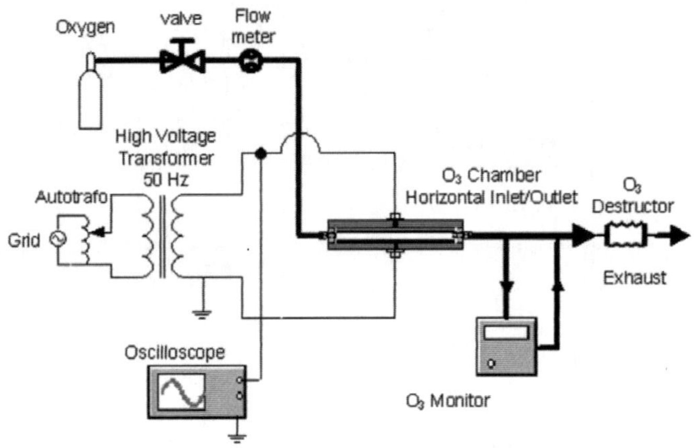

Figure 3. The experimental set up with 50 Hz AC supply

The experiment is conducted by injecting 6kVpp to the terminal of chamber. The experiment is repeated for several values of flow rate, i.e. 0.5, 1.0, and 1.5 Litre /minute. The ozone productions of different combination of electrodes are shown in Figure 4.

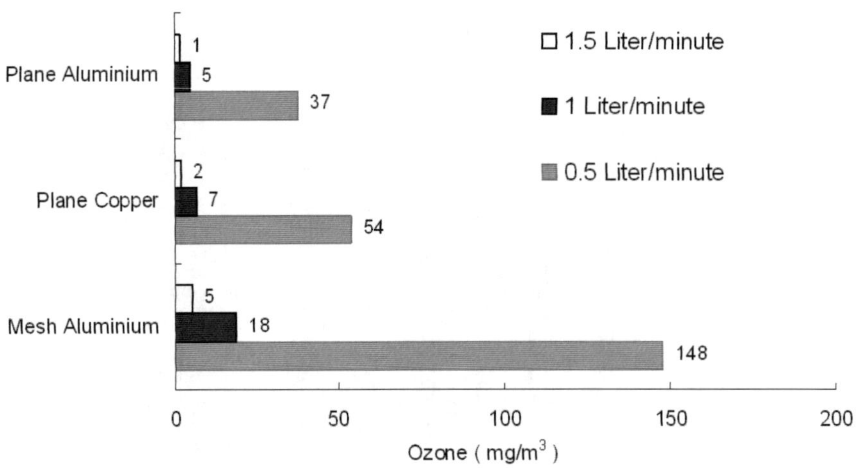

Figure 4. Comparison of the ozone production between chambers with plane electrode and mesh electrode

From Figure 4, it is clear that the mesh electrode produce the highest ozone yield compared to chamber with plane electrode for all flow rate. This significant result verified the simulation result that ionization in mesh electrode occurred in wider area than the plane electrode. The present of density current in every path of electrode provided more electrons to ionize in high electric field. These electrons created more secondary emission to initiate microdischarges inside the chamber as main process in silent discharge mechanism.

4. Conclusion

The behavior of plane and mesh electrode under AC voltage are simulated through PDE Toolbox provided in Matlab, and the result shows that mesh electrode gave better distributed electric field and current density which are needed to make secondary emission and micro discharge inside the chamber. The experimental results also verified that geometrical shape of electrode give influence to micro discharge initiation of plasma such as ozone. The present of density current in every path of electrode provided more electrons to ionize in high electric field.

5. References

[1] M.V.K. Chari and P. P. Silvester, 1980, *Finite Element in Electrical Magnetic Field Problem* (John Wiley & Sons).
[2] M. V. K. Chari and Z. J. Csendes, 1977, "Finite Element Analysis of Skin Effect in Current Carrying Conductor," *IEEE Transactions on Magnetics*, vol. Mag-13.
[3] J. R. Brauer,1982, "Finite Element Calculation of Eddy Surrent and Skin Effect " *IEEE Transaction on Magnetic*, vol. MAG-18.
[4] Facta M., Hermawan, Karnoto, Salam Z., Buntat Z., 2014, "Double Dielectric Barrier Discharge Chamber for Ozone Generation", *Proceeding The 1st International Conference on Information Technology, Computer, and Electrical Engineering,* pp 409-412
[5] Joao Pedro A. Bastos and N. Sadowski, 2003, *Electromagnetic Modelling by Finite element Methods*, (New York: Marcel Dekker Inc.).
[6] Liang Chi Shen and J. A. Kong, *Applied Electromagnetism*, 2009, (USA: Cengage Learning).
[7] M. N. O. Sadiku, 2007, *Elements of Electromagnetics*, (New York: Oxford University Press) 4 ed.
[8] A. Beiser, 2003, *Concepts of Modern Physics*, (New York: McGraw-Hill) 6 ed.
[9] S. O. Kasap, 2005, *Principles of Electronics Material and Devices*, (Mc Graw Hill) 3 ed.
[10] MATLAB, 2010, *Partial Differential Equation Toolbox 1.0.16* (The MathWorks, Inc.).

IOP Conf. Series: Materials Science and Engineering **190** (2017) 012023 doi:10.1088/1757-899X/190/1/012023

Implementation of Multipattern String Matching Accelerated with GPU for Intrusion Detection System

[1] Rangga Nehemia, [1] Charles Lim, [1] Maulahikmah Galinium and [2] Ahmad Rinaldi Widianto

[1] Department of Information Technology,
Swiss German University,
Tangerang 15339, Indonesia
[2] PT. Garuda Solusi Kreatif, Forsecnet Division,
Ruko The Icon Business Park Blok L/11
BSD, Tangerang, Indonesia

E-mail: rangga.nehemia[at]student.sgu.ac.id
charles.lim[at]sgu.ac.id
maulahikmah.galinium[at]sgu.ac.id
rinaldi[at]forsecnet.com

Abstract. As Internet-related security threats continue to increase in terms of volume and sophistication, existing Intrusion Detection System is also being challenged to cope with the current Internet development. Multi Pattern String Matching algorithm accelerated with Graphical Processing Unit is being utilized to improve the packet scanning performance of the IDS. This paper implements a Multi Pattern String Matching algorithm, also called Parallel Failureless Aho Corasick accelerated with GPU to improve the performance of IDS. OpenCL library is used to allow the IDS to support various GPU, including popular GPU such as NVIDIA and AMD, used in our research. The experiment result shows that the application of Multi Pattern String Matching using GPU accelerated platform provides a speed up, by up to 141% in term of throughput compared to the previous research.

1. Introduction

With the continuous increase of the Internet usage at the rate of 50% yearly [1], the number of security threats also increases sharply. Network Intrusion Detection System (NIDS) monitors the network traffic for harmful packets that is used as an attack vector, as described by Scarfone and Mell [2]. NIDS uses string matching for scanning and detecting patterns of threats contained in packets that match the pattern of the specified rules. Hence, to keep up with this requirement, the need of a higher speed of NIDS becomes crucial.

There are two popular Open Source NIDS, namely Snort developed by Roesch [3] and Bro developed by Paxson [4]. Compared to Snort, Bro is more flexible, customizable, and suitable for Gbps network, but is more complex to deploy. In addition, Bro's rules/signatures is also more sophisticated than the Snort ones [5]. The high flexibility and capability of Bro to handle a heavier network environment becomes the main reason for Bro to be selected to be implemented in this research. Our research aim to create an accelerated NIDS which is capable to perform efficiently in a a high speed network.

To increase the performance of NIDS, our research utilizes Graphical Processing Units for its parallel computing power to implement a parallel version of string matching algorithm. Parallel Failureless Aho Corasick (PFAC) developed by Lin et al. [6], an improvement of Aho Corasick by Aho and Corasick [7] is being implemented in our research. This research extends the research conducted by Widianto et al. [8], which showed the improved performance of the existing Intrusion Detection System by utilizing the GPGPU to allow multiprocessing of packets, effectively increasing the capacity of the IDS in high throughput environment.

Following are the organization of the paper: Section 2 summarizes the related work, Section 3 discusses our research methods, Section 4 describes our experiment environment setup and experiment results and finally section 5 concludes our research works.

2. Related Works

Many research has been performed to improve the the performance of IDS. There are three general approaches of improvements: Specialized Hardware, GPU utilization, and Algorithm Development. Specialized hardware is easy to develop but is expensive and thus not as popular as the cheaper and more efficient GPU utilization or algorithm development. GPU utilization research mostly includes the implementation of new algorithms.

Young et al [9] designed a deep packet filtering firewall on Field Programmable Gate Array to take advantage of parallelism while maintaining the programmability. Alfred et al [7], and Wu et al. [10] developed new algorithm for string or pattern matching, while others such as Kouzinopoulos et al. [11], Zha and Sahni [12], and Soroushnia et al. [13] improve the performance the pattern matching algorithms using GPU on CUDA [14]. Pyrgiotis et al. [15] perform their research based on OpenCL [16].

Vasiliadis et al. [17] developed Gnort, a GPU based Intrusion Detection System, which utilizes the computational power of GPU to handle the pattern matching operations which is costly if implemented in CPU. The implementation is using CUDA, which is NVIDIA exclusive, while our research is using OpenCL for compatibility to various machine and architecture.

This paper extends the research conducted by Widianto et al. [8] with the objective of this research is to improve the performance of the existing Network Intrusion Detection System by utilizing the GPGPU to allow multiprocessing of packets, effectively increasing the capacity of the NIDS in high throughput environment. Our research implement the Multi Pattern String Matching and compare the performance result with the performance result of the research by Ahmad Rinaldi.

3. Research Methodology

Signatures are the keywords for hinting the probability of malicious packets that NIDS uses as references to inspect packets flowing through the network. If a packet contains any of the listed signature or NIDS rule, the packet will be classified as a malicious packet. Once the packet is identified and captured, the packet is then preprocessed before the packet is sent to GPU for further inspection. Packet is inspected with PFAC algorithm to determine any suspicious or malicious pattern, that found in the NIDS rule that describe the malicious byte pattern that IDS is looking for. Using parallel computation capability offered by GPU, the search can then be executed in parallel fashion, using multiple work unit to accelerate the search.

Result of the packet inspection is sent back to CPU to determine whether the packet is malicious or not. If the packet contain malicious pattern, the reporting system to alert user for the existence of the malicious pattern. The overview of the system architecture used in our research follows the system proposed in the previous work by Widianto et al. [8] and shown in

Figure 1, which can be divided into two parts: Packet Preprocessing and Transfer on CPU, and Multipattern Matching on GPU.

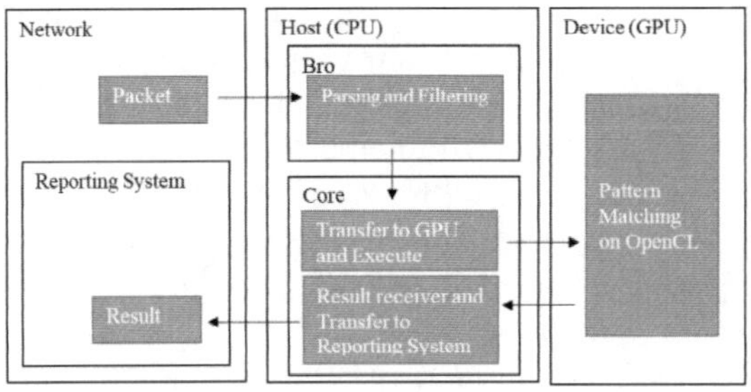

Figure 1: System Architecture Overview

In Packet Preprocessing and Transfer, the packets are captured by IDS and preprocessed before they can be inspected for malicious payload. After the packets are processed, they are sent to the core, that executes GPU specific instructions for Multi Pattern String Matching inspection based on the packets received and transfer the packets to GPU. Multipattern Matching on GPU is performed using Parallel Failureless Aho-Corasick (PFAC) algorithm by Lin et al. [6]. PFAC is an improved Aho-Corasick algorithm to used in parallel processing environment. Based on the rules, PFAC creates a trie in the initialization phase using CPU and copies it to the GPU. In the searching phase, GPU runs the PFAC in parallel fashion, using multiple threads to process different part of the packets.

Using the same method, our research also compares two different GPU platform, i.e. NVIDIA and AMD. Figure 2 shows the testing network architecture using NVIDIA GPU environment for the test subject. The test subject is the accelerated NIDS, and handle the packet capture and packet scanning. DNS query will be generated by the packet generator and sent to NIDS. The network has a single subnet where the data for testing is sent during the testing operation, with sending server connected on one end and receiving server/accelerated NIDS on the other end. After the test, the data captured by NIDS is checked with the sent data and if the two data match each other, this mean the test is successful as no packet is dropped, and the result can be measured and analyzed.

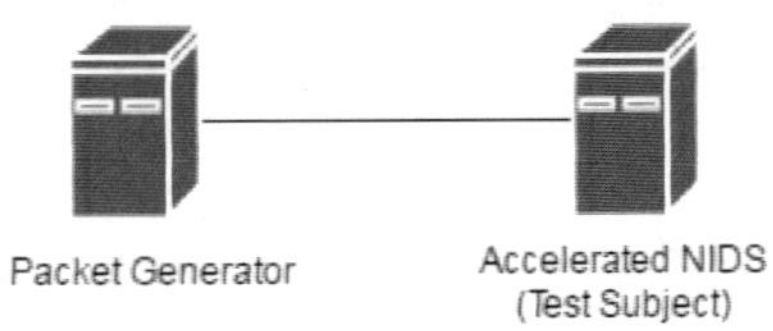

Figure 2: NVIDIA Testing Network Architecture

Figure 3: AMD Environment Testing Network Architecture

Figure 3 shows the testing network architecture using AMD GPU Environment for the test subject. The difference with NVIDIA environment is that in the NVIDIA environment, the packet capture and the service (packet scanning) is handled in the same computer, while in AMD environment the packet capture and service are handled in separate computer. Bro Control handles the packet capturing and send the packet to the test subject for packet scanning (Bro Control handles both packet capture and packet scanning in the first scenario). The criteria to check validity of the data is the same with NVIDIA environment, which the sent data and captured data are compared.

4. Experiment Results

4.1. Experiment Environment

Our research uses Intel(R) Core(TM) i7-3770 CPU equipped with NVIDIA GeForce GTX 970 GPU. Bro, an open source NIDS, is chosen since it allows developer to integrate new feature as a plugin in a more flexible way. Python programming language is used for general programming while C programming language is used for the OpenCL part of the implementation. Jmeter, developed by Apa [18] is used to send multiple packets to test the system and Bro for packet capture in Simulated Benchmark.

4.2. Initial Testing

Table 1: Initial Testing Comparison of GPU Naive and GPU PFAC

Metrics	GPU Naive	GPU PFAC	Improvement
Computation Time(sec)	3.42	2.58	25%
Requests Per Second	2196	2907	32%

Table 1 shows the result of our initial test between the system developed by the author, which is GPU PFAC, compared to the system from previous research by Widianto et al. [8], GPU system based on Naive algorithm. Given the same condition, GPU PFAC shows a computational time improvement of 25%.

4.3. Detailed Experiment Result

To further improve the performance of the system, three major optimization is conducted. The first optimization is to manage the global size to match the local size, allowing the global size to be dividable by local size. This also allows the system to be run with NVIDIA GPU. Second optimization is to perform better memory allocation to enable faster and lighter system. The last optimization is to use better handling of the result counter, reducing the computational cost for transferring results. Table 2 shows the comparison of GPU PFAC computation time before and after the optimization. Optimization of rule management allows the system to handle around 58% more rules or around 515.040 characters, limited only by the hardware. The test also included the AMD implementation of the system with the hardware that is similar to the hardware from previous system by [8]. As the hardware used in AMD is similar to the previous system, the improvement gained in the AMD implementation is the pure algorithm increased performance while the improvement gained from NVIDIA implementation, which has more powerful hardware, is the improvement gained from both algorithm and hardware. The result shows that the base performance of the algorithm has increased by up to 40% for NVIDIA GPU and 56% for AMD GPU. AMD GPU has faster computation time because of the faster GPU preparation time.

Table 2: GPU PFAC Computation Time Comparison (in Seconds)

String Length (character)	500.000	1.000.000	1.500.000	2.100.000
Unoptimized PFAC	2.08	2.11	2.14	2.2
AMD PFAC	0.94(-55%)	0.94(-55%)	0.94(-56%)	0.96(-56%)
NVIDIA PFAC	1.29(-38%)	1.30(-38%)	1.31(-39%)	1.32(-40%)

Table 3: GPU PFAC Searching Time Comparison (in Seconds)

String Length (character)	100.000	500.000	700.000
Unoptimized PFAC	0.0028	0.0048	0.0057
AMD PFAC	0.0152	0.0195	0.0242
NVIDIA PFAC	0.0071	0.0102	0.0119

A test with simulation benchmark shows that the improvements made are beneficial in real time scenario, and make the comparison between the AMD implementation of the algorithm compared to the NVIDIA implementation of the algorithm possible. The result shows computational time improvement of 51% and 58% and throughput by 106% and 141% for AMD and NVIDIA respectively compared to the GPU Naive used in the previous research [8], as shown in table 4.

Table 4: Simulated Benchmark Improvement

	Computation Time (in Second)	Throughput (Request per Second)
GPU Naive	3.42	2196
Unoptimized GPU PFAC	2.58(-25%)	2907(+32%)
AMD GPU PFAC	1.66(-51%)	4518(+106%)
NVIDIA GPU PFAC	1.42(-58%)	5282(+141%)

5. Conclusion

Existing NIDS could not cope with the increased volume of Internet traffic that flow through the organization network. This paper proposes the implementation of Multi Pattern String Matching algorithm using a GPU-based equipped computer platform. Our experiments show that the implementation of GPU accelerated Multi Pattern String Matching algorithm in NIDS, outperforms Naive String Matching algorithm, with an improvement up to 106% in term of request per second and 51% shorter computation time compared to the Naive String Matching, resulting in an NIDS that can perform twice as fast. We also prove that with a better performance GPU, such as NVIDIA GPU, the experiment shows that the performance could improve as much as 141% in term of request per second and 58% shorter computation time compared to the Naive String Matching, with a 33% more performance gain compared with AMD GPU. Due to the limitation of the string matching and scope of our research, the proposed system is suited only to scan the DNS query at the moment, among other type of possible traffics. We hope to include Regular Expression [19] in our future works to scan various type of packet that is common to real world network traffic.

References

[1] Jacob Nielsen. Nielsen's law of internet bandwidth, April 2016. URL `https://www.nngroup.com/articles/law-of-bandwidth/`.

[2] Karen Scarfone and Peter Mell. Guide to intrusion detection and prevention systems (idps). NIST special publication SP 800-94, 2007.

[3] Martin Roesch. Snort-lightweight intrusion detection for networks. In Proceedings of the 13th USENIX conference on System administration, pages 229–238. USENIX Association, 1999.

[4] Vern Paxson. Bro: a system for detecting network intruders in real-time. Computer networks, 31(23):2435–2463, 1999.

[5] Pritika Mehra. A brief study and comparison of snort and bro open source network intrusion detection systems. International Journal of Advanced Research in Computer and Communication Engineering, 1(6):383–386, 2012.

[6] Cheng-Hung Lin, Sheng-Yu Tsai, Chen-Hsiung Liu, Shih-Chieh Chang, and Jyuo-Min Shyu. Accelerating string matching using multi-threaded algorithm on gpu. In Global Telecommunications Conference (GLOBECOM 2010), 2010 IEEE, pages 1–5. IEEE, 2010.

[7] Alfred V Aho and Margaret J Corasick. Efficient string matching: an aid to bibliographic search. Communications of the ACM, 18(6):333–340, 1975.

[8] A. R. Widianto, C. Lim, and I. E. Kho. Improving performance of intrusion detection system using opencl based general-purpose computing on graphic processing unit (gpgpu). In 2015 3rd International Conference on New Media (CONMEDIA), pages 1–5, Nov 2015. doi: 10.1109/CONMEDIA.2015.7449146.

[9] Young H Cho, Shiva Navab, and William H Mangione-Smith. Specialized hardware for deep network packet filtering. In Field-Programmable Logic and Applications: Reconfigurable Computing Is Going Mainstream, pages 452–461. Springer, 2002.

[10] Sun Wu, Udi Manber, et al. A fast algorithm for multi-pattern searching. 1994.

[11] Charalampos S Kouzinopoulos, Panagiotis D Michailidis, and Konstantinos G Margaritis. Multiple string matching on a gpu using cudas. Scalable Computing: Practice and Experience, 16(2):121–138, 2015.

[12] Xinyan Zha and Sartaj Sahni. Multipattern string matching on a gpu. In Computers and Communications (ISCC), 2011 IEEE Symposium on, pages 277–282. IEEE, 2011.

[13] Shima Soroushnia, Masoud Daneshtalab, Juha Plosila, Tapio Pahikkala, and Pasi Liljeberg. High performance pattern matching on heterogeneous platform. Journal of integrative bioinformatics, 11(3):253, 2014.

[14] NVIDIA. NVIDIA CUDA Compute Unified Device Architecture Programming Guide, August 2016. [Online]. Available: `http://docs.nvidia.com/cuda/cuda-c-programming-guide/`.

[15] Themistoklis K Pyrgiotis, Charalampos S Kouzinopoulos, and Konstantinos G Margaritis. Parallel implementation of the wu-manber algorithm using the opencl framework. In Artificial Intelligence Applications and Innovations, pages 576–583. Springer, 2012.

[16] Khronos. OpenCL The open standard for parallel programming of heterogeneous systems, May 2016. [Online]. Available: `https://www.khronos.org/opencl/`.

[17] Giorgos Vasiliadis, Spiros Antonatos, Michalis Polychronakis, Evangelos P Markatos, and Sotiris Ioannidis. Gnort: High performance network intrusion detection using graphics processors. In Recent Advances in Intrusion Detection, pages 116–134. Springer, 2008.

[18] Apache jmeter, June 2016. URL `http://jmeter.apache.org/`.

[19] Yuichiro Utan, Masato Inagi, Shin'ichi Wakabayashi, and Shinobu Nagayama. A gpgpu implementation of approximate string matching with regular expression operators and comparison with its fpga implementation. In Proceedings of the International Conference on Parallel and Distributed Processing Techniques and Applications (PDPTA), page 1. The Steering Committee of The World Congress in Computer Science, Computer Engineering and Applied Computing (WorldComp), 2012.

IOP Conf. Series: Materials Science and Engineering **190** (2017) 012024 doi:10.1088/1757-899X/190/1/012024

Identification of electrical discharge on electric insulation material using acoustic emission method

J M Nainggolan[1], I Garniwa MK [2], and A Raharjo[3]

[1,2,3] Faculty of Engineering,University of Indonesia, Depok

[1] mauritus24@gmail.com

Abstract. Insulation system is essential to be equipped on high voltage equipment. One of the tests on insulation system is the electrical discharge test. In this work, the electrical discharge detected using an acoustic emission method to find out the characteristic of the discharge. Epoxy resin was taken as the insulation material, whereas one to two artificial holes were made inside the insulation. An AE sensor placed at a distance of 24 mm and 36 mm away from holes or discharge source. The main frequency component of the detected AE signal on single hole test was approximately 975 kHz. Meanwhile, test on double holes caused two main frequency component; i.e. 63 kHz and 930 Khz. For single hole test and sensor distance of 36 m, the AE signal intensity tends to be worth 40% from the AE signal at a distance of 24 mm.

1. Introduction

Electrical energy consumptions increase along with the growth of population and the development of industry and fabrication. It makes the state electric company must provide the electrical energy supply better. Therefore we need the continuity and reliability of power system in order to meet the supply of energy required by the community. A reliable electricity system requires also a good insulation system. The good insulation system in particular is very much needed in medium and high voltage system.

An insulation material used to separate two or more electrical conductor so that at the conductor, electrical spark or spark over does not happen. In insulation material, in process of time, electrical discharge or corona will occur, especially when the applied voltage exceeds dielectric strength of material [1]. Corona that take place on electrical equipment can cause damage to the appliance so that the continuity of the system will be disrupted. Therefore, the condition of insulation system must be very noted. A good insulation material must be chosen also to minimize the corona.

In general, damage in an insulation system occurs due to the high pressure of electric field. High electric field on insulation system can trigger a local discharge which is known as electrical discharge or partial discharge. Discharge is events of electrical sparks jump that occurs on one part of insulation as a result of the existence of high potential different on isolation. In addition, events of electrical discharge can also be caused due to the bad equipment construction,_continuous high pressure, and contraction of insulation material, environmental and chemical factors, and heat effects that arise because of the increase of voltage [2].

One way to know the condition and characteristics of insulation material is by performing a diagnosis using acoustic emission method. In insulation that being subjected to an aging process, an electrical discharge will occur because of high pressure of electric filed The process of electrical discharge also referred to partial ionization process. The ionization process will produce an energy that propagates in the form of elastic waves. Elastic waves will propagate from inside to the outside or at the surface of isolation. The wave then will be detected by the acoustic emission sensor [3].

2. Test Setup

The test setup is divided into two work stages.

2.1. The series of test

A series of tests were arranged as figure 1. The testing framework was created in the form of square using acrylic materials.. The top frame is made in the size of 18 cm x 20 cm as the place for placement of insulation material that will be tested. AE Sensor was installed at a certain distance from the center point of the ground electrode. The high voltage was connected to the needle electrode, while the grounding electrode connected to the ground. Sample of epoxy resin made with a thickness of 8 mm. At the middle point of the epoxy resin, one and two holes made with a diameter of 3 mm.

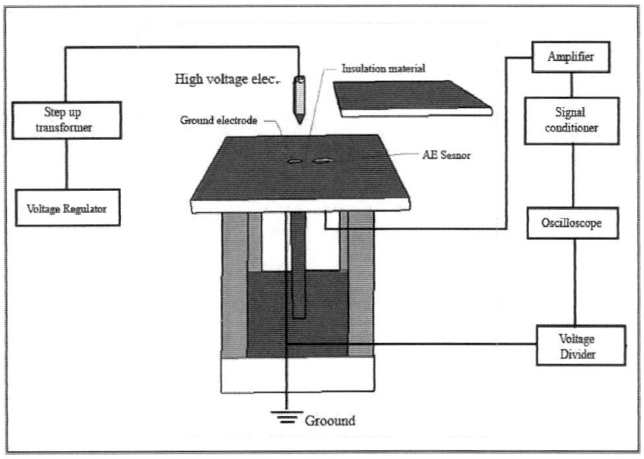

Figure 1. Test setup for identifying discharge characteristics of epoxy resin

2.2. Test and data processing

High voltage source was raised step by step until significant electrical discharge detected by AE sensor. The AE signal was detected by the AE sensor with an operating frequency 50 kHz – 900 kHz. Then, the signal was amplified by a preamplifier of 40 dB with a frequency band of 10 kHz - 2 MHz and mixed by a signal conditioner. The output of the signal conditioner was measured by a digital oscilloscope with sampling frequency of 125 MHz and recorded in a personal computer. These data will be analysed by using a Wavelet and Fast Fourier transform (FFT) to get the characteristics [4]. The discharge current was detected by a current detection circuit.

3. Result and Discussion

The identification of activities of electrical discharge on the epoxy resin conduted in two research variables, namely (1) number of holes (consist of a single hole and double holes), and (2) sensor distance from artifical holes (24 mm and 36 mm).

3.1. Tests on single hole

3.1.1. The sensor distance of 24 mm

In the test, signal of electrical discharge detected by AE sensor is shown in figure 2.

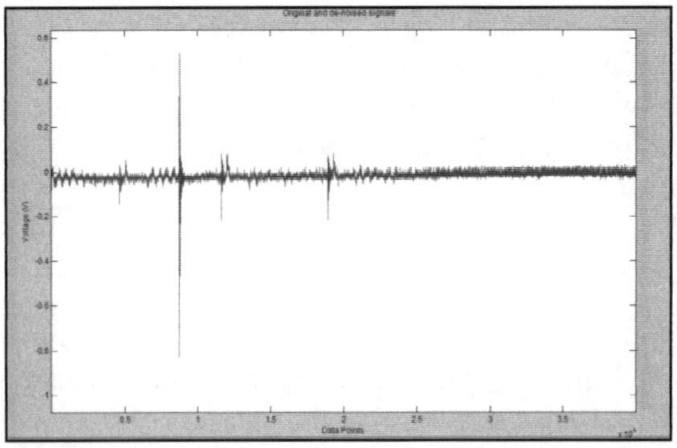

Figure 2. AE signal intensity on sensor distance of 24 mm (2,4 kV)

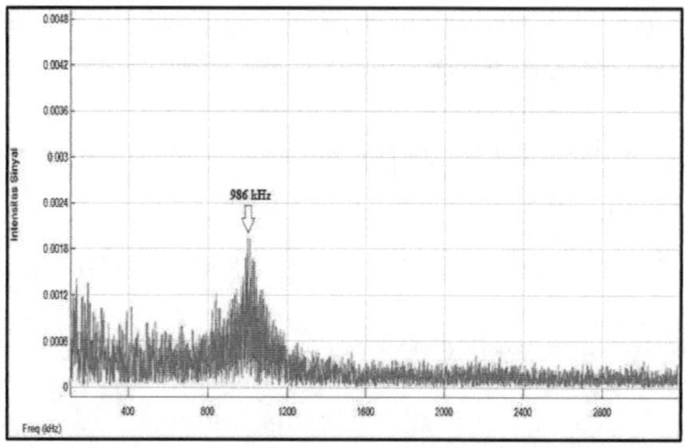

Figure 3. Frequency spectra on sensor distance of 24 mm (2,4 kV)

The signal in figure 2 was processed by using the Wavelet to reduce the noise. Spectral analysis of frequency was performed using Fast Fourier Transform (FFT), and shown in figure 3. From figure 3, it was obtained that the test had one main frequency component. Data from the test result (sensor distance of 24 mm) is shown in table 1.

Table 1. Test results on single hole with sensor distance of 24 mm

Test Voltage (kV)	AE signal intensity (V)	Discharge current (mA)	Main frequency (kHz)	Time of significant discharge current (min)
2.3	0.58	42.21	970	35
2.4	0.72	43.88	986	33
2.5	0,80	46.12	990	31

Table 1 showed the test was conducted at the voltage test of 2.2 kv, 2.3 kv, and 2.4 kv. There is an increase of AE signal instensity with a rise in voltage test. The significant discharge current is tend to increase too. Those are also illustrated better in figure 4. Figure 5 shows discharge current and AE signal intensity are directly proportional to test voltage.

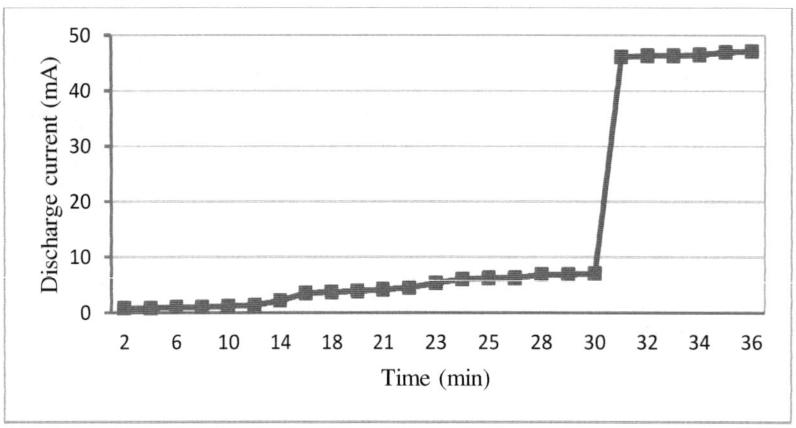

Figure 4. The increase of discharge current

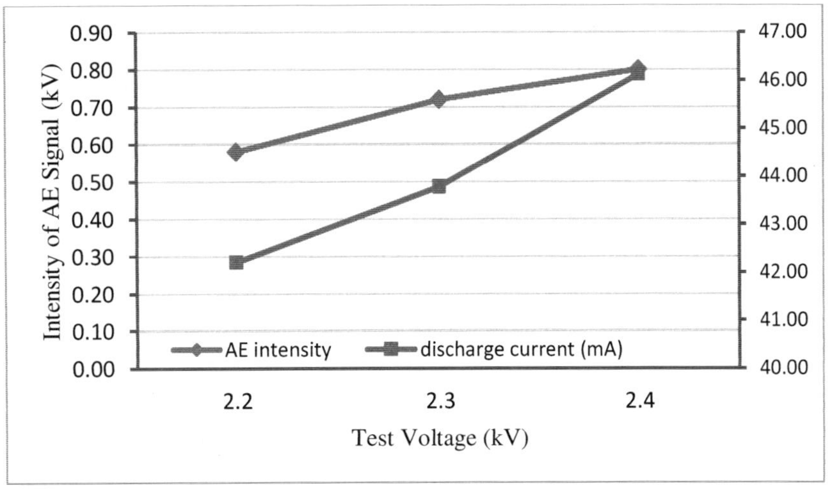

Figure 5. AE signal intensity and discharge current to voltage test at a distance of 24 mm

3.1.2. The sensor distance of 36 mm

In this test, the shape of discharge signals detected by AE sensor were almost the same with sensor distance of 24 mm that shown in figure 2. The table 2 shows the results of test. It also can be seen that with the increase of test voltage, some increase of AE signal intensity and discharge current are occur (ilustated in figure 6). Figure 3.5 is also shows the discharge current and the test voltage is directly proportional.

Table 2. Test results on single holes with sensor distance of 36 mm

Test Voltage (kV)	AE signal intensity (V)	Discharge current (mA)	Main Frequency (kHz)	Time of significant discharge current (min)
2.3	0.28	44.10	962	36
2.4	0.32	45.80	967	35
2.5	0,52	46.95	970	34

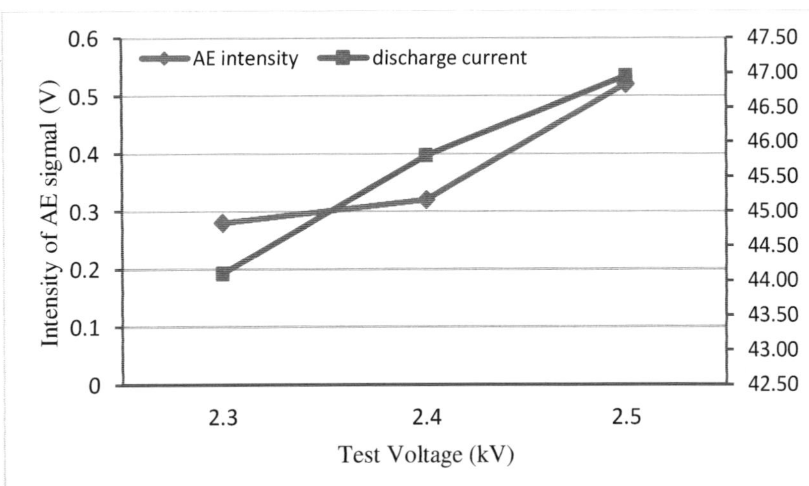

Figure 6. AE signal intensity and discharge current to voltage test at 36 mm

3.1.3. The comparison of test results between the sensor distance of 24 mm and 36 mm

(1). AE signal intensity at a sensor distance of 36 mm was smaller than the distance of 24 mm on the same test voltage. Signal intensity of AE at sensor distance of 36 mm was 40% from the distance of 24 mm. It happens because with the sensor position farther away from the hole, AE signals detected will be much smaller. The comparison is shown in figure 7.

(2). The magnitude of the discharge current at both distance of sensor was almost the same because of the same test voltage.

(3). The magnitude of main frequency component at both distance of sensor was almost the same. It happened because of the same type of material (insulation) and the same kind of test voltage. The magnitude of average of main frequency component was 975 kHz.

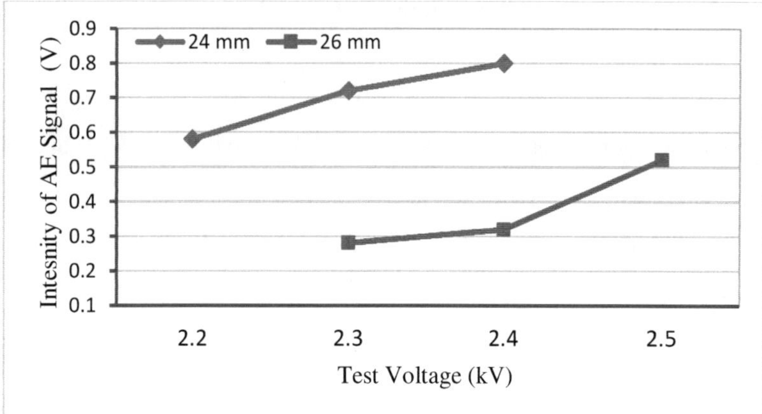

Figure 7. AE signal intensity for sensor distance of 24 mm and 36 mm

3.2. Tests on double holes

Test on epoxy resin with double holes was conducted at sensor distance of 24 mm only. Results data of the test is shown in Table 3.

Table 3. Test results on double holes

Test Voltage (kV)	AE signal intensity (V)	Discharge current (mA)	Frequency-1 (kHz)	Frequency-2 (kHz)	Time of significant discharge current (min)
2.2	0.57	45.93	62	940	33
2.3	0.60	46.71	63	930	32
2.4	0,64	47.20	63	940	30

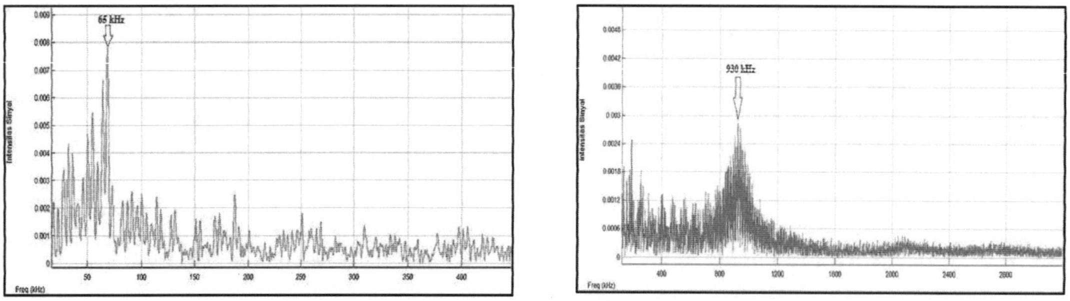

(a). Frequency 1; 63 Khz (b). Frequency 2, 930 kHz

Figure 8. Frequency spectra of test on double holes

Table 3 also shows the relationship between the discharge current and time of significant electrostatic discharge. From the table, it is apparent that the discharge current jumped up significantly to the value of 45 mA or more, started in minutes 30. From the test, two ranges of frequency component were obtained, namely 60-66 kHz and 910-950 kHz. Therefore we got two averages of main frequency; they were approximately 63 kHz and 930 kHz as they are illustrated in Figure 8).

3.3. The comparison of test results between the single hole and double holes

(1). The main frequency components acquired on single hole test at sensor distance of 24 mm and 36 mm was 975 kHz (the frequency range was 960-990 kHz). Meanwhile, test on the double holes resulted in two main frequency components; they were approximately 63 kHz and 930 kHz (frequency ranges were 60-66 kHz and 910-950 kHz). It may happen because with the existence of two cavities, electrical (electrostatic) discharge occurs twice in the duration of very short time.

(2). There was no significant difference of AE signal intensity at sensor distance of 24 mm on single hole and double holes.

(3). There was no difference significant of discharge current on both tests. There was a trend that the time of the significant discharge current on double holes test also larger than single hole test (around 30 minutes in minimum). But, further and more carefully research is required to verify the trend.

4. Conclusions

A research has been done on the epoxy resin with single and double holes by giving high voltage to find the characteristics of insulation when electrical (electrostatic) discharge happened. A single hole experiment had a main frequency component; it was approximately 975 KHz. The double holes experiment had two main frequency components; they were approximately 63 kHz and 930 KHz. On the sensor distance of 36 mm, the AE signal intensity tends to be worth 40% from the AE signal at a distance of 24 mm.

References

[1] Bonggas LT 2003 Basic of high voltage testing *Gramedia Pustaka Utama* pp 12-14.

[2] König D and Rao YN 1993 Partial discharges in electrical power apparatus *Berlin; Offenbach: VDE-Verlag* pp 15-36.

[3] Richard DF 2003 Handbook of Nondestructive Evaluation, chapter 10 – Acoustic Emission Testing *McGraw-Hill Companies pp 10-39*

[4] Jannus MN, Nakashima T, Sakoda T, Otsubo M, Kurihara S, Yarimitsu T, and Nagasato S 2010 Location of partial discharge at joint section of XLPE using AE technique *Conference on Electrical Insulation and Dielectric Phenomena* pp 442-445.

[5] Zhu D, McGrail AJ, Swingler S, Auckland DW, and Varlow BR 1994 Partial discharge detection in cable termination using acoustic emission techniques and adaptive signal processing *Conference Record of the International Symposium on Electrical Insulation*.

[6] Hsieh JC, Tai CC, Su CC, Chen CY, Huang TC and Lin YH 2006 Evaluation and comparation of on-line PD detection methods for high-voltage power cable *12th Asia-Pasific Conference on NDT* pp 84-87

[7] Rubio-Serano J, Boya, Rojas-Moreno MV, Posada J, Martinez-Tarifa JM, and Robies G 2012 Electro-acoustic detection, identitification and location of partial discharge in oil paper insulation *IEEE Trans. Dielectr. Electr. Insul.* 19(53) pp 1663-1673.

[8] Hooshmand RA, Parastegari M, and Yazdanpanah M 2012 Simultaneous location of two partial discharges sources in power transformer based on AE using the modified partial swarm optimisation algolthm *IET Science, Measurement & Technology* 7(2) pp 119-127.

IOP Conf. Series: Materials Science and Engineering **190** (2017) 012025 doi:10.1088/1757-899X/190/1/012025

Determination of Hydrophobic Contact Angle of Epoxy Resin Compound Silicon Rubber and Silica

Abdul SYAKUR[1], Hermawan[2], Heri SUTANTO[3]

[1,2] Department of Electrical Engineering, Universitas Diponegoro

3Department of Physics, Faculty of Science and Mathematics, Universitas Diponegoro., Jln. Prof. Soedarto SH Tembalang, Semarang, INDONESIA

Email : gakusei2003@yahoo.com

Abstract. Epoxy resin is a thermosetting polymeric material which is very good for application of high voltage outdoor insulator in electrical power system. This material has several advantages, i.e. high dielectric strength, light weight, high mechanical strength, easy to blend with additive, and easy maintenance if compared to that of porcelain and glass outdoor insulators which are commonly used. However, this material also has several disadvantages, i.e. hydrophilic property, very sensitive to aging and easily degraded when there is a flow of contaminants on its surface. The research towards improving the performance of epoxy resin insulation materials were carried out to obtain epoxy resin insulating material with high water repellent properties and high surface tracking to aging. In this work, insulating material was made at room temperature vulcanization, with material composition: Diglycidyl Ether Bisphenol A (DGEBA), Metaphenylene Diamine (MPDA) as hardener with stoichiometric value of unity, and nanosilica mixed with Silicon Rubber (SiR) with 10% (RTV_{21}), 20% (RTV_{22}), 30% (RTV_{23}), 40% (RTV_{24}) and 50% (RTV_{25}) variation. The usage of nanosilica and Silicon Rubber (SIR) as filler was expected to provide hydrophobic properties and was able to increase the value of surface tracking of materials. The performance of the insulator observed were contact angle of hydrophobic surface materials. Tests carried out using Inclined Plane Tracking procedure according to IEC 60-587: 1984 with Ammonium Chloride (NH_4Cl) as contaminants flowed using peristaltic pumps. The results show that hydrophobic contact angle can be determined from each sample, and RTV_{25} has maximum contact angle among others.

1. Introduction

Epoxy resin uses as an insulating material in electric power systems is still under study. The selection of epoxy resin with the base material is diglycidyl ether of bisphenol A (DGEBA), and meta phenylene diamine ripening agent as alternative isolator made by polimere insulation because it has high dielectric power (E), that is 24-25 MV/m, 10^{13}-10^{15} Ω.m volume resistance (ρ), 50/60 Hz frequency of dielectric constant (ε), 3.5-3.9 dan low power dissipation that is (35-90) x 10^{-4}. To increase epoxy resin performance to be resistant to the environmental conditions, it is be added the polydimethylsiloxane which has type is silicon rubber. Properties that polysyloxan have are high power dielectric at 20-30 MV/m, dielectric constant (ε) has frequency at 50/60 Hz, 2.5-3.2 and low power dissipation (tan δ) that is (4-25) x 10-4 and could increases the performance of water denied [1].

The addition of silicone rubber in large amounts can increase the hydrophobic properties of the surface, but the mechanical strength is reduced. To improve the mechanical strength of the epoxy resin material coupled with the filler material is silica sand (SiO2) in the hope of epoxy resin insulating material has a high mechanical strength is tensile strength and compressive strength [2]. If the silica sand is added the more the mechanical strength increases, but on the other hand will reduce the surface hydrophobic properties. Therefore, it is necessary to do a study to obtain a filler material composition of silicone rubber and silica mixture right on epoxy resin material that will be used as a high voltage insulator material. To determine the performance of surface material made of epoxy resin filler mixture of silicon rubber and silica under severe pollution, then tested the electrical tracking refers to the Inclined Plane Tracking test methods according to standards IEC 587, with duration of voltage application for 6, 12 and 18 minutes. Research on epoxy resin material performance under severe conditions has also been studied by [3] using natural and artificial aging in the laboratory for 96 hours with ultraviolet light. In this report, hydrophobic contact angle will be discussed before and after surface tracking test.

2. Fundamental Theory

2.1 Characteristics of Solid Polymer Insulation

The electron will have the electric pressure when the solid polymer insulation applied in E electric field. This electrical stress is a force that works on the electron, and raises a work. Due to the force, the electron will moves as far as dx and need an effort by W. If a different potential between two point, for example point A and point B on the solid polymer insulation material, is VAB, amount of effort required to move an electron from point A to point B has formulated by [4],

$$\int_{B}^{A} E dx = -V_{AB} \qquad (2.1)$$

Then, the value of electric field strength or electric pressure is

$$E = -\frac{\partial V_{AB}}{\partial x} \qquad (2.2)$$

2.2 Epoxy Resin Insulating Materials

Epoxy resin is a plastic which including thermoset category, but it is hard as phenolic and polyester. Epoxy resin offers the incredible function, low shrinkage rate, good chemical resistant and good sticking power.

The strength of epoxy resin is located on the hardness, resistance to friction and chemical resistance. Solid resin at medium molecular weight can be combined with similar polymerization or non-similar polymerization through the epoxy group. The property of epoxy resin adhesive which is perfect, malleable, good mechanical strength and strong chemical resistance are the most important benefits from the solid resin [5]. Surface degradation caused by a reaction which causes broken chain of the main molecular bond and the change of polymer base material happens. Surface degradation of polymer insulation material as insulator material on electricity power system, caused by high pollutant which came from the water sea pollution around the beach which contains salt, industrial pollution is chemical pollution because of the soot, industrial fog; desert contains sand and dust, volcanic dust pollution, and cement factory's dust pollution. Pollutant which sticks on polymer insulation material as high voltage insulator material, influenced by the electric field and tropical climate, those are temperature, humidity, air pressure, wind velocity, rain and ultraviolet ray, causes aging of insulation, that is aging of physically aging, chemical, electrical, and combination of mechanical and electrical [6]

2.3 Metaphenylene Diamine (MPDA)

Ripening of the epoxy resin can be divide into three main groups, there are: Hydroxyl group (R-OH), Amine group and Acid Anhydride group. Each group has so many types. Every type has different level on the mixture process. Ripening reaction occurs through the epoxy groups from base material and reactive groups from reinforcement material. Reaction between base-material of epoxy resin with ripening agent of acid anhydride a group is a more complex reaction than the ripening agent of amine group. The test uses ripening agent from amine group that is Metaphenylene-diamine (MPDA) because MPDA included the most chemical that is used as ripening agent of epoxy resin [5].

2.4 Silicon Rubber (SiR)

Silicon rubber is a material resistant to high temperature; usually it is used to cable insulation and high voltage insulator material. Physical property of this thing can be improved by mix the filler material, is like silica sand. Silicon rubber is safe to use at -55°C – 200°C temperature. It has good barrier to the ozone, corona, and water. This has well resistant against the alcohol, salt and oil. Silicon rubber ($[CH_3]_2SiO$) is formed by cyclic siloxane monomers that created polydimethylsiloxane. The hydrophobic property of it gives a high surface obstacle in the humid condition and contaminated. It decreases the leakage current. Uncontrolled leakage current can increase the intensity of spark in dry condition, when the crack occurred, and erosion (flashover) in the polymer material. Silicon rubber has very good insulation property, there are: has tan $\delta = 3\text{-}30.10^{-3}$, dielectric constant $\varepsilon r = 2\text{-}4$, resistivity $\rho = 1015\ \Omega m$ (without filler material), resist to light (metil group absorbs light in $\lambda > 300$ nm area) and it can be stabled until the temperature is 250^{0}C by resists the elasticity in the low temperature because PDMS has glass transition temperature around -120^{0}C (long thermal stability).

2.5 SiliconDioxide (SiO_2)

Silica sand or quartz sand is excavated materials composed of crystals - crystalline silica (SiO2) and contains a compound impurities carried during the deposition process. Silica sand or quartz having a combined composition of SiO2, Fe2O3, Al2O3, TiO2, CaO, MgO and K2O translucent white or other colours depending on the compound impurities. Combination of silicon with oxygen to form crystal formation depends on the temperature of the called silica. Silica is an insulating material that is often used in ceramics. Packaging atoms enclosed in the bonding of silicon and oxygen, making the structure tends to be stable in general does not decline to environmental changes. This means that the insulator is not easily damaged by ultraviolet, electrical treatment such as providing a voltage gradient on the insulator and environmental conditions such as temperature, humidity and so forth. Insulator will be more resistant to aging.

3. Experimental Set Up

3.1 Materials

Materials composition used in this research are Diglycidyl Ether Bisphenol A (DGEBA), Metaphenylene Diamine (MPDA) as hardener with stoichiometric value of unity, and nano-silica mixed with Silicon Rubber (SiR) with 10% (RTV21), 20% (RTV22), 30% (RTV23), 40% (RTV24) and 50% (RTV25) variation.

Table 1 Code and the composition of the sample test materials research

Test sample code	DGEBA (% pbw)	MPDA (% pbw)	Filler (% pbw)		
			Silicon Rubber	Silica sand	Filler Total
RTV_{21}	45	45	5	5	10
RTV_{22}	40	40	10	10	20
RTV_{23}	35	35	15	15	30
RTV_{24}	30	30	20	20	40
RTV_{25}	25	25	25	25	50

3.2 Contact Angle Measurement
Equipment for research performance hydrophobic contact angle include: conductivity meter, pipette, lights, cameras, and computers as shown at Figure 1.

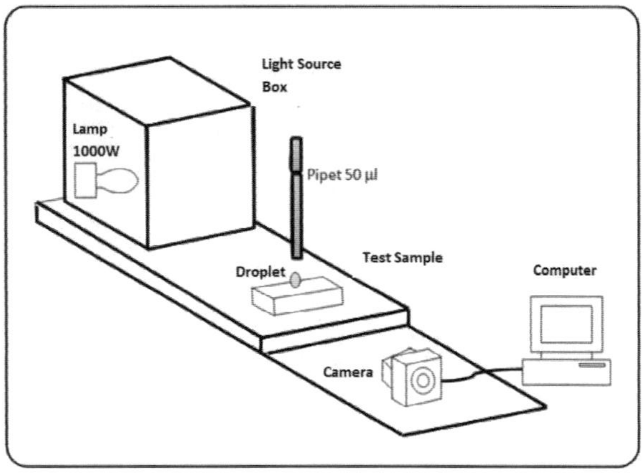

Figure 1 Equipment for surface contact angle measurements

3.3 Leakage Current Measurement
Equipment for surface leakage current performance study, include: transformer test with 220 Volt input voltage and output 50 kV with a capacity of 5 kVA, table controls to adjust the applied voltage, high voltage probes, voltmeter, voltage divider, arresters, oscilloscope, peristaltic pumps, hoses, measuring cups, digital camera, flash and computer as shown at Figure 2.

Figure 2 Equipment for measuring of leakage current

4. Results and Discussion
Contact angle measurements aims to determine the surface properties of materials, water-repellent (hydrophobic) or absorb water (hydrophilic). Hydrophobicity of the surface of the test material is influenced by the type of material and the composition of the filler constituent [7,8]. The hydrophobic properties can be determined by measuring the contact angle between the tests materials with distilled water dripped on the material surface. Value of surface contact angle of the droplets of liquid material obtained by direct observation through the digital camera scene which is then stored on the computer. The photographs processed using software to obtain the contact angle on the right side and the left side of the test sample is measured. Epoxy resin material contact angle increased after the addition of filler material silicon rubber and silica. Surface hydrophobicity properties of the epoxy resin material

increases with silicone rubber. This happens because the more a methyl group (CH3) on the hydrophobic silicon rubber. Methyl group is derived from the polysiloxane [3]

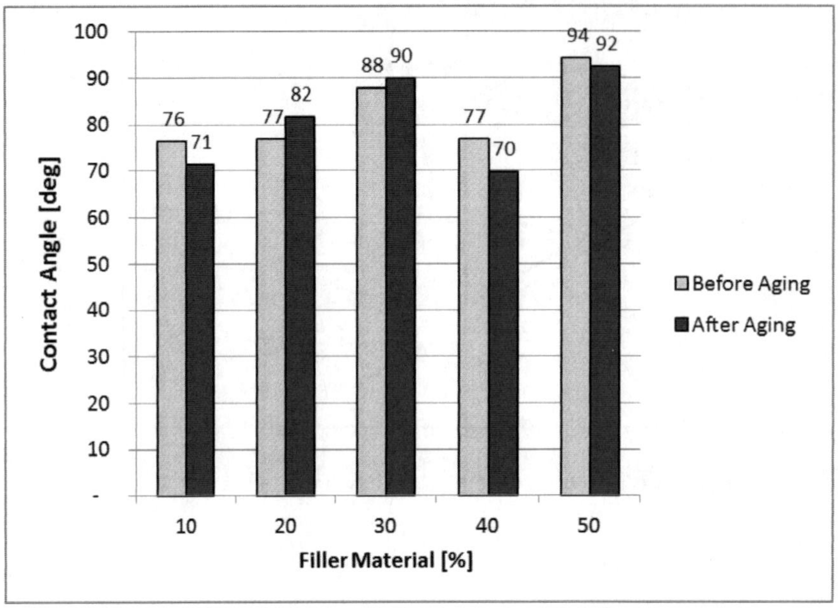

Figure 3 Correlation graph for contact angle of epoxy resin

Surface hydrophobicity of epoxy resin composition is influenced by the chemical structure of polysiloxane more dominated by the polydimethyl functional group (n [CH3]2) are hydrophobic. The presence of methyl groups CH3 rotating freely and Si-O bond polarizability causing siloxane chains are able to straighten them and produce on their surface hydrophobicity.

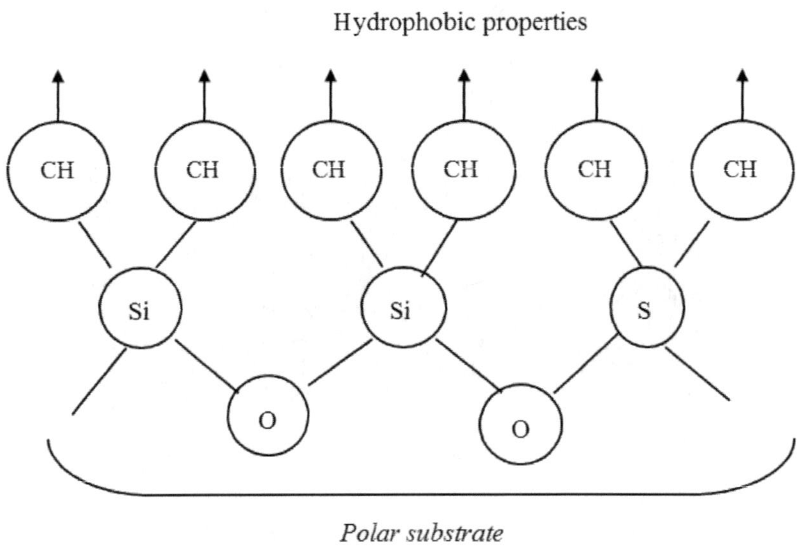

Figure 4 Hydrophobic properties of freely rotating methyl group

Figure 4 shows that Polydimethylsiloxsane molecules randomly shaped zigzag. Chain configuration can curl because the valence angle is relatively large silicon (Si-O-Si ~ 143o and 109o ~ O-Si) and the radius of the elements Si = 1.46 Angstroms larger than the carbon element radius 0.91 Angstroms so that the distance between the axis of methyl compound Si-O-Si greater. If the configuration chain structure curl then reoriented methyl group (CH3) and siloxane bond in opposite directions. When a methyl group oriented to the surface, then the surface will be hydrophobic [9]. The increasing the contact angle of an epoxy resin material after aging for 6 minutes because the surface of the resin material and RTV23 and RTV22 able to withstand sparks during an electrical process tracking, the surface does not have a fire. However, as a result of sparks on the surface, causing the surface to become rough and formed pores filled with air.

When the surface is rougher, then the cohesive forces between water molecules also greater than the adhesion force between water molecules at the surface of the insulating material, thus making the surfaces more hydrophobic insulation material. The more rough the surface of the insulating material contact angle greater hydrophobic [10].

5. Conclusion

The results of the contact angle measurements before aging, show that the test sample RTV_{21} to RTV_{24} have contact angle below 90° means that the test sample is normally hydrophilic, only RTV_{25} test sample that has reached the contact angle 94°, meaning that the test sample is normally hydrophobic. The contact angle of 76° epoxy resin in the composition of the mixture of silicone rubber and silica RTV_{21} to be the contact angle 94° after the addition of filler material silicon rubber and silica mixture of RTV_{25} showed an increase in the hydrophobicity properties of epoxy resin material.

References

[1] Raju G G 2003 Dielectrics in Electric Fields *Marcel Dekker Inc.*

[2] Ghunem R A, Jayaram S H and Cherney E A 2012 Erosion of ATH and silica filled silicone rubber in the DC inclined plane test *2012 IEEE 10th Int. Conf. Prop. Appl. Dielectr. Mater.* 1–4

[3] Jaya A and Berahim H 2011 The hydrophobicity improvement of high voltage insulator based on epoxy polysiloxane and rice husk ash compound *Proceedings of the 2011 International Conference on Electrical Engineering and Informatics* (IEEE) pp 1–6

[4] Arismunandar A 1982 *Teknik Tegangan Tinggi Suplemen* (Jakarta: Ghalia Indonesia)

[5] Lee H and Neville K 1967 *Handbook of Epoxy Resin* (New York: McGraw-Hill Book Company)

[6] Dissado L A and Fothergill J C 1992 *Electrical Degradation and Breakdown in Polymers* (London: Peter Peregrinus Ltd.)

[7] Rao U M, Majeed S S M S A, Venkataseshaiah C and Sarathi R 2002 Investigations of surface characterization of silicone rubber due to tracking phenomena under a.c. and d.c. voltages *Bull. Mater. Sci.* **25** 473–5

[8] Du B, Guo Y, Liu Y and Tian L 2014 Effects of adding nanofiller on DC tracking failure of epoxy/MgO nano-composites under contaminated conditions *IEEE Trans. Dielectr. Electr. Insul.* **21** 2146–55

[9] Tumiran T, Jaya A, Berahim H and Rochmadi R 2012 Accelerated Aging Effect on Epoxy-polysiloxane-Rice Husk Ash Polymeric Insulator Material *TELKOMNIKA (Telecommunication Comput. Electron. Control.* **10** 655–66

[10] Hazlet R D 1990 Fractal Application: Wettability and Contac Angle *J. Colloid Interface Sci.* **137** 527–33

IOP Conf. Series: Materials Science and Engineering **190** (2017) 012026 doi:10.1088/1757-899X/190/1/012026

Statistical Analysis of Reducing Biochemical Oxygen Demand (BOD) on Industrial Rubber Wastewater using Dielectric Barrier Discharge Plasma

Abdul SYAKUR[1], Badrus ZAMAN[2], Dias Yunita NURMALIAKASIH[3]

[1]Department of Electrical Engineering, Faculty of Engineering, Universitas Diponegoro.
[2,3]Department of Environmental Engineering, Faculty of Engineering, Universitas Diponegoro.
Jln. Prof. Soedarto SH Tembalang, Semarang, INDONESIA
Email: syakur@undip.ac.id

Abstract. Dielectric Barrier Discharge plasma (DBD) is one of type non-thermal plasma (non-equilibrium plasma) or can be referred to as cold plasma. In this research, DBD plasma be utilized to reduce organic compounds like Biochemichal oxygen demand in the wastewater rubber processing. In the environment field DBD plasma has been used as a treatment for reducing air pollutants such as gas COx, NOx and HC. In addition DBD plasma have been developed to processed wastewater as an alternative technology in wastewater treatment. DBD plasma appears when the electrode is given a high voltage so that, it will form electric field in the area of the electrodes which allows the ionization and the presence of high-energy electrons in the area. The presence of these electrons will ionize molecules of H_2O into active species such as OH•, H• and H_2O_2. The active species that can oxidize into CO_2 and H_2O so, BOD that can be degraded. In this research for wastewater treatment used high voltage are 10kV, 11kV, 12kV and 13kV and variations of processing time for 5, 10, 15, 20, and 25 (minutes). By increasing the voltage and extend the contact time then the speed variation of electrons to ionize the greater and more active species to be formed to degrade the pollutants to the maximum. This research used quantitative analysis with statistical analysis using SPSS software.

1. Introduction

Dielectric-barrier discharges plasma (DBD), also referred to silent discharges have found amount of interesting industrial applications in addition to the historical ozone generation[1]. Originally it was called the silent discharge in contrast with a spark discharge. From a fundamental point of view, the main difference between DBD and spark discharges is the presence of the dielectric barrier which precludes AC operation of the DBD[2]. Configuration of DBD plasma is a discharge having two dielectric barrier boundaries has many similarities with discharges operated between metal electrodes[3]. Plasma occurs when the electrode is given a high voltage so that the area around the electrodes form an electric field to ionize and the formation of electrons with great energy. On development of plasma technology, DBD is currently widely used as a treatment of wastewater to reduce polluter [4]-[7]. Characteristic of DBD can be provided atmospheric pressure. In this paper DBD plasma is used for treating industrial wastewater. Biochemical oxygen demand is a parameter that be studies researched to decreaced of amount on the variation of voltage and duration of time application.

2. Fundamental Theory

2.1 Dielectric Barrier Discharge Plasma

Dielectric Barrier Discharge (DBD) Plasma or usually called as silent plasma is gas ionized, when given by high voltage on the atmospheric pressure. DBD plasma can be generated between two electrodes. Electrodes are located outside and inside the dielectric barrier. The electrode can be termed as active electrode to be energized with High-voltage AC in specific capacity, and beyond the electrode is located on outside of dielectric barrier affairs called passive electrode surrounding of Pyrex-glasses as a dielectric barrier. When given two of electrode high voltage will be generated electricity that can form a plasma discharge between the electrodes around the room. The electrodes materials made from metal which is good conductor such as iron, copper, aluminum etc. The voltage source typically used to generate High voltage derived from the AC voltage of 1-100 kV [8]. The following Figure 1 is a geometric configuration DBD plasma reactor with forms of coplanar.

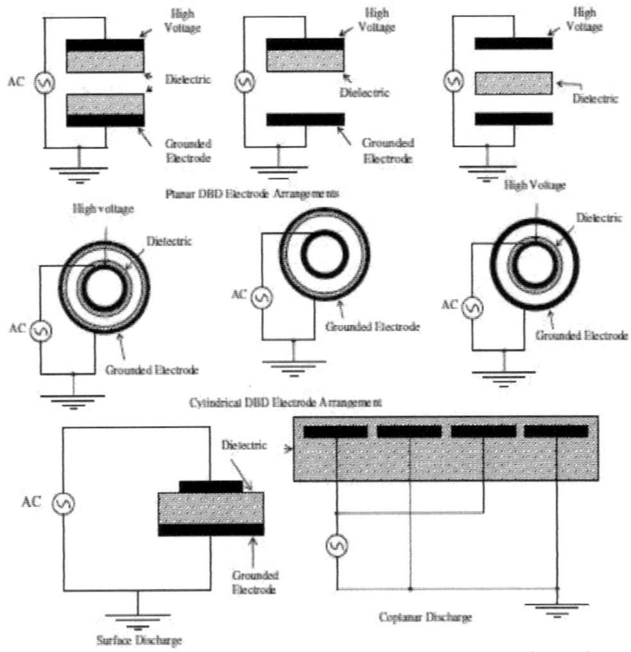

Fig. 1 Typical electrode arrangements of DBD configurations

On Figure 1 DBD plasma type is shaped coaxial with having two electrodes. The electrodes are cylinder or coaxial shaped. Inside electrodes could be replace as active electrode will be energized with high voltage and beyond a passive electrode will be streamed on ground.

2.2 Geometry Horizontal Roughing Filtration

Horizontal Roughing Filtration (HRF) is a physical filter that has the ability to keep the solids in large enough quantities. Solids are filtered on filter media that will settle to the bottom filter. On the HRF filters used are typically made of a filter which has pores that are large and broad media surface, so that it can absorb a maximum of solids.

The solids are absorbed will be deposited on the basis of a filter that can be optimized to filtrated of suspended solid. HRF flowed with the flow horizontal or straight stream that passes through the filter 3-4 compartment [9].

2.3 Biochemical Oxygen Demand

Biochemical oxygen demand (BOD) or organic matter is the amount of oxygen that needed by micro-organisms to break down organic matter in the water. BOD in water will break down into carbondioxide and NH_3. Organic substances contained in the water because of industrial activity in the disposal of waste in water bodies.

Disposal of waste with high levels of BOD or organic substances that are high enough to be causing problems to water due the amount of oxygen dissolved in water decreases. It can harm the ecosystem living things in the water [10].

3. Experimental Set Up

3.1 Materials

Configuration materials of DBD Plasma reactor are: pyrex glass, active electrode is alumunium (Al), passive electrode is copper wire,natural rubber wastewater, beaker glass, pipette and measuring cup.

3.2 DBD Plasma Reactor

Dimension and size of Pyrex glass as barrier between high voltage electrode and ground electrode as shown in Fig. 2. Diameter of Pyrex is 16 mm. At the centre of Pyrex, there is aluminium as high voltage electrode. The radius of aluminium is 3 mm.

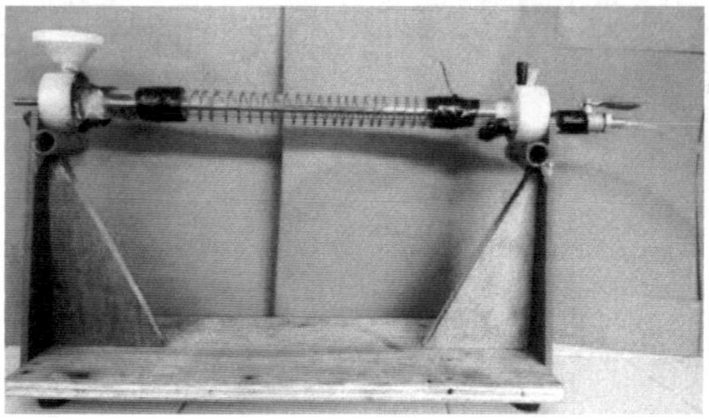

Fig 2. Geometry Plasma DBD Reactor

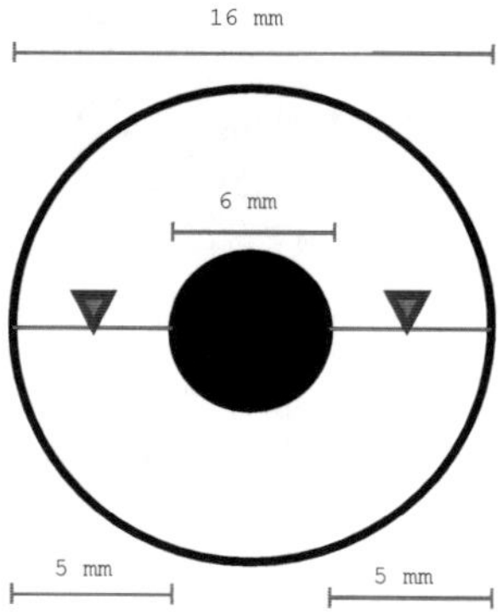

Fig 3. Geometry Plasma DBD Reactor

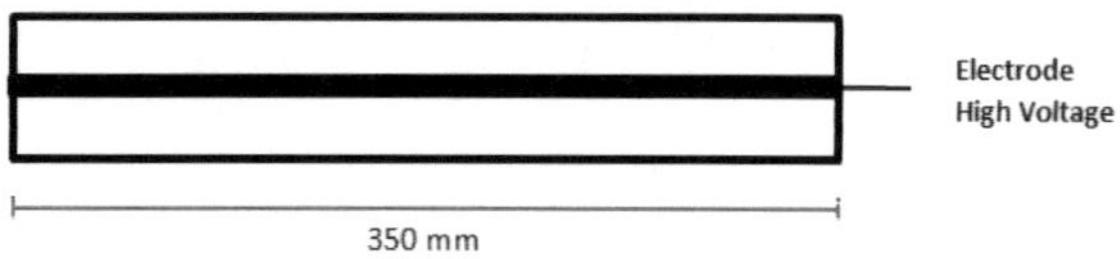

Fig 4. Geometry of Plasma DBD Reactor

3.3 Test Circuit

Plasma is generated from high voltage application in aluminium electrode in the Pyrex tube. While the copper wire was wrapped around the Pyrex tube. The test circuit for generating dielectric barrier discharge plasma as shown in Fig 5 as follow.

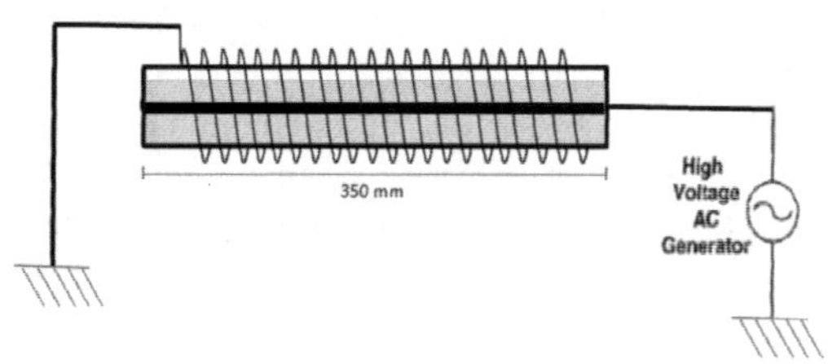

Fig 5. Test Circuit for Generating DBD Plasma

3.4 Titration Test

Standard specifies a method for the determination of the biochemical oxygen demand (BOD) of wastewater following SNI 6989.72:2009. The principle of Biochemical Oxygen Demand (BOD) test sample are added diluent solution saturated oxygen has been added a solution nutrients (1 ml buffer phosphate, 1 ml $MgSO_4$, 1 ml CaCl, and 1 ml $FeCl_3$) and microbial seed, then add demineralised water to 1 litter and mix until homogenous. incubated in the dark at a temperature of 20 °C ± 1 °C for 5 days. value BOD calculated based on the difference in the concentration of dissolved oxygen 0 days and 5 days.

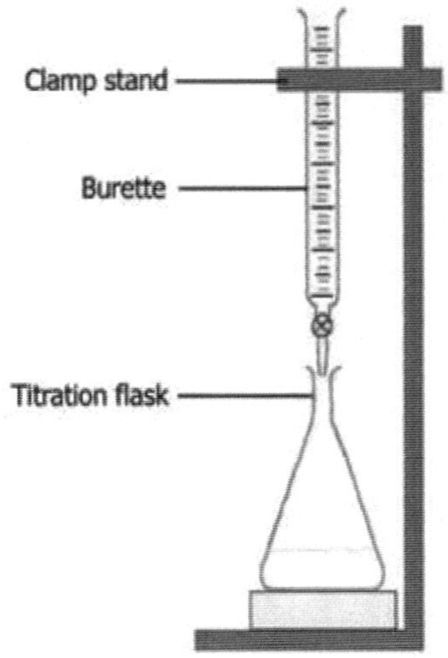

Fig 6. BOD Titration Test

3.5 SPSS Analysis

In this paper used the software SPSS 17 for testing the quantitative analysis. SPSS performed including normality test to ensure that the normally distributed data, correlation to know the how much the relationship between independent variables and the dependent variable, and regression test to determine how much influence the dependent variable to the independent variables and give equality to achieve the test expected

4. Results and Discussion

Measurement results of the BOD concentration after treatment wastewater of rubber industry with DBD plasma will be discussed. Becomes treatment with DBD plasma, wastewater of rubber industry performed pre-treatment first with Horizontal Roughing Filtration (HRF) to reduce the work DBD plasma reactor because a lot of suspended solid contains BOD is an indicator of contaminants that could pollute the environment. The previous study showed a reduction of BOD concentration after DBD plasma treatment. The following graph shows a decrease in the concentration of BOD after treatment.

On this research, the treatment of wastewater variation of times contact and voltages DBD plasma. BOD measurement is an indicator of the presence of organic compounds that can interfere with the environment. The Organic compounds contained in the rubber industrial wastewater coming from the washing process the raw material rubber (latex). Reduction of BOD concentrations were shown in Figure 7 and 8. Decrease of BOD because of the presence of active species such as OH•, H• and H_2O_2 contained in DBD plasma due to the ionization of H_2O crushed by electrons.

Increasing the voltage and variation of duration contact times in this paper due to being a different amount of electrons and energy electrons generated with increased voltage. In the high voltage electrons produced will be more [11] that shows the process of ionization, dissociation and excitation in the plasma. Increasing number of electrons, the active species such as OH •, H • and H_2O_2 formed will be more rapid and substantial to oxidize organic compounds such as BOD in wastewater.

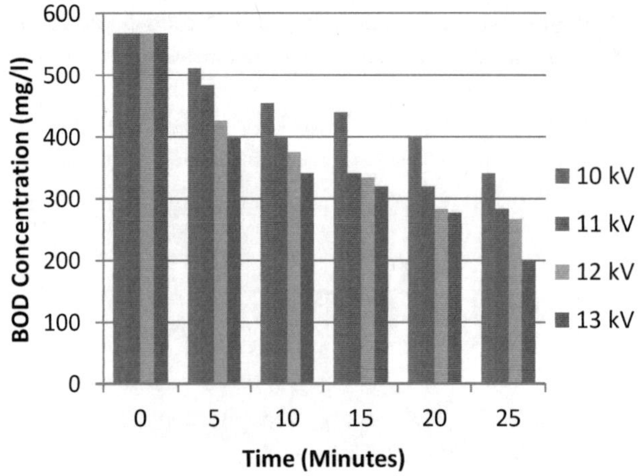

Fig 7. Graph of reducing concentration of BOD with Plasma DBD for time variation.

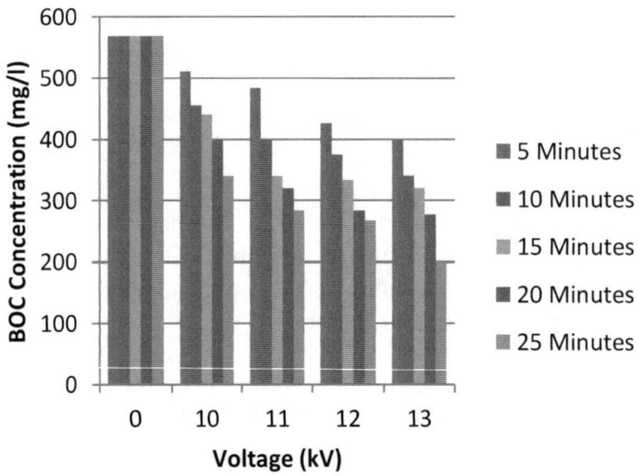

Fig 8. Graph of reducing concentration of BOD with Plasma DBD for Voltages variation.

Table 1 One-Sample Kolmogorov-Smirnov Test

		Unstandardized Predicted Value
N		20
Normal Parameters [a,b]	Mean	365.8125000
	Std. Deviation	71.05676834
Most Extreme Differences	Absolute	.093
	Positive	.093
	Negative	-.093
Kolmogorov-Smirnov Z		.415
Asymp. Sig. (2-tailed)		.995

Quantitative analysis was performed by statistical tests. To know normally distributed data or not, it is necessary to test for normality. BOD normality test results shown in column Kolmogorov-Smirnov Z obtained a value of 0.415 (>0.05) so that BOD distribution is the norm.

Table 2 Correlations BOD with Voltages

		BOD1	Voltage 1
BOD	Pearson Correlation	1	-.499*
	Sig. (2-tailed)		.025
	N	20	20
Voltages	Pearson Correlation	-.499*	1
	Sig. (2-tailed)	.025	
	N	20	20

*. Correlation is significant at the 0.05 level (2-tailed).

Table 3 Correlations BOD with Contac Times

		BOD	Contact Time
BOD	Pearson Correlation	1	-.599**
	Sig. (2-tailed)		.005
	N	20	20
Contac Times	Pearson Correlation	-.599**	1
	Sig. (2-tailed)	.005	
	N	20	20

**. Correlation is significant at the 0.01 level (2-tailed).

The relationship test for determine how much influence the voltage and the contact time of the BOD concentration. Based on the results of correlation on Table 2 and 3, it can be seen that the correlation between the independent variable voltage and contact times with BOD concentration had a correlation coefficient of -0.499 and -0,559 it means to have a medium relationship. Figures negative correlation coefficient indicates the opposite relationship; in this case, the greater the voltage applied and contact times with BOD concentration will decrease. Values of Sig. (2-tiled) 0.005 < 0.05 its means that the voltage and contact times independent variable relationship with the BOD concentration significantly, meaning that the variable voltage and contact times give great on influence decreasing of BOD.

To determine the influence of the voltage and the contact time of wastewater by the DBD plasma concentrations of BOD then do regression testing. Table 4 shows that the influence of the independent variable voltage and the contact time of the BOD concentration. The test results show regression R^2 of 0.887. That is, the variable contact time has an influence of 88.7 % of the BOD concentration. The other 11.3% is influenced by other variables that are not performed in this study. Tabel 5 is regression test produces the following equation:

$$y = 1087{,}946 - 51{,}354(a) - 9{,}130(b)$$

where:

y = concentration of BOD

a = the Voltage

b= the contact time

IOP Conf. Series: Materials Science and Engineering **190** (2017) 012026 doi:10.1088/1757-899X/190/1/012026

Table 4 Model Summary

Model	R	R Square	Adjusted R Square	Std. Error of the Estimate
1	.942[a]	.887	.868	36.380

Table 5 Coefficients

Model		Unstandardized Coefficients		Standardized Coefficients	t	Sig.
		B	Std. Error	Beta		
1	(Constant)	1087.946	91.763		11.856	.000
	Voltage	-51.354	7.877	-.632	-6.519	.000
	Time	-9.130	1.253	-.706	-7.288	.000

5. Conclusion

Based on the results of laboratory tests and statistical analysis can be concluded that when higher voltage and the duration of the contact were applied to the treatment of wastewater using DBD plasma has the effect of reducing the BOD concentration of 88.7%. The combination of the two independent variables resulted in a decrease in the concentration of BOD significantly. BOD normality test results shown in column Kolmogorov-Smirnov Z obtained a value of 0.415 (>0.05) so that BOD distribution is the norm. The correlation between the independent variable voltage and contact times with BOD concentration had a correlation coefficient of -0.499 and -0,559 it means to have a medium relationship.

References

[1] O. Dmitry F, 2012, "Dielectric Barrier Discharge Plasma Actuator For Flow Control," Princeton University, Princeton, New Jersey.

[2] K. Ulrich, 2000, "Dielectric Barrier Discharge: Their History, Discharge Physics, and Industrial Application," *Plasma Chemistry and Plasma Processing*, Switzerland, Vol 23, No 1, 2003.

[3] U. Kogelschatz, "Fundamentals and Applications Of Dielectric Barrier Discharges," ABB Corporate Research, Switzerland.

[4] W. B. Zimmerman, E. Karunakaran, W. S. Abdul Majeed, 2015, "Develompent of Wastewater Treatment System Based On Cascade Dielectric Barrier Discharge Plasma Atomizers," Departemen of Chemical and Petrochemical Engineering, University of Nizwa, Nizwa, Oman.

[5] L. Heon-Ju, J. Jin-Oh, M. Youg Sun, 2008, "Dielectric Barrier Discharge Plasma Induced Photocatalysis and Ozonation for The Treatment of Wastewater," *Plasma Science and Technology*, Vol 10, No 1.

[6] M. M. Kuraica, B. M. Obradovic, D. Monojlovic, D. R. Ostojic, J. Puric, 2006, "Application of Coaxial Dielectric Barrier Discharge For Potable and Wastewater Treatment," Faculty of Physics and Chemistry, Center for Science Technology Development, Serbia and Montenegro.

[7] M. Tichonovas, E. Krugly, V. Racys, R. Hippler V. Kauneliene, I. Stasiulaitiene, D. Martuzevicius, 2013, "Degradation of Various Textile Dyes as Wastewater Pollutants Under Dielectric Barrier Discharge Plasma Treatment," Chemical Engineering Journal, Vol. 229, pp. 9-19.

[8] V. Nehra, A. Kumar, H. K. Dwivedi, 2008, "Athmospheric Non-Thermal Plasma," *International Journal Of Electrical*, India, 2008.

[9] Wegelin M, 1996, "Surface Water Treatment by Roughing Filters", *A Design, Construction and Operation Manual*, Swiss Federal Institute for Environmental Science And Technology (EAWAG) and Departement Water and Sanitation In Developing Countries (SANDEC).

[10] M.R. Penn, J.J. Paeur, and J.R. Mihelcic., 2006, "Biochemical Oxygen Demand". *Environmental and Ecological Chemisrty*, Vol II.
[11] M. Nifuku, M. Horvath, J. Bornar, G. Zhang, T. Tanaka, E. Kiss, G. Woynarovich, and H. Katoh, 1997 "A Study on the Decomposition of Volatile Organic Compounds by Pulse Corona."*Journal of Electrostatic* , pp. 40-41.

IOP Conf. Series: Materials Science and Engineering **190** (2017) 012027 doi:10.1088/1757-899X/190/1/012027

Developing Control System of Electrical Devices with Operational Expense Prediction

Siti Sendari[1], Heru Wahyu Herwanto[2], Yuni Rahmawati[3], Dendi Mukti Putranto[4], Shofiana Fitri[5]

[1, 2, 3, 4, 5]Department of Electrical Engineering, Faculty of Engineering, Universitas Negeri Malang, Jalan Semarang No. 5, Malang, Jawa Timur, Indonesia,

siti.sendari.ft@um.ac.id

Abstract. The purpose of this research is to develop a system that can monitor and record home electrical device's electricity usage. This system has an ability to control electrical devices in distance and predict the operational expense. The system was developed using micro-controllers and WiFi modules connected to PC server. The communication between modules is arranged by server via WiFi. Beside of reading home electrical devices electricity usage, the unique point of the proposed-system is the ability of micro-controllers to send electricity data to server for recording the usage of electrical devices. The testing of this research was done by Black-box method to test the functionality of system. Testing system run well with 0% error.

1. Introduction

In recent years, the use of electrical energy has become the basic human needs in daily activity. However, the electrical energy needs every year has increased in line with population, wealthy, and the overall economic growth. One important factor to support economic growth, i.e., the rapid growth of electricity demand of Indonesia's consumers for product and services. Deputy Minister of Energy and Mineral Resources, Rudi Rubiandini and Indonesia's state utility company, PT Perusahaan Listrik Negara's (PLN) Director of Planning and Risk Management said that Indonesia's annual electricity usage is expected to rise by 9% in 2013, while The National Energy Council (DEN) expects a broader threefold increase in energy demand by 2030 [1]. The awareness of society towards electricity usage of electrical devices is less. This behavior make users tend to behave wasteful. So, the use of electricity energy excessively in the household lead to wastage of electricity expense every month. [2], it showed that the amount of unused energy from electrical devices, which are connected to a voltage source, is about 5%-10% from its normal used. It can affect in financial loss and decrease the efficiency of electricity usage.

In order to improve the efficiency of electricity usage in household, control system has been developed rapidly. In the modern era, integrating new technologies in control system could improve the quality of human life.The aim of this paper is to develop a control system which can help users to control the electricity usage and also to develop a system for monitoring and controlling home electrical devices. Currently for monitoring electrical quantities, it is mostly done by installing electrical measuring instrument at the electrical panel before going to its electrical devices. However, this method has some lacks, that is difficult to identify which electrical devices waste the use of electrical power, it also does not have data record for monitoring electrical power usage [3].

This paper proposes a solution that is advantageous for (1) providing information of electricity status without checking directly at the location where the measuring instrument is installed, (2) providing recorded information for monitoring electrical power usage for analysis, (3) providing an expense prediction of electricity usage for each electrical devices, and (4) providing controlling facility for electrical devices remotely, and setting usage permissions of electrical devices.

The developed system requires module controllers, and web-based application in server. The module controllers that is installed to electrical devices, use Arduino Nano as microcontroller [4], while ESP8266-01 as WiFi module [5]. Current sensor of Electronic Brick TA12-100 Itead Studio [6] is used to inform the electricity usage, while relay driver is used to switch the electrical devices either "On/Off". The data of electrical current that has been read by microcontroller is sent by WiFi module to the server so that the measurement result can be stored in database. The server in home network can be connected to internet, so the monitoring process can be done remotely.

The rest of this paper is organized as follows. Section II presents the proposed method to develop the system. In Section III describes the result and discussion about the performance of proposed method. The last, Section IV present the conclusion.

2. Method

The proposed method that used to develop this system is Systems Development Life Cycle (SDLC). [7], i.e., a methodology that used to develop a system in the field of engineering, information system, and software system. This method is described in several processes, i.e. planning, analysis, design, implementation and maintenance.

2.1. System

The designed system is shown in Fig 1. The system is sent by detect the status of electrical devices which is shown as "On/Off" based on the information from current sensor.

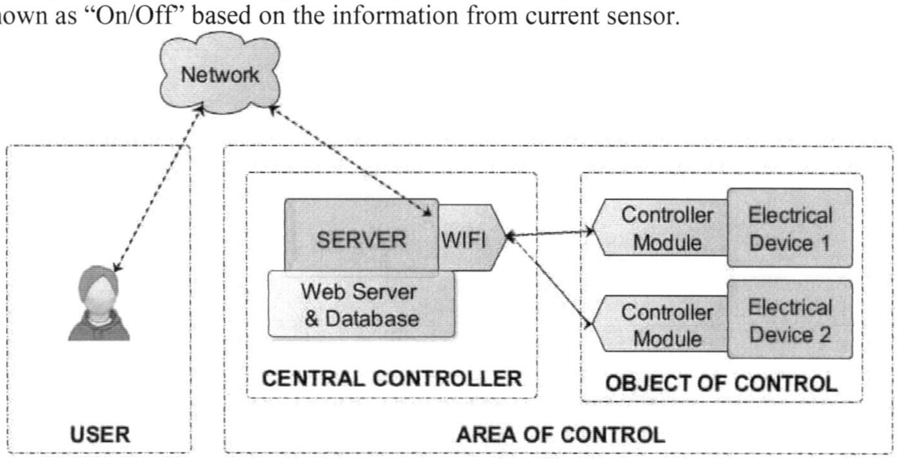

Figure 1 The Designed System

Each electrical device has its own module controller, which given action control from microcontroller based on information of current sensor. The current sensor transfers its current data to microcontroller Arduino Nano to be processed by action control. The information of electrical devices status is sent via micro-controller to server. The server's duty is to receive an input and control module through wireless network. In order to the system can be accessed by users from several devices, the application has been designed in web based.

On the server, collected data in database will be processed to be useful information and presented to the user which include data of electricity usage and prediction of operational expense daily, monthly or in the specific time. The designed system works as the flowchart shown in Figure 2.

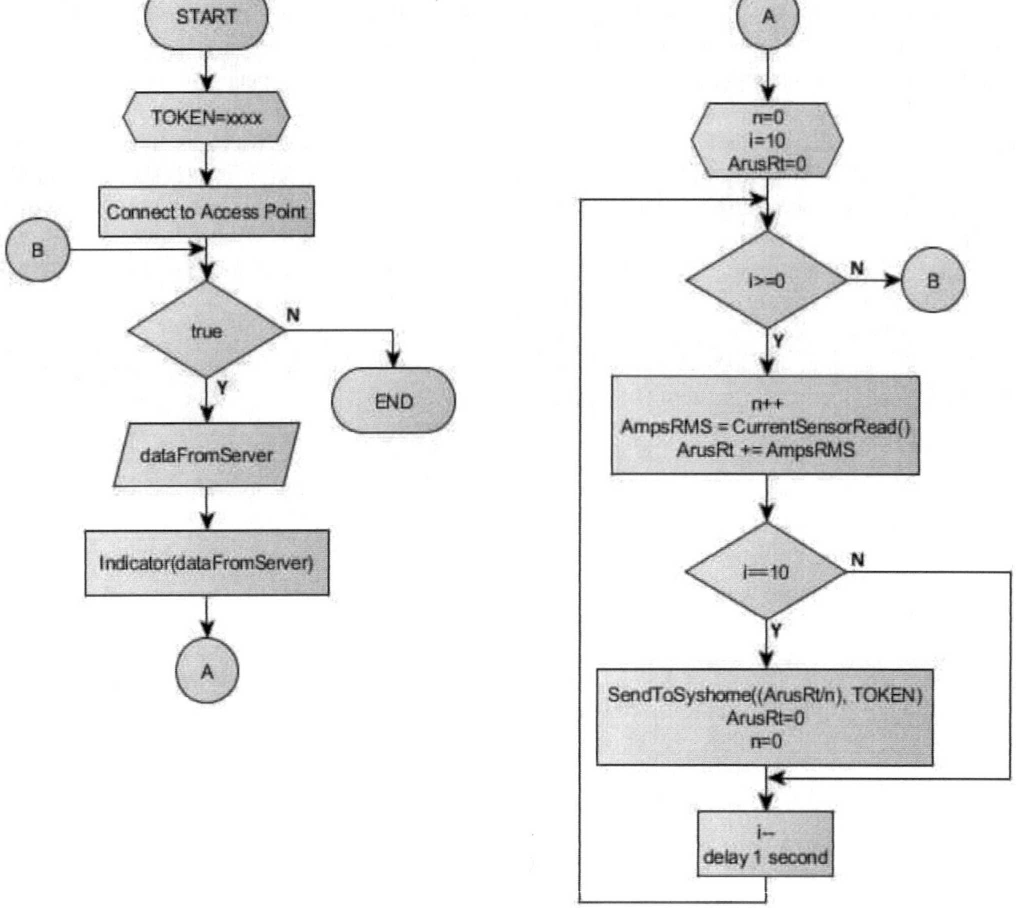

Fig 2. Flowchart System

2.1.1. Module Controllers

Module controllers are devices that used to monitor and control electrical devices. Module controllers has five main parts, i.e. microcontroller (Arduino Nano), WiFi module (ESP8266-01), current sensor (TA12-100 Itead Studio), indicator, and driver circuit using relay for controlling the status of "On/Off" in electrical devices. Fig 3 shows the designed module controllers.

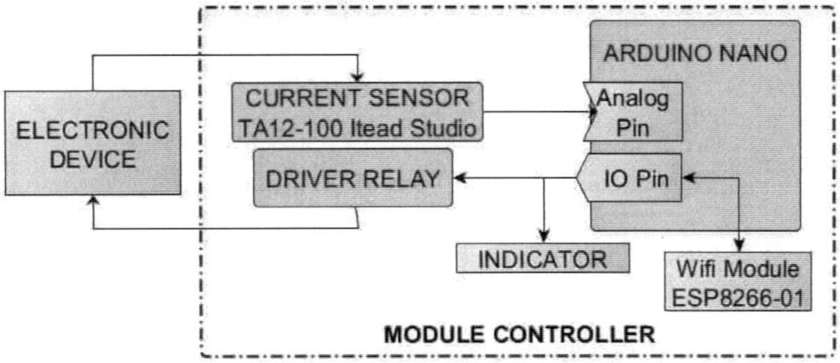

Fig 3. Designed-controller module

The Microcontroller that used has twelve digital pin and eight analog pin. Current sensor will be connected to an analog pin. Relay access driver will be handled by digital pin. While indicator circuit as an status information of module controllers uses eight digital pin.

2.1.2. Web-based Application

Website, i.e., part of system that handle all process in server and as a controlling interface by users. Here is some steps to design website :

- Design database and analyze web application to be divided into several part.
- Design website layout.
- Web programming in each part of website, that include : (1) *Log in* process page; (2) *Form* page to manipulate master data (*retrieve*, *input*, *update*, and *delete*) of module controller, electrical devices, and electricity expense / kWh; (3) *Form* page to define module controller; (4) *Form* access control setting page; (5) Notification page; and (6) *Report* page.
- Evaluate each part of software

Data Flow Diagram (DFD) is graphical representation of system that describes components. Fig 4 shows data flow diagram (DFD) level 0 of developed-system.

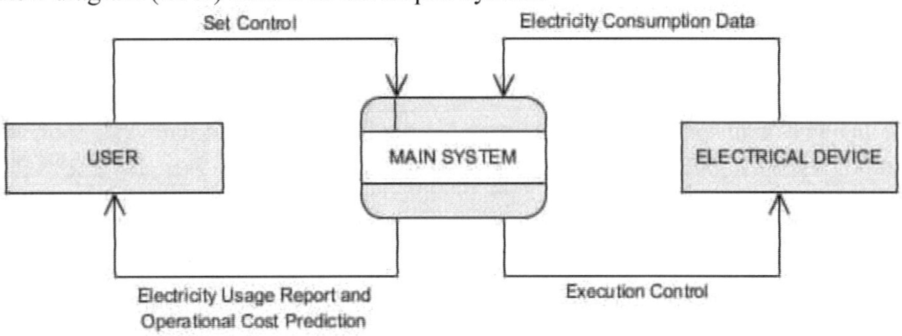

Fig 4. The Data Flow Diagram Level 0 Context Diagram of system

This diagram describes the basic process, where user can input data of control time setting to the system as a system reference to Main System. In the Main System data will be proceed to execute the controlling in electrical devices.

At the same time, the system receives an input of electricity data usage from electrical devices to be processed in the Main System. Then the data will proceed as an output record of home electrical device's electricity usage and its operational expense prediction.

2.1.3. Implementation

The implementation process is done by creating module controller and website application based on the earlier design process. The system design is translated into codes using programming language i.e., Arduino for hardware programming, PHP for web server programming, HTML, Java script, and CSS for web programming, also MySql as database management system. The program that has been developed will be tested per unit and overall using Black-box method.

In order to observe the implementation of the entire system components, the testing has been held. The test was carried out by testing per sub-system to know the performance of each sub-system and the compliance with design specification. The further testing is done by combining all sub-system to know the assessment result of overall system.

This designed system use case diagram, i.e., to describe a set of actions (use cases) as representation of user's interaction which consists of actors, use case and their relationships are shown in Fig 5.

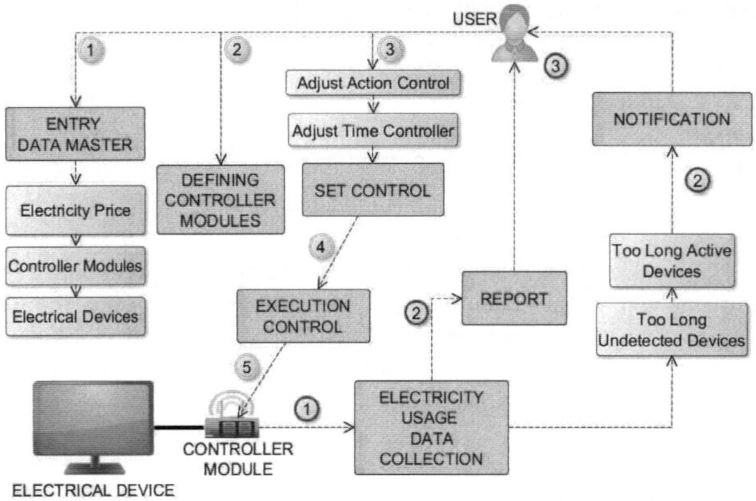

Figure 5 Use Case Diagram of system

The next step was system maintenance. The aims of this step is to ensure the system is always in good performance and also evaluate every error that might be occur. So that system maintenance can improve system performance and keep the system from being obsolete.

2.1.4. Testing Design

The functionality is tested using Black box testing method [8]. The testing was carried out on module controller, software (website application) and overall system. The test is used to find errors that occur on the system. The errors were found by analyzed the equation below [9].

$$E = 100 \times (1 - (1 - L)^{n}) \quad (1)$$

Where:
E = Errors found (%)
n = Number of testers
L = Problems found (%)

3. Result and Discussion

3.1. Module Controller Testing

The aim of testing controller module is to avoid any errors when the system is being executed. The testing was carried out by testing each part of module controllers, which consist of testing the running of (1) microcontrollers, (2) WiFi modules, (3) current sensors, (4) indicators, and (5) relay drivers. The testing results presented that all parts of module controller was valid and running well.

3.2. Software Testing

The software testing is done to avoid errors when the system is executed by testing the functionality each part of website. The testing was done to check the page of (1) *Log in* process page; (2) *Form* page to manipulate master data (*retrieve*, *input*, *update*, and *delete*) of module controller, electrical devices, and electricity expense / kWh; (3) *Form* page to define module controller; (4) *Form* access control setting page; (5) Notification page; and (6) *Report* page. The testing result showed the 0% error percentage. So it can be concluded that the software is valid, running well and ready to be used for system development

3.3. Overall System Testing

The overall system testing is done to ensure that all sub-systems can work together, interconnected and shows good performance. The result of overall system testing generates 0% error percentage which can be concluded that the development of system is valid and overall can run well. Here, the graph on Figure 6 shows the electricity usage daily with each detailed usage for 24 hours.

MODUL 1 - KKSZG7Q731										
No.	Electrical Devices	Time	Duration	Hour	Cost	Current (A)	Power (W)	Energy (WH)	Energy (KWH)	Total Cost
1	Charger Laptop Toshiba 15 inch (60 Watt)	00:00:24	00:00:27	0.00750	1550	0.59	129.8	0.97350	0.00097	Rp. 1.50893
2	Charger Laptop Toshiba 15 inch (60 Watt)	00:00:51	00:00:28	0.00778	1550	0.56	123.2	0.95822	0.00096	Rp. 1.48524
3	Charger Laptop Toshiba 15 inch (60 Watt)	00:01:19	00:00:27	0.00750	1550	0.59	129.8	0.97350	0.00097	Rp. 1.50893
4	Charger Laptop Toshiba 15 inch (60 Watt)	00:01:46	00:00:27	0.00750	1550	0.57	125.4	0.94050	0.00094	Rp. 1.45778
5	Charger Laptop Toshiba 15 inch (60 Watt)	00:02:13	00:00:28	0.00778	1550	0.59	129.8	1.00956	0.00101	Rp. 1.56481
6	Charger Laptop Toshiba 15 inch (60 Watt)	00:02:41	00:00:27	0.00750	1550	0.53	116.6	0.87450	0.00087	Rp. 1.35548
7	Charger Laptop Toshiba 15 inch (60 Watt)	00:03:08	00:00:27	0.00750	1550	0.58	127.6	0.95700	0.00096	Rp. 1.48335
8	Charger Laptop Toshiba 15 inch (60 Watt)	00:03:35	00:00:28	0.00778	1550	0.59	129.8	1.00956	0.00101	Rp. 1.56481
9	Charger Laptop Toshiba 15 inch (60 Watt)	00:04:03	00:00:27	0.00750	1550	0.57	125.4	0.94050	0.00094	Rp. 1.45778
10	Charger Laptop Toshiba 15 inch (60 Watt)	00:04:30	00:00:27	0.00750	1550	0.54	118.8	0.89100	0.00089	Rp. 1.38105
11	Charger Laptop Toshiba 15 inch (60 Watt)	00:04:57	00:00:28	0.00778	1550	0.56	123.2	0.95822	0.00096	Rp. 1.48524
12	Charger Laptop Toshiba 15 inch (60 Watt)	00:05:25	00:00:27	0.00750	1550	0.54	118.8	0.89100	0.00089	Rp. 1.38105
13	Charger Laptop Toshiba 15 inch (60 Watt)	00:05:52	00:00:27	0.00750	1550	0.56	123.2	0.92400	0.00092	Rp. 1.43220
14	Charger Laptop Toshiba 15 inch (60 Watt)	00:06:19	00:00:28	0.00778	1550	0.56	123.2	0.95822	0.00096	Rp. 1.48524
15	Charger Laptop Toshiba 15 inch (60 Watt)	00:06:47	00:00:27	0.00750	1550	0.53	116.6	0.87450	0.00087	Rp. 1.35548
16	Charger Laptop Toshiba 15 inch (60 Watt)	00:07:14	00:00:27	0.00750	1550	0.55	121	0.90750	0.00091	Rp. 1.40663
17	Charger Laptop Toshiba 15 inch (60 Watt)	00:07:41	00:00:27	0.00750	1550	0.56	123.2	0.92400	0.00092	Rp. 1.43220
18	Charger Laptop Toshiba 15 inch (60 Watt)	00:08:08	00:00:28	0.00778	1550	0.55	121	0.94111	0.00094	Rp. 1.45872
19	Charger Laptop Toshiba 15 inch (60 Watt)	00:08:36	00:00:27	0.00750	1550	0.55	121	0.90750	0.00091	Rp. 1.40663
20	Charger Laptop Toshiba 15 inch (60 Watt)	00:09:03	00:00:27	0.00750	1550	0.54	118.8	0.89100	0.00089	Rp. 1.38105
21	Charger Laptop Toshiba 15 inch (60 Watt)	00:09:30	00:00:28	0.00778	1550	0.54	118.8	0.92400	0.00092	Rp. 1.43220
20	Charger Laptop Toshiba 15 inch (60 Watt)	00:09:03	00:00:27	0.00750	1550	0.54	118.8	0.89100	0.00089	Rp. 1.38105
21	Charger Laptop Toshiba 15 inch (60 Watt)	00:09:30	00:00:28	0.00778	1550	0.54	118.8	0.92400	0.00092	Rp. 1.43220

(a) The electricity usage table report

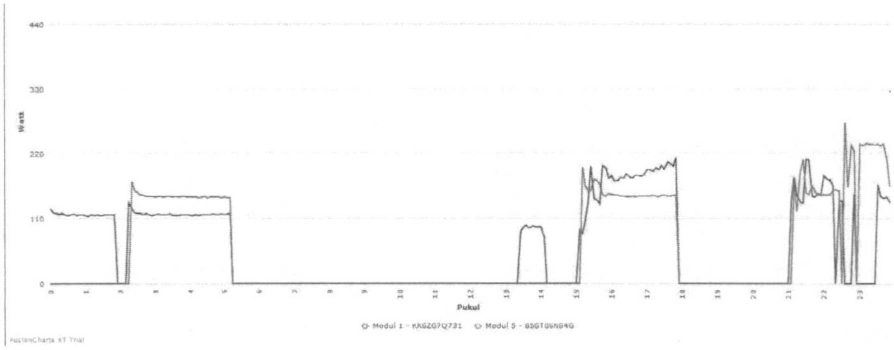

(b) Electricity usage graph record

Figure 6 Example of electricity daily usage

4. Conclusions

In this paper, we have developed a control system with aims to monitor and record home electrical devices electricity usage. This research can be concluded that the system has ability as the proposed-system, while it also can predict operational expense daily, monthly or in specific time. Furthermore, this system can control electricity usage of electrical devices remotely with usage permission setting for users.

The testing has been done with functionality test using Black box method. The testing result shows that the system worked properly with 0% error percentage, confirming that the control system of electrical devices and its operational expense prediction can work well and ready to be implemented.

5. Acknowledgement

We would like to express our sincere appreciation to all the members of our team, for their encouragement and advices. We also want to express our gratitude to the Minister of Technology and Higher Education for supporting the research fund.

6. References

[1] Noname. The Swelling Demand for Electricity: Commercial Opportunities in Indonesia's Power sector, (Online), https://photos.state.gov/libraries/indonesia/890636/pdf_001/business_power-sector.pdf)

[2] Sripan, M., Lin, X., Petchlorlean, P.,dan Ketcham, M. 2012. Research and Thinking of Smart Home Technology. International Conference on Systems and Electronic Engineering (ICSEE'2012).

[3] ElShafee, A., and Hamed, K. A. 2012. “Design and Implementation of a WiFi Based Home Automation System”, International Journal of Computer, Electrical, automation, Control and Information Engineering, Vol. 6, No. 8, pp.1074-1080.

[4] Arduino Manual. Getting Started with Arduino. (Online), (https://www.arduino.cc/en/Guide/HomePage).

[5] ESP8266EX Datasheet, (Online), (https://espressif.com/sites/default/files/documentation/0a-esp8266ex_datasheet_en.pdf/), diakses 18 Juli 2016.

[6] TA12-100 Itead Studio Data Sheet. Electronic Brick of Electricity Meter Sensor (online). ftp://imall.iteadstudio.com/Electronic_Brick/IM120710018/DS_IM120710018.pdf.

[7] Blanchard, B. S., dan Fabrycky, W. J. 2006. Systems engineering and analysis (4th ed.) New Jersey: Prentice Hall.

[8] Nidhra, S., and Dondeti, J., Black Box and White Box Testing Techniques - A Literature Review. International Journal of Embedded Systems and Applications (IJESA), Vol. 2, No. 2, pp. 29-50, June 2012.

[9] Myers, G. J.; Sandler, C.; and Badgett, T. 2012. The Art of Software Testing. Canada: John Wiley & Sons, Inc.

Social Media Success for Academic Knowledge Sharing in Indonesia (Conceptual Model Development)

Setiawan Assegaff

STIKOM Dinamika Bangsa, Jambi, Indonesia

setiawanassegaff@stikom-db.ac.id

Abstract. The aim of this study is to investigate how success is the social media as a tool for knowledge sharing among scholars in Indonesia. To evaluate the success of social media we develop a model base on Delone and McLeane IS Success Model. In this article, we would like discuss the process of developing the research model. In developing the model, we conduct literature review from knowledge management, social media and IS Success Model area from previous study. This study resulted in the social success model for academic knowledge sharing in Indonesia.

1. Introduction

The research community is recognized as one important element in creating reliable and resilient researches in universities [1]. Nowadays, many universities in Indonesia faced with various problems that the program to create a reliable researcher could be inhibited [2]. The small number of experienced researchers in certain fields followed by the uneven distribution of their presence on college only available in a few major colleges and famous only which can affect the reliable researchers [2]. Universities in Indonesia spread from Sabang to Marauke with most colleges have faculties with qualification lower than doctorate lecturers. All conditions above resulted in a remarkable gap among educational institutions in Indonesia. This resulted in a slowdown in creating the reliable researchers in Indonesia.

Virtual community is one of potential solution to create research community among academician in Indonesia. Using information technology many features could be exploiting to enhance academic collaboration and networking [3]. One of potential application is social media. Social media is powerful applications that enable their member in sharing, chatting, collaborating and sharing document via internet [4]. However how success social media is in facilitating academic in Indonesia to sharing their experience and knowledge in virtual communities is still do not identify yet. This study is interested to investigate the phenomenon. This article focus on develop model that in future could apply in investigate the success of social media for knowledge sharing.

2. Literature Review

2.1. Social Media

Koplan and Haenlein define social media as a collection of internet-based application that provides a web 2.0 interactive services among individuals and communities to interact, discuss, create and modify content [4]. Social media is now believed to be a tool that provides tremendous potential for the organization to improve the performance of the company [5]. To get the maximum benefit from the potential of social media, the company developed a variety of strategies for the use of social media in their business activities. Hanna et al [5] proposed a concept to develop a strategy for the utilization of social media in an organization that they are familiar with the term "social media ecosystem". In the concept of this, organization need to understand that the use of social media in supporting the activities organized needs to be aligned, by appropriately combining social media means in achieving organizational goals

2.2. Knowledge Sharing

Knowledge sharing is the main activity in knowledge management activities. Knowledge sharing allows individuals in an organization to work together to exchange information, ideas, suggestions, ideas and experience in the end creating the formation of a new knowledge.

Knowledge sharing is a mechanism of the spread of knowledge of the organization to all members of an organization. One of the leading theories relating to knowledge creation comes from Nonaka [6]. In theory, Nonaka argues that knowledge is created only through interaction between people or between organizations. Barthold and Srivastava [7] suggested that knowledge sharing as sharing expertise, information, advice, and ideas between individual to individual within an organization

2.3. IS Success Model

The use of information systems in an organization is expected to have a positive impact. In order to ensure that any implementation of the information system is a success, then knowing the elements that influence the success of information systems becomes very attractive to researchers.

Measurement of the success of an information system has received wide public attention from researchers since long. DeLone and McLean has developed a model that can be used as a guide in order to successfully implement the IS organization [8]. They propose six elements related to IS success is: "quality system", "information quality", "quality services", "use", "user satisfaction" and the "net benefit". DeLone and McLean conduct a revisit for their successful model for the implementation of IS [9]. The model is tested on E-commerce applications. And found that the model of IS success was successfully tested in the context of e-commerce [10].

3. Research Method

In develop the research model we do some stage. We conduct literature review to investigate the current state of social media, knowledge sharing and is success model. Some reputable database journal such as Sciendirect, Emerald, ProQuest, and Google Scholar we use in this literature review process. Result from literature review is became foundation for us to develop the research model.

4. Result and Analysis

The use of information systems in an organization is expected to have a positive impact. In order to ensure that any implementation of the information system is a success, then knowing the elements that influence the success of information systems becomes very attractive to researchers.

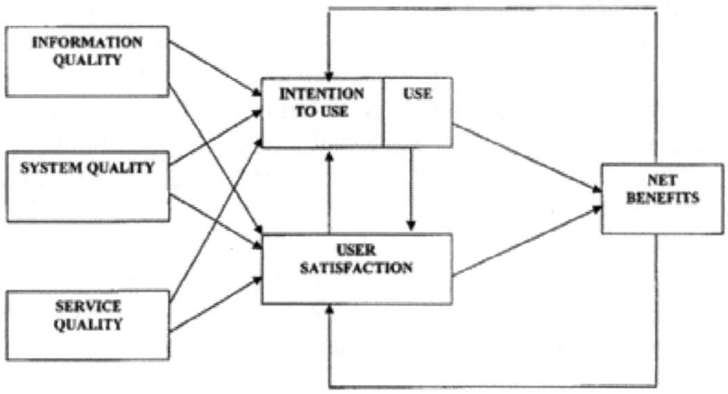

Figure.1. IS Success Model

IS success model has been used successfully and tested empirically in the context of various information systems such as e-commerce, e-government, and e-banking. But until now the development of IS success model in social media in the context of knowledge management is still very little in to be explored. Research related to the successful model of social media to share knowledge among researchers still rarely performed. Sucess Model of information system developed by DeLone and McLean [8] and [11] is the basis of this research to be developed into a model that can be specifically used as a reference in the utilization of social media for knowledge sharing.

Research in the field of social media and KS that has done this time includes conceptual models, case studies, empirical studies (e.g., [5, 12-15]) One of the studies related to the social media and KS by Razmerita et al. [14]. They argue that there are principles in common between social media and KS [14]. Social Media has shown a positive role in KS. Some researchers have successfully identified the benefits obtained through the use of social media in KS [14, 16]. However, although the principle believed social media synergy with KS but researchers also found the constraints faced in the implementation of social media in KS [17]. KS in social media can be achieved through a variety of tools that allows for creating, codify, organize, and share knowledge, but also to socialize and improve personal network and collaborate in order to organize and create new knowledge[14]. Taking into account the characteristics of social media and knowledge sharing, the successful model of information system [11] will be developed.

Research Model

A research model developed by applying the theory of IS Success Model [11] as a base. In develop the research model we do enrich the conceptualization of variable in indicators. The aim of reconceptualization the definition of variable and the indicator is to make sure the variable and indicators that measure it would appropriate with our research goal. We hope that the indicators in each variable so that it accommodates the issues and needs of the user and the system in the context of the use of social media. In addition we also incorporate a user characteristic variable that is recommended by Petter, DeLone and McLean [9] to be explored.

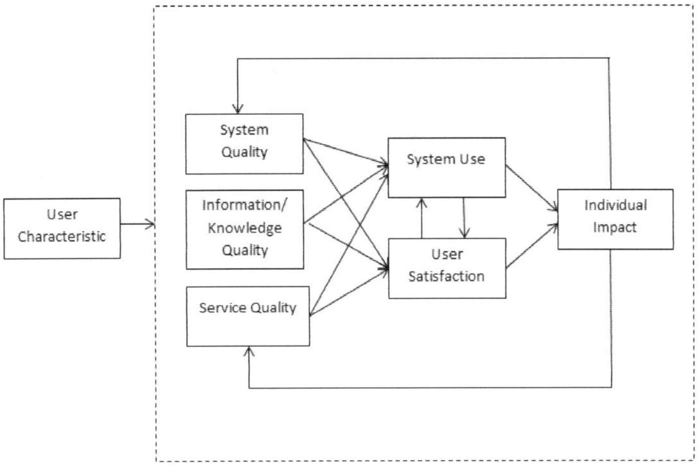

Figure 2. Research Model

The explanation of all variable in our research is describes as follow:

A. Information / Knowledge

Information and knowledge becomes the most important commodity for academics and researchers in Indonesia to develop them. When used as a medium of social media to share knowledge and information, the number of issues arise, such as whether the information, knowledge can be trusted? Is it relevant? Whether delivered as a whole? Is it easy to understand? Is it easy to access? and others. The issue is not the same if the application used is an internal application that is built within an organization, which has no validation, filter, mechanisms to ensure that the information and knowledge that is consumed by the user is guaranteed by the institution. IS Success Model theory has to accommodate the variable quality of the information / knowledge but the indicators used has not significantly describe user needs for information and knowledge in the context of social media applications. So, we enrich by adding new indicators as contained in Table 1.

Table 1. Indicator for Information/Knowledge Quality

Indicators
KI1: Information / knowledge is delivered in social media must be reliable (Believability)
KI2: Information / knowledge is delivered in social media must be objective (Objectivity)
KI3: Information / knowledge is delivered in social media should be accurate (Accuracy)
KI4: Information/knowledge presented in social media should be reputable (Reputation)
KI5: Information/knowledge presented in social media should have the value of benefits (Value-added)
KI6: Information/knowledge presented in social media should be relevant (Relevancy)
KI7: Information/knowledge presented in social media should be on schedule needs (Timelines)
KI8: Information/knowledge presented in social media should be intact (Completeness)
KI9: Information/knowledge presented in social media should be packaged in small amounts of relative data (Appropriate amount of data)
KI10: Information / knowledge is delivered in social media should be interpreted properly (Interpretability)
KI11: Information / knowledge is delivered in social media should be easily understood (Ease of Understanding)
KI12: Information / knowledge is delivered in social media should be presented consistently (Representational consistency)
KI13: Information / knowledge is delivered in social media should be exposed briefly (Concise Representation)

B. User characteristics

Users become the main actors in the use of social media applications. In contrast to the use of information systems in an organization that binds the user should adhere to a mechanism of information systems they build, then the social media applications users are not tied to specific rules, so that utilization would be very influence by user characteristic. In research relating to the IS success model information systems do Petter, DeLone and McLean [18] found that the role of the user characteristics was affecting the pattern of information system success. In this study, we will also explore whether user characteristic and also has a role in the utilization of social media by academics in Indonesia. The indicators are describing as follow:

Table 2. Indicator for User Characteristic

Indicators
UC1: I find the use of Social media application is the right choice (Attitude Toward Technology)
UC2: I feel enjoy utilizing Social Media (Enjoyment)
UC3: I believe the Social media application that I use (Trust)
UC4: Social media application that I use meet my expectations (User Expectation)

C. System Quality

System quality is defined as the desirable characteristics of an Information System quality [19]. In measure individual impact we develop indicator that refer to Gable, Sedera and Chan [19]. The indicators are describing as follow:

Table 3. Indicator for System Quality

Indicators
SMQ1: Social media application should be easy to use
SMQ2: Social media application should be easy to learn
SMQ3: Social media should be easily accessible application
SMQ4: Social media application should be able to meet the need of users
SMQ5: Social media application must have features that support information sharing activity
SMQ6: Social media application must be supported by a system That is accurate
SMQ7: Social media application should be accessible on various devices (Flexibility)
Social media should be reliable application (Reliability)
SMQ8: Social media application should be efficient when used (Efficiency)
SMQ9: Social media application should be effective when used (Effective)
SMQ10: Social media application should be able to satisfy user (Sophistication)
SMQ11: Social media application must be integrated with other applications that support (Integration)
SMQ12: Social media application should be customized (Customization)

D. Service Quality

Quality of the service or support that system users receive from the IS organization and IT Support personnel in general or for a specific IS [8]. The indicators are describing as follow:

Table 4. Indicator for Services Quality

Indicators
SQ1: Social media application should have a fast response when used (Responsiveness)
SQ2: Social media applications should have good accuracy (Accuracy)
SQ3: Social media application is always running and operate properly while in use (Reliability)
SQ4: Social media application developers should be supported by a competent team (Technical Competence)
SQ5: Social media application developers should be supported by staff who empathize with the user (Empathy of personal Staff)

E. System Use

System Use is define as degree and manner in which staff and customers utilize the capabilities of an IS [20]. In measure individual impact we develop indicator that refer to Wu and Wang [20]. The indicators are describing as follow:

Table 5. Indicator for System Use

Indicators
SU1: I use social media to communicate knowledge and information (Communicate Knowledge and Information)
SU2: I use social media to share knowledge and general information (Share General Knowledge and Information)
SU3: I will use social media to share specific knowledge and information; (Share Specific Knowledge and Information)

F. User Satisfaction

Individual impact is referring to users' level of satisfaction with the IS. In measure individual impact we develop indicator that refer to Delone and Mclean [8]. The indicators are describe as follow:

Table 6. Indicator for User Satisfaction

Indicator
US1: I am satisfied with the Social media application that I use (general satisfaction)
US2: I am satisfied with the knowledge and information I got (knowledge/information satisfaction)
US3: I am satisfied with the features available on Social media application (System feature satisfaction)

G. Individual Impact

Individual impact is refer to the extent to which IS are contributing to the success of individuals [19]. In measure individual impact we develop indicator that refer to Gable, Sedera and Chan [19]. The indicators are describing as follow:

Table7. Indicator for Individual Impact

Indicators
II1: Social Media is one of my forums in learning (Learning)
II2: Social Media increases my awareness on certain issues (Awareness/ Recall)
II3: Utilizing social media to make decisions which I take to be more effective in working (Decision Effectiveness)
II4: Utilizing social media makes me productive at work (Individual Productivity)

5. Conclusion

Result of this study is the research models that appropriate to use in our investigation. Our next agenda is to develop the research instrument and validate the instrument in order to ensure the instrument meet the criteria in validity and reliability test.

Acknowledgments

This work was supported by MENRISTEK DIKTI under Hibah Fundamental Grant 2016.

References

[1] P. G. Altbach and J. Balán, *World class worldwide: Transforming research universities in Asia and Latin America*: JHU Press, 2007.

[2] Q. Wang, Y. Cheng, and N. C. Liu, "Building World-Class Universities," in *Building World-Class Universities*, ed: Springer, 2013, pp. 1-10.

[3] W. W. Ding, S. G. Levin, P. E. Stephan, and A. E. Winkler, "The impact of information technology on academic scientists' productivity and collaboration patterns," *Management Science,* vol. 56, pp. 1439-1461, 2010.
[4] A. M. Kaplan and M. Haenlein, "Users of the world, unite! The challenges and opportunities of Social Media," *Business horizons,* vol. 53, pp. 59-68, 2010.
[5] R. Hanna, A. Rohm, and V. L. Crittenden, "We're all connected: The power of the social media ecosystem," *Business horizons,* vol. 54, pp. 265-273, 2011.
[6] I. Nonaka and H. Takeuchi, "Organizational knowledge creation," *Creative Management, SAGE, London,* pp. 64-82, 2001.
[7] K. M. Bartol and A. Srivastava, "Encouraging knowledge sharing: The role of organizational reward systems," *Journal of Leadership & Organizational Studies,* vol. 9, pp. 64-76, 2002.
[8] S. Petter, W. DeLone, and E. McLean, "Measuring information systems success: models, dimensions, measures, and interrelationships," *European journal of information systems,* vol. 17, pp. 236-263, 2008.
[9] W. H. Delone and E. R. Mclean, "Measuring e-commerce success: Applying the DeLone & McLean information systems success model," *International Journal of Electronic Commerce,* vol. 9, pp. 31-47, 2004.
[10] W. H. DeLone and E. R. McLean, "Information systems success revisited," in *System Sciences, 2002. HICSS. Proceedings of the 35th Annual Hawaii International Conference on*, 2002, pp. 2966-2976.
[11] W. H. Delone and E. R. McLean, "The DeLone and McLean model of information systems success: a ten-year update," *Journal of management information systems,* vol. 19, pp. 9-30, 2003.
[12] H. H. Chang and S.-S. Chuang, "Social capital and individual motivations on knowledge sharing: Participant involvement as a moderator," *Information & management,* vol. 48, pp. 9-18, 2011.
[13] L. Prusak and T. H. Davenport, "Knowledge after the Knowledge Creating Company: A Practitioner Perspective," in *Towards Organizational Knowledge*, ed: Springer, 2013, pp. 255-262.
[14] L. Razmerita, K. Kirchner, and T. Nabeth, "Social media in organizations: leveraging personal and collective knowledge processes," *Journal of Organizational Computing and Electronic Commerce,* vol. 24, pp. 74-93, 2014.
[15] G. Von Krogh, "How does social software change knowledge management? Toward a strategic research agenda," *The Journal of Strategic Information Systems,* vol. 21, pp. 154-164, 2012.
[16] P. Bharati, W. Zhang, and A. Chaudhury, "Better knowledge with social media? Exploring the roles of social capital and organizational knowledge management," *Journal of Knowledge Management,* vol. 19, pp. 456-475, 2015.
[17] G. Schiuma, V. Vuori, and J. Okkonen, "Knowledge sharing motivational factors of using an intra-organizational social media platform," *Journal of knowledge management,* vol. 16, pp. 592-603, 2012.
[18] S. Petter, W. DeLone, and E. R. McLean, "Information systems success: The quest for the independent variables," *Journal of Management Information Systems,* vol. 29, pp. 7-62, 2013.
[19] G. G. Gable, D. Sedera, and T. Chan, "Re-conceptualizing information system success: The IS-impact measurement model," *Journal of the association for information systems,* vol. 9, p. 377, 2008.
[20] J.-H. Wu and Y.-M. Wang, "Measuring KMS success: A respecification of the DeLone and McLean's model," *Information & Management,* vol. 43, pp. 728-739, 2006.

IOP Conf. Series: Materials Science and Engineering **190** (2017) 012029 doi:10.1088/1757-899X/190/1/012029

Information Framework of Pervasive Real Time Monitoring System: Case of Peat Land Forest Fires and Air Quality in South Sumatera, Indonesia

Siti Nurmaini[1], Reza Firsandaya Malik[2], Deris Stiawan[2], Firdaus[1], Saparudin[1], Bambang Tutuko[1]

[1] Robotic and Control Research Lab. Computer Engineering Department, Universitas Sriwijaya
[2] Computer Networking Research Lab. Computer Engineering Department, Universitas Sriwijaya

sitinurmaini@gmail.com, rezafm@unsri.ac.id, deris.stiawan@gmail.com, virdauz@gmail.com, saparudinmasyarif@gmail.com

Abstract. The information framework aims to holistically address the problems and issues posed by unwanted peat and land fires within the context of the natural environment and socio-economic systems. Informed decisions on planning and allocation of resources can only be made by understanding the landscape. Therefore, information on fire history and air quality impacts must be collected for future analysis. This paper proposes strategic framework based on technology approach with data fusion strategy to produce the data analysis about peat land fires and air quality management in in South Sumatera. The research framework should use the knowledge, experience and data from the previous fire seasons to review, improve and refine the strategies and monitor their effectiveness for the next fire season. Communicating effectively with communities and the public and private sectors in remote and rural landscapes is important, by using smartphones and mobile applications. Tools such as one-stop information based on web applications, to obtain information such as early warning to send and receive fire alerts, could be developed and promoted so that all stakeholders can share important information with each other.

1. Introduction

Indonesia peat land and forest fires occur during the dry season on an almost annual basis, burning vast areas of peat land and forest across the archipelago. When peat layers are dry, fires can burn roots and underground organic matter for up to months at a time, and spread below the surface very quickly. From the data of terrestrial ecosystems of Southeast Asia, around 25 million hectares of peat land area and 70% of those found in Indonesia. Peatlands are the most important area for the storage of carbon, equal to all other terrestrial biomass and twice as much of all forest biomass. In case of peatland fires in Indonesia, is estimated cause by emissions by 2 billion tones of CO2 per year, equivalent to almost 4% of the emissions caused by global fossil fuels. Indonesia air pollution was extremely poor in late 2000, due to haze from massive peat land fires in South Sumatera and Kalimantan [3]. Peatland forest fires in South Sumatera are contributing to the extremely poor air quality and spewing greenhouse gases into the atmosphere. Not only do they pose an immediate threat to human health and wildlife, haze from the fires also disrupts transport and the economic activities of millions of people. The results are devastating loss and irreparable damage to the environment and atmosphere, due to 30 % of carbon dioxide is released to the atmosphere [2].

One of the biggest threats in this century, is the declining the environmental quality in the world, due to it can reduce the level of human well-being because of disruption in human health [1]. The most important problems for stopping forest fires is the difficulty to find the exact location of the fire and no evidence to determine who is responsible for it. Moreover, the fire control is hard, because the process of detection and prevention occurred when a fire is widespread. Thus it is impossible to be done at that time. Early warning and initial data from monitoring system about pollution levels in central cities is key to provide adequate information to citizens and take actions to reduce it.

To overcome such problems and to produce the impact of policies that have been made by governments and public institutions regarding air pollution, it can be achieved through increased public awareness of the need for clean air and healthy. It is because the impact will be felt bad for human health if air pollution continues to rise. The slightest increase in air pollution within a certain time can cause various symptoms of the disease and lead to various health risks. Due to smoke particles from burnt wood are almost entirely < 1 μm in size and mostly ranged from 0.15 to 0.4 μm[6]. Real-time air pollution monitoring including concentration of particulate matter (PM_{10}), sulphurdioxyde (SO_2), nitrogen dioxide (NO), carbon monoxide (CO), and ozone (O_3) very necessary to provide updated data on the daily air quality. If the data can be monitored in real time, the air pollution control will be easier to be done. All of the process to overcome bad impacts from air pollution, in order to protect human beings and other living things.

In 2015, Indonesia 's Pollutant Standards Index (PSI) which incorporates PM10 , SO2 , CO, O3, NO increased above 2,000. Any score above 350 is considered hazardous to human health. Indonesia's National Disaster Mitigation Agency reports that 135,000 Indonesians are suffering from respiratory problems. For long time the peat fires have had negative impacts on economy, human health, environment, and climate. Greenpeace notes that Indonesia's fires in 2015 will release more carbon to the atmosphere. The estimation between 0.81 and 2.57 Giga Tonnes of carbon emissions were released. Peat land fires produce large amount of dangerous haze and deteriorate air quality; the dense haze also causes various health problems.

Currently, there are fast developments in pollution monitoring technique. Intelligent sensors and satellite imageries are being applied worldwide. Free-access sensing images provide a big opportunity to track the change and support ground–based observations effectively and continuous monitoring capacity makes sensors become very efficient in supporting government decision making. Besides, the involvement of citizens in monitoring is receiving more attention since local knowledge and aspiration can ensure the sustainability of projects. This paper proposes the information framework from the decision making planning until the platform technology to solve various problems concerning primary and secondary pollutants in the air. Primary pollutants emitted directly on the pollution sources and secondary pollutants formed when other pollutants interact with the atmosphere or referred to as a major pollutant.

2. Peat Land Forest Management

The frequency and extent of peat and forest fire events in Indonesia implies that effective fire-prevention measures and early-control systems are not yet in place. This is partly attributed to the absence of village-level management bodies, which are tasked to prevent and mitigate devastating forest and peat land fires. Furthermore, peat land's geographical conditions and characteristic terrain, coupled with the lack of access roads, often pose considerable challenges in conducting patrols and the early suppression of fires. Low awareness about the causes and impacts of peat land fires and the lack of coordination among local government authorities, business players and local communities also undermine efforts at proper land management.

It must make a commitment to cumulate the management effects which focus on the outcome achievements, understand the effects of multiple development including the new and the existing pressures, risk assessment, collaborative work that shared responsibility for action, and improve the economic integration and the environmental and social considerations. It follows an adaptive control version in which selection-makers research from eexperience and new facts and adapt to changing social expectations and needs. Overall performance control, with pollutants prevention principles, is important to supplying information on environmental situations and identifying the need for any changes and modifications on an ongoing foundation.

To make changes to the management of cumulative impacts, required a control framework. Next will be a concept that became the basis of the Management framework in terms of joint ownership, accountability, clear management in a dynamic information database. All of these principles should be used to establish measurable criteria so that all the goals that have been made can be achieved in accordance with the desired. The illustration of the interconnectedness of national, provincial, and regional air quality and peat and forest fire plans is presented through Figure 1.

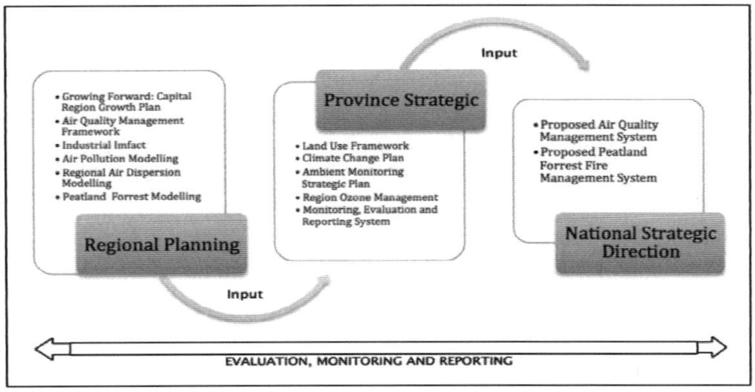

Figure 1. Proposal of strategic framework for inter-region planning

The aim of the strategic framework to assist stakeholders in carrying out their responsibilities at the time, and the right conditions. The process of such framework to align each policy will be developed or revised at the regional, provincial or national. This framework is proposed to assist stakeholders to develop the capacity in order to perform their responsibilities in a timely manner. It became a process to align framework updates with other policies being developed or revised at a regional, provincial or national level.

Previous study about the air pollution and forest fires with many tools are proposed, from ground-base sensors to data collected by satellites [4-8]. Many, identification of potential hot-spot and fire zones has been done by remote sensing strategy. Everyday satellites have been monitoring the hotspot area. In general, all the forest fires in South Sumatra detected using equipment that only measures in near real time by Moderate Resolution Imaging Spectroradiometers (MODIS) [8][9]. The MODIS instruments capture of about all of Earth's apparent every day and can ascertain hot spots acquired by fires [8]. However, not easy for detecting fire from space. The Smoke or clouds can blemish an instrument's adeptness to sense fires on the surface. Clouds over south Sumatera are absolutely close, the acclimate can still affectation as obstacle. Moreover, low-temperature or smouldering, underground fires, which are accepted in South Sumatera, sometimes cannot be detected at all. Fires in burn area, from surface until bake is not only in the surface, but as well in the underground. Acclimate altitude accept cogent influences on blaze behavior. Figure 2(a), shows the hotspot sample in Indonesia, while peat land forest fire in 2015.

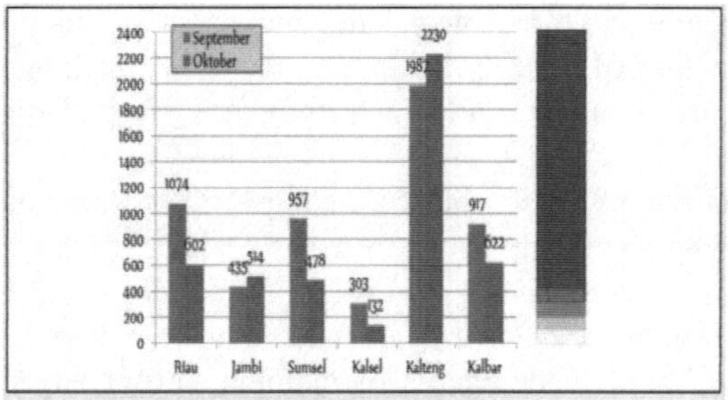

(a) Hotspot data in 2015

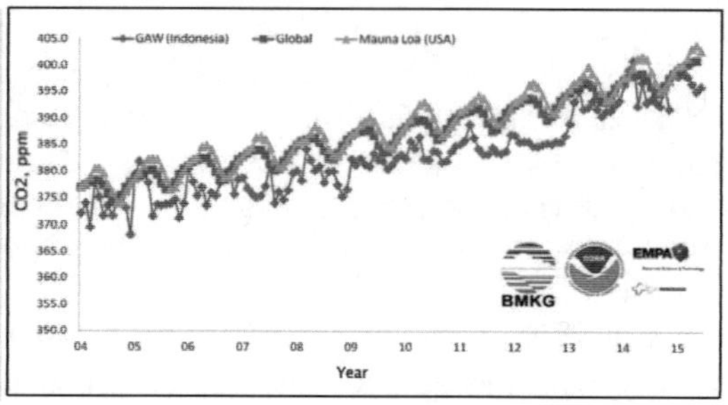

(b) Comparison of CO2 concentration

Figure 2. Indonesian Hotspot and Air Pollution

The sample of carbon emission detection in 2015 comparison between Indonesia (GWW), global and Mauna Lua (USA) can be shown in Figure. 2 (b). The value can cause various health problems. Only a few organizations collect the data of such high quality. Even so, this limits the amount of environmental monitoring data that is often available for exposure and health assessments. Hence, getting the right air pollution sensors are also critical in this part of the air pollution management research.

3. Framework Design

Currently, Indonesia only have 43 conventional stations measurement for daily air quality monitoring, but only 17 in active condition and 2 of which are in South Sumatera. However, static air quality measurement stations in a city is expensive cost of building and maintaining. These monitoring systems are usually complex, inflexible, and time consuming, and they provide spatial resolution in the order of several hundred kilometres [10]. Indonesia doesn't have ground-base sensor especially for detection peat land forest fire, only use a satellite data. Such problem makes the air quality management issues in South Sumatera become more complex, because the purpose of the amount and the ability of environmental monitoring devices increase continuously, with the specifications of the device technology is reliable and sturdy. The monitoring system must be able to detect and analyse indications of peat fires and the impact on air quality. One monitoring technology being developed today are wireless sensor networks and monitoring with unmanned aerial vehicles (UAVs).

Ground-base sensor technology is kind of static sensor that is able to send environmental data without using wires is expected to become one of the development of applied technology that can support forest conservation program [11]. In the WSNs detection, many sensors are deployed in the target area to measure environmental data such as, SO_2, CO, O_3, NO, PM_{10}, and $PM_{2.5}$ for daily air quality. Rapid developments in technology led to the production of small, low-cost air pollution sensors to symbolize the future of air quality monitoring.

However, many potential issues which arise as a result of these studies including,

1. Mobile technologies for monitoring air pollution levels and source of pollution;
2. Real time air pollution application on a smart phone, the data pollution will be shares and interacted to the public;
3. Mathematical models are used to combine spatial and temporal data through the integration of monitoring data with the data fusion method;
4. Utilization of remote sensing satellite technology to improve the process of forecasting and estimation of air quality and exposure assessment;
5. Real time monitoring through a network to facilitate the detection distance and large area of which are covered.

4. System Architecture

For overcoming such issues in environmental monitoring, we propose low cost pervasive wireless sensor network management as shown in Figure 3. It means the sensors have characteristic simple and

IOP Conf. Series: Materials Science and Engineering **190** (2017) 012029 doi:10.1088/1757-899X/190/1/012029

cost effective through protocol 802.15.4 or Zigbee. Such module ensures data communication at all times while allows to send SMS alerts according to defined thresholds. The sensors use the battery as a power supply, for making the sensor network in a longtime operation and ensuring a virtually perpetual operation, solar panels can recharge the batteries. The sensors on board have several gas sensors, it produces a heterogeneous sensor motes. For monitoring environmental pollution, and peat land forest fires, the sensor network operation based on the same network.

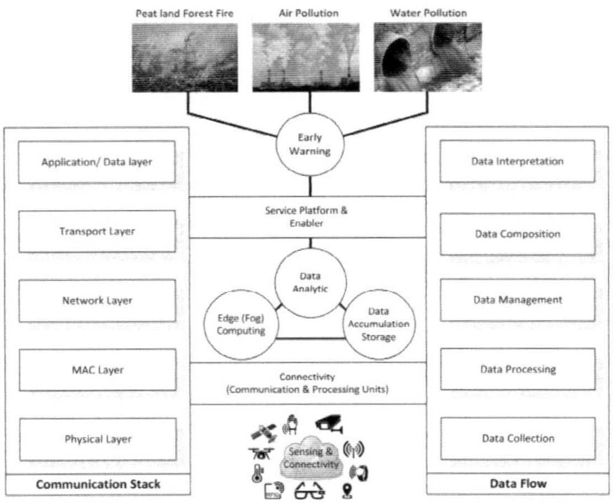

Figure 3. Proposed of information framework of pervasive monitoring system

In Figure 3, describes a process of implementation of air quality sensor network applications that will be used. One type of sensor platform that is ground-base sensor. These sensors will take air quality data directly at the location of peatland and forest fires. Detection system will record the results of air quality measurements into a set of data in certain intervals based on the level of pollutant contamination. The air quality data that has been generated is sent to the base station via wireless communication. It would be very useful to make a situation analysis in the burn area. In the proposed platform, ground-base sensor consisting of air quality sensors, heat sensor and infrared camera to monitor the condition that occurs before, during and after the fires.

Ground-base of the sensor is required since the sensor system with remote sensing method with satellite technology is difficult to obtain accurate data, because of fires difficult to monitor. Clouds and smoke often cover the spread of fire in the fire. So, the estimation data on the dot summer forest fires and air pollution would occur in combination. To produce more accurate data.in this project WSNs with mobile technology for air pollution monitoring, unmanned aerial vehicle (UAVs) with infra-red camera for light and fire detection to give early detection of peat land forest fires and satellite for mapping the hotspot in the burning area are combined. The sources of data observations are used to measure air quality in real-time and forecast the peat land fire data. Analysis of all data, by fusing them with data obtained from statistical or deterministic air pollution model. The combination satellite and ground observations are integrated into a computer model to predict the future location of pollution and peat land fire. The information fusion processes at the heart of system framework to exploit the capabilities offered by geographic information systems (GIS) and for the assimilation of the three data sources such as ground-base analytical measurements, satellite observation, and remote sensing by UAVs. The framework of monitoring system is proposed in Fig. 4.

4. Data Fusion Approach

Currently, many researchers have been developed a method that is in the process combining observational data in the form of an objective mathematical model. Such a method is called data fusion approach. In the context of wireless sensor network, it will produce a high-density value. Due to the data coming from various sources detection and sensor devices are different will be combined in order to find the most optimal of data with the smallest probability error value and generate high robustness. In observation of environmental quality can improve the accuracy of the value of all the observational data that has been done, because in the process of data fusion use monitoring data from various sources including the data history. The results can be able to improve their skills in mapping the value of the air quality in an environment, because all the data is the data from various sources contribute significantly. In this paper source of data from four sources including ground-base sensor, remote sensing with UAVs, data from satellite and historical data as presents in Figure. 4.

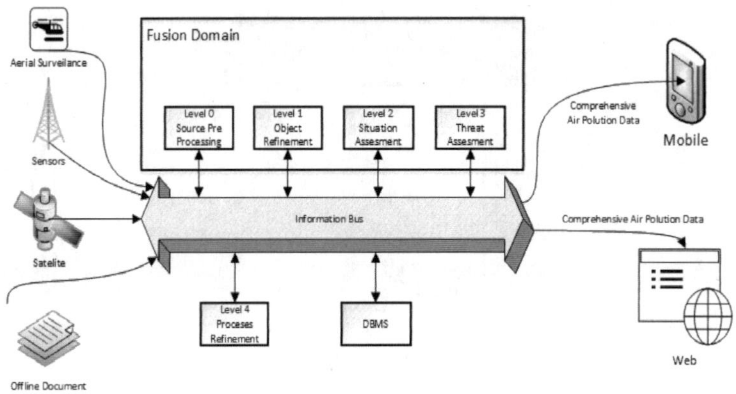

Figure. 4. Data sources for monitoring system

The available data fusion approach can be classified into four nonexclusive categories, data smoothing, data analysis, modelling and data control. The fusion level is a part of all process is divided 4 levels and all the process are presented in Figure 5.

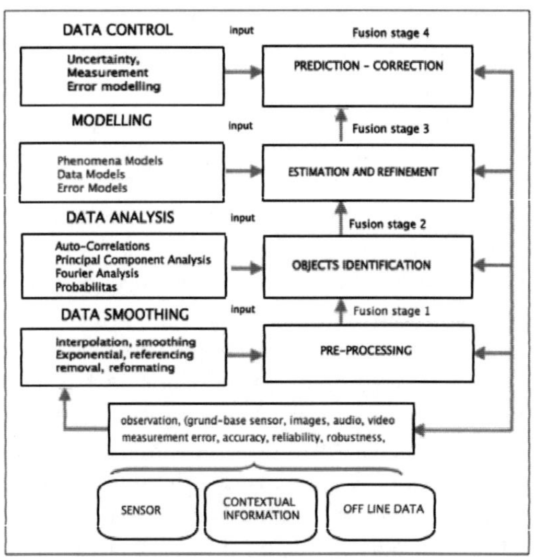

Figure 5. Data Fusion Approach

5. Conclusion

Based on some of the resumes that have been done in the previous section, the peat land and forest fires seriously produce impacts on the climate change. The impact that occurs is not only in the decrease in human health, but also affects in the economic, social, cultural and even cause increased the Indonesian mortality. Nowadays reducing peatland and forest fires is a high priority for the Indonesian government. However, every steps that must be taken by the government should not only in the form of regulations on paper, but also depend on its implementation at the local level and provincial areas. Effective measures should involve the community. With increasing public awareness of its dangers how the impact caused by the peatland and forest fires, the expected implementation of the working steps which have been collated the government will produce a significant reduction in the rate of fire. The propose framework of monitoring system is expected to present the data in the form of a peat land fire indication for the vast land even, early warning and daily air quality in South Sumatra. The propose framework encompasses the complete information system, from the sensory level and networking support structure through to data management and Cloud based integration of respective systems and services. The information is used to disaster early warning, to increase public awareness of environmental pollution and health effects, and as the data analysis for the government to make policy.

Acknowledgments
The authors thank the Ministry of Technology Research and Higher Education (Kemenristek-Dikti), Indonesia, and Universitas Sriwijaya for their financial support in this Competitive Grants Project.

References

[1]. Page, S.E., Siegert, F., Rieley, J.O., Boehm, H.D., Jaya, A., Limin, S., 2002. The amount of carbon released from peat and forest fires in Indonesia during 1997. NATURE Vol. 420. Issued November 7, 2002

[2]. Hooijer, A., Silvius, M., Wösten, H. and Page, S. 2006. PEAT-CO2, Assessment of CO2 emissions from drained peatlands in SE Asia. Delft Hydraulics report Q3943 (2006).

[3]. CONRAD, ESTHER. "Managing Peatland Fire Risk in Central Kalimantan, Indonesia.World Resources Report

[4]. Hooijer, A, Page, S., Canadell, J. G., Silvius, M., Kwadijk, J., Wösten, H., and Jauhiainen J. 2010. Current and future CO2 emissions from drained peatlands in Southeast Asia. Biogeosciences 7, 1505–1514.

[5]. Ceccato, P. N., Jaya, I. N. S., Qian, J., Tippett, M. K., Robertson, A. W., &Someshwar, S. (2010). Early warning and response to fires in Kalimantan, Indonesia.

[6]. Hayasaka, H., Noguchi, I., Putra, E.I., Yulianti, N., and Vadrevu, K. (2014). Peat-fire-related air pollution in Central Kalimantan, Indonesia. Environmental Pollution xxx, pp.1-10.

[7]. Murdiyarso, D., Lebel, L., 2007. Local to global perspectives on forest and land fires in Southeast Asia. Mitig. Adapt. Strateg. Glob. Change 12 (1), 3-11.

[8]. Vadrevu, K.P., Ellicott, E., Badarinath, K.V.S., Vermote, E., 2011. MODIS derived fire characteristics and aerosol optical depth variations during the agricultural residue burning season, north India. Environment Pollution 159 (6), 1560-1569.

[9]. Lentile, L. B., Holden, Z. A., Smith, A. M., Falkowski, M. J., Hudak, A. T., Morgan, P., & Benson, N. C. (2006). Remote sensing techniques to assess active fire characteristics and post-fire effects. International Journal of Wildland Fire, 15(3), 319-345.

[10]. Ma, Y., Richards, M., Ghanem, M., Guo, Y., &Hassard, J. (2008). Air pollution monitoring and mining based on sensor grid in London. Sensors, 8(6), 3601-3623.

[11]. Sader, S. A., Hayes, D. J., Hepinstall, J. A., Coan, M., &Soza, C. (2001). Forest change monitoring of a remote biosphere reserve. International Journal of Remote Sensing, 22(10), 1937-1950.

Designing Albaha Internet of Farming Architecture

A Alahmadi

Department of Computer Sciencee and Computer Engineering, College of Computer Science & IT, Albaha University, Albaha P.O. Box 1988, Saudi Arabia
aalahmadi@bu.edu.sa

Abstract. Up to now, most farmers in Albaha, Saudi Arabia are still practicing traditional way, which is not optimized in term of water usage, quality of product, etc. At the same time, nowadays ICT becomes a key driver for Innovation in Farming. In this project, we propose a smart Internet of farming system to assist farmers in Albaha to optimize their farm productivity by providing accurate information to the farmers the right time prediction to harvest, to fertilize, to watering and other activities related to the farming/agriculture technology. The proposed system utilizes wireless sensor cloud to capture remotely important data such as temperature, humidity, soil condition (moisture, water level), etc., and then they are sent to a storage servers at Albaha University cloud. An adaptive knowledge engine will process the captured data into knowledge and the farmers can retrieve the knowledge using their smartphones via the Internet.

1. Introduction

Al-Baha region is a tourist destination in Saudi Arabia rich with natural resources and is known for its cool weather and natural diversity. Its seasonal crops are pomegranates, almonds, bananas, and kadi patta. Most farmers in Albaha are still practicing traditional way, which is not optimized in terms of water usage, quality of product, etc., whilst modern agriculture provides farmers with new innovations, research and scientific advancements to produce safe, sustainable and affordable food. Intensive scientific research and robust investment in modern agriculture during the past 50 years has helped farmers double food production while essentially freezing the footprint of total cultivated farmland.

In the context of our work, we have specified a farm management system that takes advantage of the new characteristics that ''Future Internet'' offers, we propose a smart Internet of farming system to assist farmers in Albaha to optimize their farm productivity by providing accurate information on the right time to harvest, to fertilize, to watering and other activities related to the farming/agriculture technology. The proposed system utilizes wireless sensor cloud to capture remotely important data such as temperature, humidity, soil condition (moisture, water level), etc., and then they are sent to a storage server. An adaptive knowledge engine will process the captured data into knowledge and the farmers can retrieve the knowledge through their smartphone via the Internet.

2. Related Works

To enhancing crop production, we need to monitor the crops constantly using sensors. Wireless Sensor Network in agriculture is showing progress [1-3]]. Jiou, et al. [4] has designed a farm environmental monitoring system based on the Internet of Things (IoT) and to realize the automate management of agriculture and the implementation of precision production. The system is made up of three layers which are sensor layer, transmission layer and application layer, respectively. Kaloxylos, et al., discussed in [5] the implementation of an innovative cloud based Farm Management System, It provides a framework that allows the interconnection among services developed by different service providers. Furthermore, the work utilized domain independent software modules called generic enablers, developed within the FI-WARE project.

Research works have been done on WSN implementation in agriculture/farming. Malik et al. [6] implemented energy aware WSN to monitor soil of palm-oil farm using PEGASUS routing protocol introduced by Lindsey and Raghavendra [7]. Perera and Collins [8] introduced ZigBee wireless soil moisture sensor design for vineyard management system. Mohanan et al. [9-10], proposed a holistic network selection algorithm to provide personalized always best connection (ABC) in 4G communication. The proposed algorithms consider dynamic parameters in achieving the ABC to the nodes in the network.

3. Proposed Design

The architecture of the whole system is exhibited in Figure 1. The system consists of 4 main components as follows.

1. Albaha University (ABU) Cloud
2. Adaptive Knowledge Engine
3. Mobile Applications/Services
4. Wireless Sensor Cloud

The ultimate goal of the system is to create a knowledge of the farming domain from data that collected via sensors deployed in the farm. The farmers need to procure the data acquisition components which will be deployed in their farms and then via Internet they can monitor the situation in their farm remotely using their smart phones.

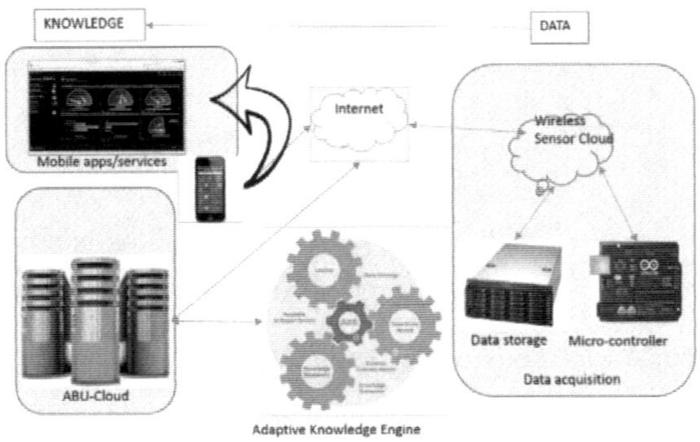

Figure 1. Overall design of the proposed design.

3.1. Albaha Univesity (ABU) Cloud

Albaha University (ABU) Cloud is a simple enterprise cloud designed and developed by the researcher in the College of CSIT. It consists of two application servers and two data storage servers, cloud communication servers, including: cloud controller, cluster controller, data storage (Walrus) controller, and node controller. ABU Cloud uses UEC (Ubuntu Enterprise Cloud), a free Linux-based cloud computing with Eucalyptus middleware and Windows server 2013 as for storage and application servers. Figure 2 illustrated the ABU cloud.

3.2. Adaptive Knowledge Engine (AKE)

Machine learning methods will be elaborated in this project with the intention of producing new Algorithms for AKE. To create knowledge on farming from the sensed data needs adaptive capability of the AKE due to the nature of farming and weather data that changing from time to time. Besides, accuracy is also needed to provide accurate prediction as well as suitable recommendation to the farmers.

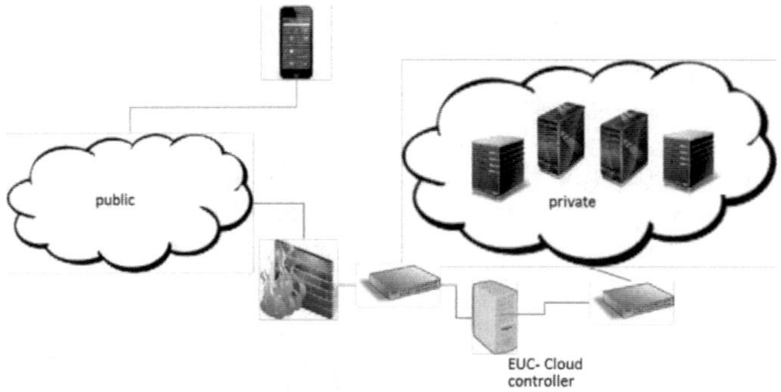

Figure 2. ABU Cloud

3.3. Mobile Applications
Overall architechture design of the mobile Baha farming cloud is illustrated in Figure 3, whereas its backend architecture is shown in Figure 4.

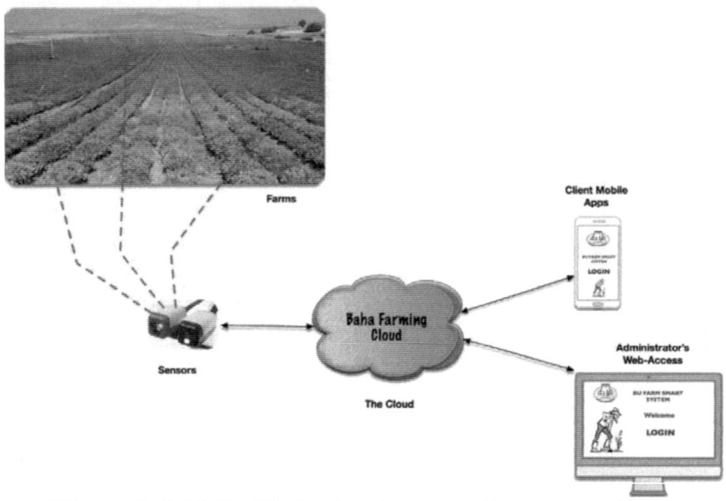

Figure 3. Mobile Albaha farming overall system architecture

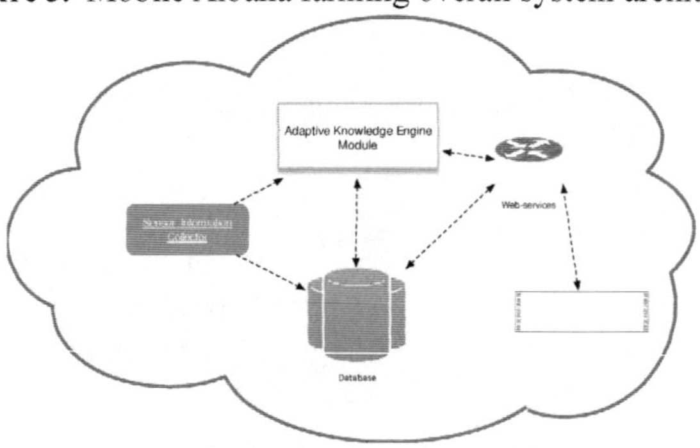

Figure 4. The backend engine architecture

The sensed and collected data/information are stored in a database, the AKE module will process some of the data/information to produce a useful knowledge to the farmers. Web application run at background and web service is employed to support interoperable machine-to-machine interaction over a network.

The farmers, using their smart phones will be able to access services with a graphical user interface (GUI) (see Figure 5). The services include the following:

- Real time farm monitoring system; Remotely or locally
- Recommender system; assist the farmer in emergency, extreme weather situation, etc.
- Storage cloud services; storing the farming data resulting from sensing the farm
- Farming management systems; to optimize the farm production
- Forum; sharing knowledge, experience, and other activities

Weather prediction system; short term and long term prediction.

Figure 5. The GUI dashboard design.

3.4. The wireless sensor cloud

The wireless sensor cloud (WSC) for data sensing is designed according to the requirements of specific farm. Geographically, Albaha region is a hilly area. In addition, some area in Albaha have extreem wheather.

3.4.1. The wireless sensor network (WSN) design. A wireless sensor network (WSN) is a wireless network consisting of spatially distributed autonomous devices that use sensors to monitor physical or environmental conditions. These autonomous devices, known as routers and end nodes, combine with a gateway to create a typical WSN system. The distributed measurement nodes communicate wirelessly to a central gateway, which acts as the network coordinator in charge of node authentication, message buffering, and bridging from the IEEE 802.15.4 wireless network to the wired or mobile Ethernet network.

The design uses three layers: network layer is responsible for communication with the data sink/server; Reliable distributed storage layer is responsible to collect the data using coordinator node (sn); Sensor node layer is responsible to sense the data. The sensor nodes (sn) can be deployed in two different ways using grid topology or random topology as shown in Figure 6.

3.4.2. The Always Best Connection (ABC) network. The proposed WSN system is built on an IEEE 802.15.4 wireless mesh network. The 802.15.4 radio in each WSN device provides for lowpower communication of measurement data across a large network of devices. A software built on top of that to provide reliable communication from the host PC or Programmable Automation Controller (PAC) to the node devices.

The access network selection mechanism must also be flexible in adapting to fast changing environment. Current access network selection mechanism use single or limited criteria to identify the context on which the selection occurs and this will not scale up to the more challenging 4G environment. Current access network selection methods tend to allot static weights to the criteria.

Predetermined weights that are fixed will not correctly represent a scenario that has changed. A new algorithm to select the best connection is proposed by considering a dynamic weight to each available connectivities based on the most active applications that are using the connections.

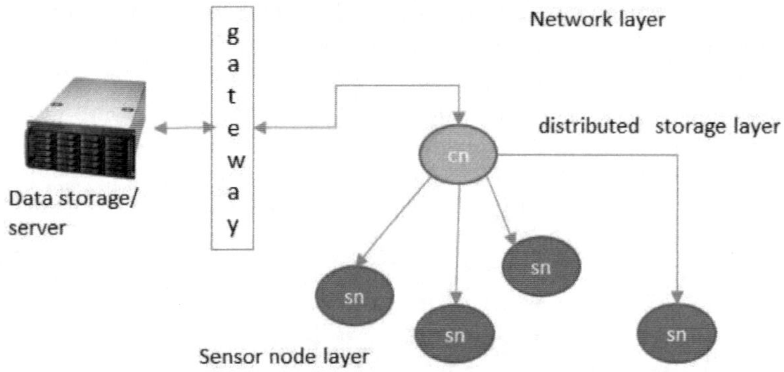

Figure 6.The sensoring components

3.4.3. The sensor design. Providing an always best connection (ABC) using wireless technologies in a region such Albaha is a big challange. Thus, we attempt to design a wireless sensor with the capability to provide ABC feature through a multiple communication protocols running in a sensor unit. Besides, in order to come up with a low cost sensor design we put multiple sensors in one microcontroller. The challange here is to deal with an efficient battery power usage. Overall block diagram of the sensor design is shown in Figure 7.

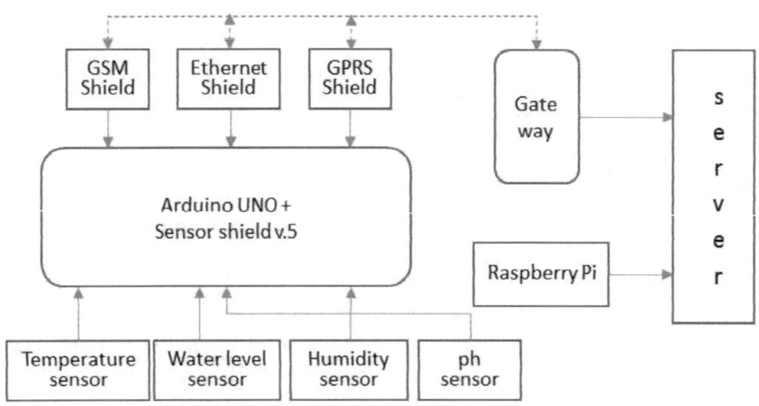

Figure 7. Block diagram of the sensor

3.4.4. Initial experiment and the results. For the initial experiment purpose, the WSN hardware is designed using 3 units of microcontroller Arduino Uno Rev5 with additional Raspberry Pi unit. Four sensors are attached (soil temperature, soil water level, soil pH, humidity soil humidity). Only GSM communication protocol is used in this simple experiment. The data sensing capability and end-to-end delay are measured during the *experiment.* Figure 8 (a) and Figure 8 (b) exhibit the end-to-end delay for WSN with 2 nodes and 3 nodes, respectively. Delay average for 3-nodes WSN is 30.31 ms., while delay average for the 2-nodes WSN is 24.19 ms. It is shown that WSN nodes, able to transmit sensed data to the server through its gateway with a reasonable delay. We do not apply any WSN routing protocols in this simple experiment.

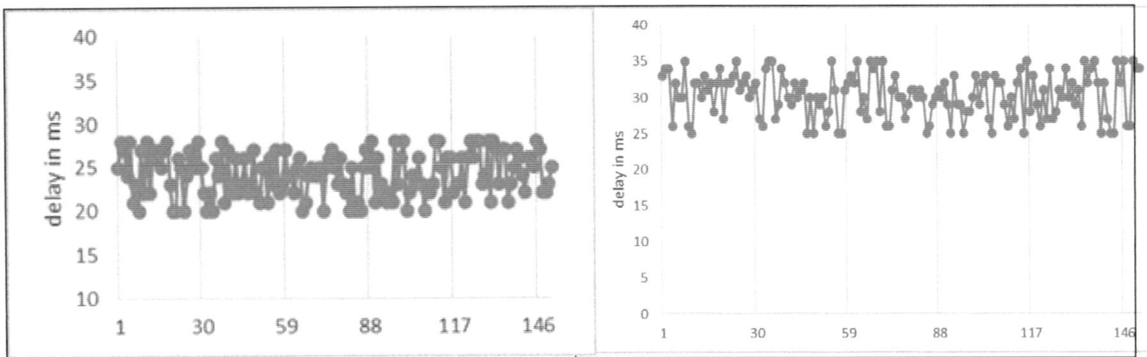

(a) WSN with two nodes	(b) WSN with three nodes

Figure 8. End-to-end delay of the WSN

4. Conclusion

In this paper, the author shared an experience on the development of Internet of Farming system for rural area in Albaha province, Saudi Arabia. The project is at the early stage. The next stage is to detail the design and to implement the four components at laboratory scale. Having done some experiments and measurements, the pilot project will begin to deploy the real system in 10-15 farms as a pilot project.

Acknowledgment

The author would like to thank Albaha University for the research funding under research contract number 1437/02.

References

[1] Kshitu S, Arti N, Neelam S, and Raghuvir S 2010 *International Journal of Engineering Science and Technology*, **2** 8 pp 3955-3963.
[2] Chaudhary D D, Nayse S P, and Waghmare L M 2011 *Journal of Wireless & Mobile Networks (IJWMN)* **3** 10 pp 140-149.
[3] Syed D, Masood U, Farwa U and Khurram M 2015 *Journal of Computer Engineering and Applications* **4** 3 pp 153-163.
[4] Jiao J, Ma H, Qiao Y, Du Y, Kong W, and Wu Z-C 2014 *Advance Journal of Food Science and Technology* **6** 3 pp 368-373.
[5] Kaloxylosa A, Groumasb A, Sarrisb V, Katsikasb L, Magdalinosb P, Antoniouc E, Politopoulouc Z, Wolfertd S, Brewstere C, Eigenmannf R and Tero C-M, *Journal of Computers and Electronics in Agriculture* 100 pp 168–179.
[6] Malik R F, Hafiz M, Nopransyah A, Zalbina MR and Septian L M, 2015 *Proc. of Int. Conf. on Electrical Engineering, Computer Science and* Informatics (*Palembang, Indonesia*).
[7] Lindsey S and Raghavendra C S, 2002 *IEEE Aerospace Conference* (*Montana, USA*) **3** 3 p 1125.
[8] Perera T A and Collins J 2010 *Auckland: Auckland University of Technology. Thesis*.
[9] Mohanan V, Budiarto R, Zainon W M N W 2012 *Proc. 18th Asia-Pacific Conference on Communications (Jeju island, Korea)* pp 53-58.
[10] Mohanan V, Budiarto R 2013 *Proc. 19th Asia-Pacific Conference on Communications (Bali, Indonesia)* pp 196-201.

Detection Learning Style Vark For Out Of School Children (OSC)

Ali Amran[1], Anita Desiani [2], MS Hasibuan[3]

[1, 2] Math and Science Faculty, Sriwijaya University
[3] Department of Electrical Engineering & Information Technology, UGM

[1] Ali_42@yahoo.com, [2] Anita.arhami@gmail.com, [3] said.s3te15@mail.ugm.ac.id

Abstract. Learning style is different for every learner especially for out of school children or OSC. They are not like formal students, they are learners but they don't have a teacher as a guide for learning. E-learning is one of the solutions to help OSC to get education. E-learning should have preferred learning styles of learners. Data for identifying the learning style in this study were collected with a VARK questionnaire from 25 OSC in junior high school level from 5 municipalities in Palembang. The validity of the questionnaire was considered on basis of experts' views and its reliability was calculated by using Cronbach's alpha coefficients (α=0.68). Overall, 55% preferred to use a single learning style (Uni-modal). Of these, 27,76% preferred Aural, 20,57% preferred Reading Writing, 33,33% preferred Kinaesthetic and 23,13% preferred Visual. 45% of OSC preferred more than one style, 30% chose two-modes (bimodal), and 15% chose three-modes (tri-modal). The Most preferred Learning style of OSC is kinaesthetic learning. Kinaesthetic learning requires body movements, interactivities, and direct contacts with learning materials, these things can be difficult to implement in eLearning, but E-learning should be able to adopt any learning styles which are flexible in terms of time, period, curriculum, pedagogy, location, and language.

1. Introduction

From the data published by UNESCO Institute for Statistics (UIS) in 2011, there were 122 million children and teenagers who have never gone to school. But in the year of 2013, the number has been increasing to 124 million children and teenagers. Their ages were between 6 and 15 years old, and around 65 million of those live in Asia [1].

According to UIS Indonesia was the 5th of the countries that had large numbers of children and teenagers who had been dropped out of school and rest of them were children who never attended to school, its estimated around 1.3 million (figure 1).

The increasing number of OSC in Indonesia becomes a serious problem, since children and adolescents are the future generation has the right to access proper education. Education and Sport departments had attempted to reduce the OSC number by 13%, but the target was achieved only 1%[2]. Nowadays is the Internet (multimedia) generation, where there are many internet based applications for various purposes, the problem of OSC can be reduced by providing Internet-based learning applications (E-learning). E-learning is internet application that can be accessed anytime and anywhere, so out of school children can still have the access to education.

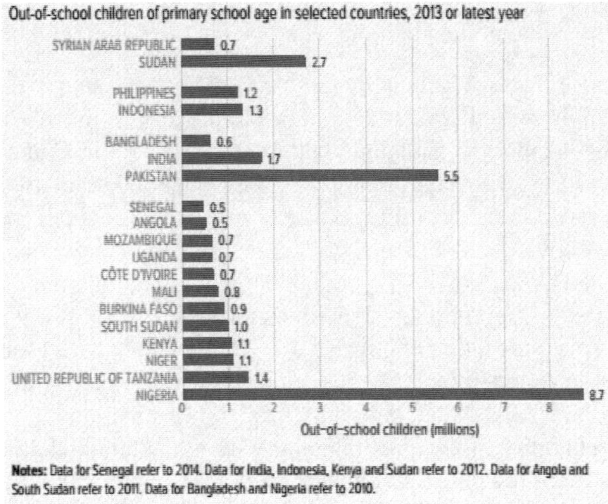

Figure 1. Number of Out of School Children in the World [1]

E-learning is established as all forms of electronic support from the learning process, which is oriented toward the user's character and aims to build systems that refer to the reference contraction of knowledge, experience and training of the user or learner [3]. An E-learning should have instructional contents and provide a learning experience through electronic technology [4].

However, the development of E-learning is not quite enough to think about how to design websites and what are website content. The development of ELearning so far is deficient in providing the proper materials, especially for OSC. OSC do not have teachers as an assistant in process of learning, so it should be considered which learning style that fits OSC individually. Learning style is defined as the various types of signals of different students. The differences are sometimes referred to as a "lifestyle" and is sometimes interpreted as a type of personality, so learning style called as a part of Intellectual styles [5]. Learning styles also often means the differences of academic ability, skills and interests [6]. Learning style is also interpreted as an individual's preferred method for approaching learning and obtaining knowledge [7].The interesting approach to observe the learning styles that is proposed by Fleming and Mills is VARK (Visual, Aural, Read / Write, Kinaesthetic) [8]. VARK has been used widely in many current studies [8]–[14]. The concept of this approach is to consider the type or style of student interest in receiving the provided information's.

VARK questionnaire was developed by Lincoln University of New Zealand in 1998. The questionnaire was based on three principles, which are as follows: 1., everyone can learn academic issues; if not, everyone has their own style of learning 2. Learners' Motivation increases when the learning styles of students are adopted to the learning content and 3. The educational concept of learning through the use of different senses and different perceptions [15], [16].
Based on this perspective, people gain knowledge of the environment through four modalities sensory: visual, auditory, reading / writing and kinaesthetic[16]. Four models of the sensors has been described by Fleming on VARK. V (Visual) students prefer to obtain information in the form of images, making them easier to learn by using maps, graphs, charts, and more. A (Aural) students are more easily adopt information from the hearing. R (Reading/Writing) students who can assimilate new information by reading and writing. K (kinaesthetic) students take advantage of the physical experience to learn new things using many sensory modalities like the somatosensory, auditory, smell, taste and visual [17]. Some individuals may have more than one modality learning style, they can be unimodal or multimodal[17], [18].

The purpose of this study is to find the learning style of out of school children, in order to enhance the education of OSC. E-learning help OSC to learn like in the classroom, even though there are no teacher or tutor who help and accompany them. This study took 25 out of school children from five municipalities in Palembang.

2. Research Method

1. Data

   Data for this research were gathered with VARK questionnair from 25 total respondent of OSC in junior high school level. Data was taking place in five OSC shelters in the area Rambutan, plaju, Jakabaring, Sukajaya and Gandus district in Palembang. Respondents were just taken for children who had attended school before. The difficulty of this sample is the data cannot be categorized according to age or sex. Because there were just 4 number of girls and 21 number of boys.
2. Instrument for Data Collection

   Data were collected using interviews and questionnaires. This study used VARK 7.8. questionnaires. This questionnaire consisted of 16 questions where each question had four possible answers. respondents could choose more than one answer according to their needs.
3. Data Processing

   at this step, the reliability of the questionnaire with the obtained data would be tested using Cronbach alpha. If the results of Cronbach alpha coefficient was reliable, then data would be processed in accordance with the rules of VARK. if the result wasnot reliable, the data would be taken over by adding the number of respondents
4. Conclusion

   The conclusions was based on the results of questionnaires processing and interviews to determine the learning styles of OSC.

3. Discussion

Everyone has different learning styles as well as for OSC in Indonesia, out of school children dropouts can still attend formal education by following an equivalency test in accordance with their education level for elementary school level, they can take the equivalency test for package A, for junior high school level can follow the package B, and Senior High School level can take the packet C. The equivalence test results can be used as formal evidence that the student has successfully completed their formal education. The faced problems OSC is the lack of learning facilities for them. OSC should be able to learn independently without teacher or assistant because they did not attend formal school like formal student, although there have been several shelters that provide learning facilities for OSC, but it is inadequate because their capacity, teacher and learning facilities are limited.

One of the alternative solutions is E-learning, that can be accessed anytime and anywhere. But to develop an e-learning is not easy because ELearning is not only intended to store and provide a variety of learning materials, but also E-learning is able to increase the interest of children to study especially for OSC. So e-learning can help to enhance learning ability of OSC. A good E-learning is it should be able to accommodate the various learning styles of each person who uses the e-learning.

To optimize the development of E-Learning, this study took several samples of OSC in Palembang. Data were collected by visiting several shelters in the some district of Palembang. The problem is in data collecting. Data collecting is difficult to trace the children who did not attend to school, and determine areas in Palembang which has a greater number of OSC. So This study only took 25 respondents in five district in Palembang. These districts were area Rambutan, Plaju, Jakabaring, Sukajaya and Gandus.

VARK questionnaire was used to detect the learning styles of the OSC. Questionnaires VARK presents 16 questions where each respondent can give more than one answer depends on their answer. The validity and reliability of the questionnaire were calculated using the Cronbach alpha. In this study, Cronbach alpha coefficient is $\alpha = 0.68$, which means VARK questionnaire could be used as a reliable tool for collecting data and were able to reveal the actual information of data. Overall it can be concluded there were 55% of respondents have one learning style (uni modal), where 33.3% prefer kinaesthetic learning styles, 27.76 prefer audio learning styles, 23.13% chose a visual learning style and 15.78% selecting learning styles Reading and writing (fig 2). There were 45% of respondents chose more than one learning style, where 30% they chose two learning styles at once (bi-modal) and 15% they choose 3 styles of learning styles (trimodal).

From the data processing, it could be concluded that the most dominant learning styles for OSC was uni modal learning. The first thing should adopt in e-learning was kinaesthetic learning style, the second was the learning style and the third was audio visual learning style, and the last were reading and writing style.

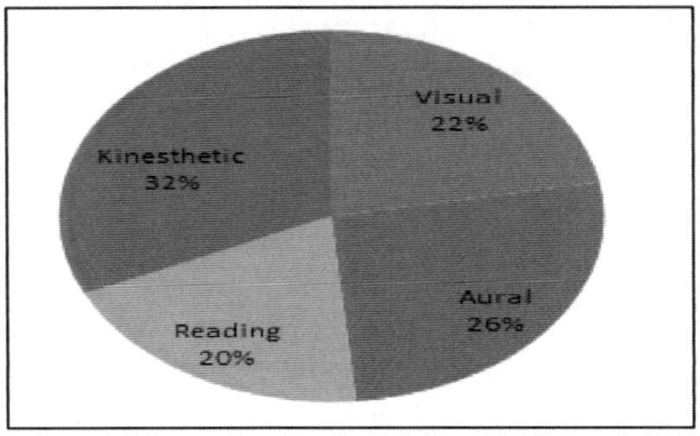

Figure 2. Unimodal Learning Style Of Out Of School Children (OSC)

Bimodal learning style showed in Fig.3, where the dominant learning style was still kinaesthetic even it combined with another learning style. The first learning styles in bimodal were kinaesthetic combined with aural, the second was kinaesthetic and reading, the third were visual and reading, and kinaesthetic and visual, the last were combination of visual and aural

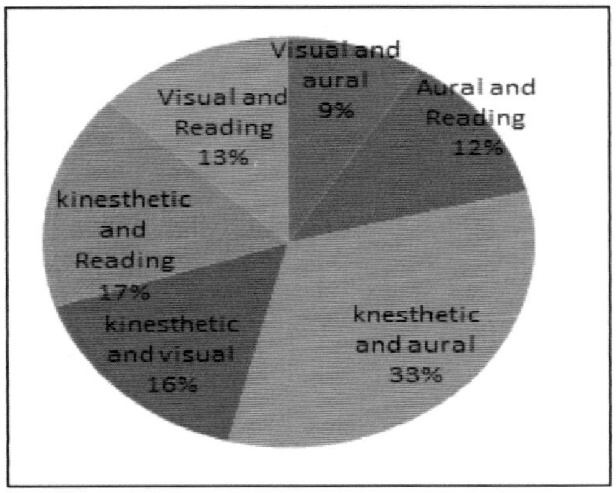

Figure 3. Bimodal Learning Style Of Out Of School Children (OSC)

In trimodal learning style, there were consist of three combination of learning styles. The first combination were Kinaesthetic, Aural and Visual learning style, the second place were combination of Kinaesthetic, aural and reading learning style, and the third were kinaesthetic, visual and reading (Fig.4).

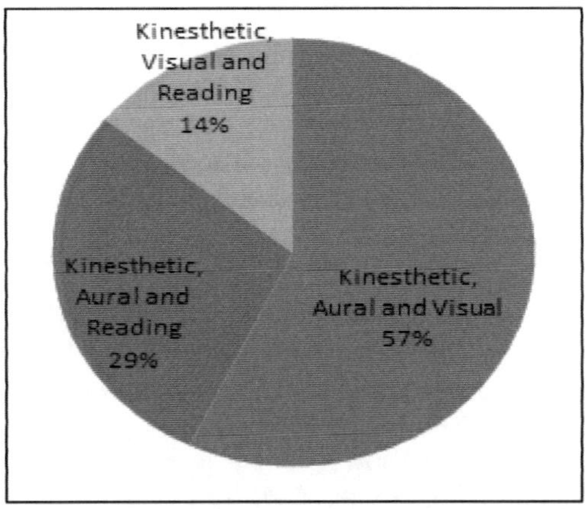

Figure 4. Trimodal Learning Style Of Out Of School Children (OSC)

From the obtained results, there were several points to be considered in developing an e-learning for OSC. E-learning should be able to adopt four learning styles to meet the needs of OSC, especially for kinaesthetic learning style. But this is not easy because kinaesthetic learning require movement, interactivity, and direct contact with the learning materials, so e-learning needs some strategies in order to Kinaesthetic learning can be adopted in e-learning.

4. Conclusion

In the development of e-learning is not only concerned with technology. But it has to be able to provide various needs of users with different characters and ways of learning that called learning styles. Based on the data processing, it can be concluded that the most suitable method for OSC are kinaesthetic learning style, even there were some respondents chose other learning styles such as audio and visual. It means E-learning should be as a medium of learning for OSC that flexible in terms of time, period, curriculum, pedagogy, location, and languages, so they can still learn and study according to their learning styles, especially kinaesthetic learning style which is quite difficult to be adopted directly into an e-learning

References

[1] UNESCO Institute, “A growing number of children and adolescents are out of School as aid fails to meet the mark,” 2015.
[2] Official Palembang. of Education. Youth. and Sport., “Work Indicator.” [Online]. Available: http://disdikpora.palembang.go.id/indikator_kinerja. [Accessed: 28-Sep-2016].
[3] D. P. Nagarajan and J. Dr.G.Wiselin, “ONLINE EDUCATIONAL SYSTEM (e- learning) ,” Int. J. u- e- Serv. Sci. Technol., vol. 3, no. 4, pp. 37–48, 2010.
[4] C. S. Ong, J. Y. Lai, and Y. S. Wang, “Factors affecting engineers’ acceptance of asynchronous e-learning systems in high-tech companies,” Inf. Manag., vol. 41, no. 6, pp. 795–804, 2004.
[5] L. F. Zhang and R. J. Sternberg, “A threefold model of intellectual styles,” Educ. Psychol. Rev., vol. 17, no. 1, pp. 1–53, 2005.
[6] R. M. Felder, G. N. Felder, and E. J. Dietz, “The effects of personality type on engineering student performance and attitudes,” J. Eng. Educ., vol. 91, no. 1, pp. 3–17, 2002.
[7] N. D. Fleming and C. Mills, “Not another inventory, rather a catalyst for reflection,” To Improv. Acad., vol. 11, no. 1, p. 137, 1992.
[8] K. M. A. Rogers, “A preliminary investigation and analysis of student learning style preferences in further and higher education,” J. Furth. High. Educ., vol. 33, no. 1, pp. 13–21, 2009.
[9] T. A. Meehan-Andrews, “Teaching mode efficiency and learning preferences of first year nursing students,” Nurse Educ. Today, vol. 29, no. 1, pp. 24–32, 2009.
[10] K. Becker, J. Kehoe, and B. Tennent, “Impact of personalised learning styles on online delivery and assessment,” Campus-Wide Inf. Syst., vol. 24, no. 2, pp. 105–119, 2007.
[11] T. Sinnerton, L. Leonard, and K. M. A. Rogers, “Using Learning Style Preferences to Enhance the Education and Training of Allied Health Professionals,” Internet J. Allied Heal. Sci. Pract., vol. 12, no. 1, pp. 1–5, 2014.
[12] A. Banu, G. K. Huilgol, and J. Rajkumar, “Learning Style Preferences Among Medical and Nursing Students in a Medical College in South India,” J. Educ. Res. Med. Teach., vol. 2, no. 3, pp. 21–24, 2014.
[13] N. Othman and M. H. Amiruddin, “Different perspectives of learning styles from VARK model,” Procedia - Soc. Behav. Sci., vol. 7, no. 2, pp. 652–660, 2010.
[14] L. Ambrose, M. Mcgregor, D. Sabiston, and S. Vriliotis, “VARK learning styles and student performance in principles of micro vs macro-economics,” J. Econ. Econ. Educ. Res., vol. 15, no. 3, pp. 113–121, 2014.
[15] D. . W. Mills, “Applying what we know: Student learning styles,” 2002. [Online]. Available: http://www.adesignmedia.com/onlineresearch/Applying What We Know - Student Learning Styles.htm. [Accessed: 28-Sep-2016].

[16] H. Peyman, J. Sadeghifar, J. Khajavikhan, M. Yasemi, M. Rasool, M. Y. Yaghoubi, M. Mohammad Hassan Nahal, and H. Karim, "Using VARK approach for assessing preferred learning styles of first year medical sciences students: A survey from Iran," J. Clin. Diagnostic Res., vol. 8, no. 8, pp. 1–4, 2014.
[17] N. Fleming and D. Baume, "Learning Styles Again : VARKing up the right tree !," Educ. Dev., vol. 7, no. 4, pp. 4–7, 2006.
[18] T. F. Hawk and A. J. Shah, "Using Learning Style Instruments to Enhance Student Learning," Decis. Sci. J. Innov. Educ., vol. 5, no. 1, pp. 1–19, 2007.

Study of Smart Campus Development Using Internet of Things Technology

Marti Widya Sari[1], Prahenusa Wahyu Ciptadi[2], R. Hafid Hardyanto[3]

[1, 2, 3] Informatics Engineering, Universitas PGRI Yogyakarta, Yogyakarta, Indonesia

[1] widya@upy.ac.id, [2] nusa@upy.ac.id, [3] hafid@upy.ac.id

Abstract. This paper describes the development of smart campus using Internet of Things (IoT) technology. Through smart campus, it is possible that a campus is connected via online by the outside entity, so that the teaching approach based on technology can be conducted in real time. This research was conducted in smart education, smart parking and smart room. Observation and literature studies were applied as the research method with the related theme for the sake of system design of smart campus. The result of this research is the design of smart campus system that includes smart education development, smart parking and smart room with the sake of Universitas PGRI Yogyakarta as the case study.

1. Introduction

In the developed countries, Information and Communication Technology (ICT) has been used as the unseparated parts to increase the quality of higher education. ICT can be used to fix and increase the quality of learning process, research, library, information services and university management. One of the ICT implementations is the using of internet technology that is integrated to all of things of daily life, that is called as Internet of Things (IoT). IoT is a structure in which objects, people are provided with exclusive identity and the ability to relocate data over a network without requiring two-way handshaking between human-to-human or human-to-computer interaction [1]. IoT technology has been widely used for the development of smart home, smart campus, smart building and smart city. The concept of smart campus already being developed in developed countries several years ago. Smart campus is a trendy application in the paradigm of the IoT. The concept of constructing a smart campus implies that the institution will adopt advanced ICTs to automatically monitor and control every facility on campus [2]. The design and the implementation of smart campus is different with others, depends on the campus needs. The infrastructures to build a smart campus is costly. However, when it is implemented, all the campus activities will be effective and efficient. To build a smart campus, it needs to build the digital infrastructure inside campus that can give services so that it will be beneficial for surrounding citizens. IoT which bases on the internet, uses a variety of information sensing identification device and information processing equipment, such as RFID, GPS, GIS, JIT, EDI and other devices to combine with the internet to form an extensive network in order to achieve information and intelligence for entity [3]. In this research, the study of smart campus development with the discussion of smart education, smart parking and smart room that are located in Universitas PGRI Yogyakarta (UPY).

2. Literature Review

The research by [4] proposes the solution of RFID design structure integrated by using modern technology from cloud computation, supported by good quality technology and economy. This technology supports things to increase the campus security, asset track record, valuable things, student record, security of paper exam and original certificate. The future class appears to increase pedagogy, in which the students can participate more in the learning process through interaction and collaboration. To make environment that supports future classroom, it needs integrated mechanism of any related things being needed. In this research, the design of structure of Internet of Things was built, in which application domain is being made as system unity related to the internet [5]. To fasten and develop Intelligent Campus Internet of Things (ICIoT) more efficiently, the approach based on the runtime model to manage campus was applied as the result of the use of Wireless Sensor Network (WSN). It is hoped that all of management job description can be conducted by executing program that the model has been made appropriate. In the learning part, experiment was conducted and compared by using the traditional method. This method can increase the management effectiveness of campus facilities, save energy as much as 16,7% [6]. In the study conducted by [7], the concept of internet based parking guide by using QR code was introduced. This system runs on cellular platform by visualizing the available parking lot for the customers so that they can make order.

3. Smart Campus Infrastructure

Smart Campus is one of the innovations that will be developed in UPY. Adopted from the existing smart city system, one of aspects that being considered in developing smart campus is the infrastructure. The infrastructure is the main key of the campus smart program. If the infrastructure has been well developed, the information related to the campus can be accessed from mobile phone or others gadgets. Some parameters of the smart campus that being adaptated from smart city are:

a. Smart education consists of: eLearning, Personalized Learning, Virtual Classroom.
b. Smart parking, a parking system that provides the information about available parking lot and the information that the parking lot has been full.
c. Smart room, system that provides information related to the classroom that is being used or vacant.

4. Proposed Work

4.1. Existing Condition

The existing condition has not covered a smart campus. Between spaces, there is no connection that provides information regarding the total amount of present students or vacant classrooms. The system of parking lot is still manual, no information regarding the available parking lot. In UPY there are three parking lot, those are in unit 1, 2 and 3.

4.2. Proposed Method

The communication in smart campus consists on IoT unit that uses radio with the capability of receiving and sending wireless connection. Besides wireless, bluetooth is possible to be used as the connection. The component of IoT consists of hardware, software, cloud service. The design of hardware platform is built based on the needed. The hardware consists of microcontroller board that equipped by many sensor module, wireless, and other connection. The sensor module that being used is based on the needed. Generally, sensor module connected by microcontroller board through wireless. Sensor module being used is RFID, PIR sensor, IR sensor, camera, ultrasonic. Information being sent from the sensor, then being processed and saved in cloud that transmitted to android application. The application provides information to the users about condition and spatial condition, parking lot.

4.3. System Design

The system design of smart campus which described in Figure 1 below limited on (1) Smart education that consists on: eLearning, Virtual classroom, (2) Smart parking, a parking system that provides information related to the available parking lot, and also provides information when the parking lot is full. There are 3 parking lot. (3) Smart room: system that provides information related to the vacant room.

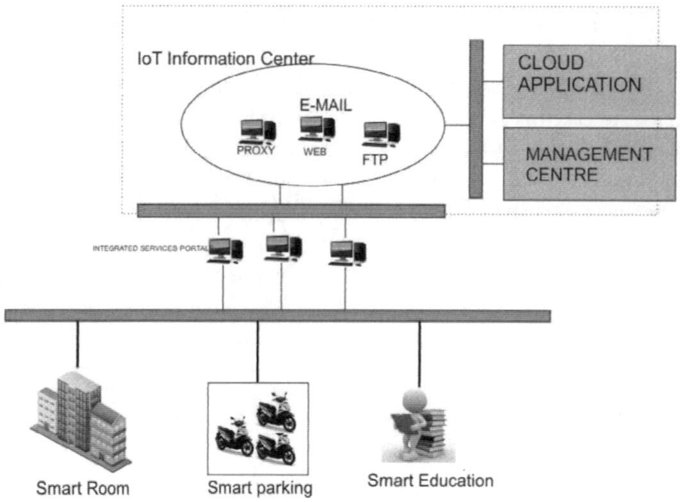

Figure 1. General scheme of system design

4.3.1. *Smart education.*

In smart campus sector, the learning process is conducted through e-Learning system, that makes it possible for students to be able to join learning from anywhere, anytime with the internet connection. E-learning eqquips with video conference facility so that it is possible for students to face the teacher from different place. Besides, virtual class feature can help simulation for students to solve problem in learning. Virtual class can be used for practicum lessons.

4.3.2. Smart parking.

A parking system that provides information related to the available parking lot, and also provides information when the parking lot is full. There are 3 parking lot. Sensor is put in the parking lot to scan the vehicle that enters the parking lot. The total amount of the vehicles that in the parking lot are revealed on the board. Next, the information will be processed by the system that provides information to the users about the available parking lot. The scheme is as described in Figure 2 below.

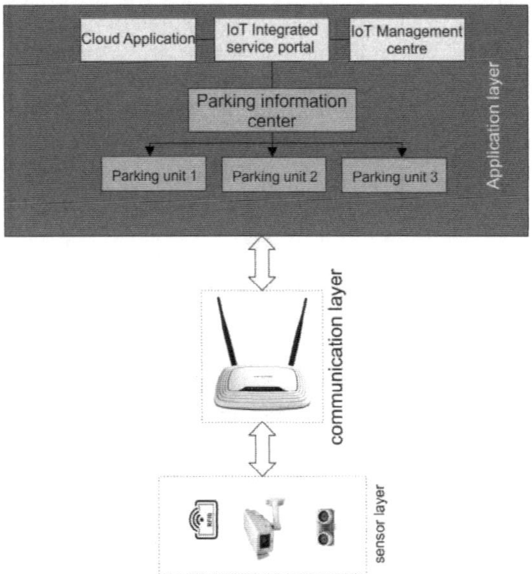

Figure 2. The scheme of smart parking

4.3.3. Smart Room

It uses the sensor of PIR, RFID, and camera. The concept of smart room is giving information about the vacant room, and the amount of the students being present on it. The data about amount and the name of the students are saved on the database. In this system, the students use RFID, so that the data of the students are in database. Smart system is being used in smart room. By using the sensor of PIR, the lights will be off automatically if there is no human in the room. On the other hand, the lights will be automatically on if there is or there are people in it.

5. Conclusion

IoT technology can be developed in any field. One of which is in the development of smart campus. Smart campus is an emerging and challenging concept for the technology to bring it in reality. The design of the system has created a scheme for implementation of smart campus limited on smart education, smart parking and smart room. This paper described the study of the concept that can be helpful in building the smart campus. The result of this research is the design of smart campus system that includes smart education development, smart parking and smart room with the sake of Universitas PGRI Yogyakarta as the case study.

References

[1] Burange, A.W, Misalkar, H.D. 2015. Review of Internet of Things in Development of Smart Cities with Data Management & Privacy. International Conference on Advances in Computer Engineering and Application (ICACEA). IEEE.

[2] Wang, H. 2013. Toward a Green Campus with the Internet of Things – the Application of Lab Management. Proceedings of the World Congress on Engineering 2013 Vol II.

[3] Fan, P., Zhou, G. 2011. Analsysis of The Business Model Innovation of the Technology of Internet of Things in Postal Logistics. IEEE : 978-1-61284-449-7 page 532 – 536.

[4] Mirza, Z.R., Brohi M.N. 2013. An In-Depth Analysis On Integrating Campus Radio Frequency Identification System On Clouds For Enhancing Security. Journal of Computer Science, Vol. 9, No. 12, hh. 1710-1714.

[5] Chang, F.C., Chen D.K. 2015. Future Classroom with the Internet of Things A Service-Oriented Framework. Journal of Information Hiding and Multimedia Signal Processing, Vol. 6, No. 5.

[6] Zhang, P., Wang, J. 2015. Management of Intelligent Campus Wireless Sensor Networks Based on Runtime Model. Journal of Computer and Communications, Vol. 3, hh. 22-31.

[7] Shinde Smita N. Et al. 2015. An Android Application for Parking Management and Dissemination System. International Journal of Advanced Research in Computer Engineering & Technology, Vol. 4, No. 3.

IOP Conf. Series: Materials Science and Engineering **190** (2017) 012033 doi:10.1088/1757-899X/190/1/012033

Model of load balancing using reliable algorithm with multi-agent system

M F Afriansyah[1], M Somantri[1], and M A Riyadi[1]

[1]Department of Electical Engineering, Diponegoro University, Semarang, 50275, Indonesia

[1]afriansyahfaizal@gmail.com

Abstract. Massive technology development is linear with the growth of internet users which increase network traffic activity. It also increases load of the system. The usage of reliable algorithm and mobile agent in distributed load balancing is a viable solution to handle the load issue on a large-scale system. Mobile agent works to collect resource information and can migrate according to given task. We propose reliable load balancing algorithm using least time first byte (LFB) combined with information from the mobile agent. In system overview, the methodology consisted of defining identification system, specification requirements, network topology and design system infrastructure. The simulation method for simulated system was using 1800 request for 10 s from the user to the server and taking the data for analysis. Software simulation was based on Apache Jmeter by observing response time and reliability of each server and then compared it with existing method. Results of performed simulation show that the LFB method with mobile agent can perform load balancing with efficient systems to all backend server without bottleneck, low risk of server overload, and reliable.

1. INTRODUCTION

Server becomes a very important thing to be aware of on the era of technology. Server become a services center of all users. The more users lead to the more complex of the server infrastructure. Development of server infrastructure must be able to accommodate a number of requests coming from users. Load of the server will be high in accordance with the number of user of the services provided by the server. High server load can affect the speed of process that will be slowing down and can make server stop working. The appropriate technology is needed to overcome this problem. Distributed systems are used in the infrastructure to distribute incoming requests from users to servers to be processed. In addition to distributed systems, the application of load balancing mechanism can be able to handle that problem reliably. Load balancing in distributed system has been proposed by a number of scientists [1]–[4]. Load balancing is necessary to improve the reliability of services with many users on a large scale. In distributed systems, load balancing is a method to share the load equally across all unit components (server) [5]. Load balancing is needed on a large scale process. The load balancing mechanism aims to distributing resources to each user's demand or request on the servers so that no node server is overloaded or remained idle. Load balancing focused on cloud computing environment using mobile agent has been analyzed by Singh [6]. Load balancing method using a multi-agent system can make reliable and efficient system.

Multi-agent system is a software-based distributed system using the agent with special abilities and autonomous (independent) on a network [3]. Software agents will continuously perform tasks that have been assigned by the user in a specific environment [7]. Agent has the ability or characteristic such as cooperative, independent, able to communicate between agents, and mobility [8]. Agents can move from one node to another node freely according to a given task.

Previously, many methods of developing load balancing using a mobile agent. Mobile agent method is an implementation of load balancing server dynamically using the mobile agent, the agent collects information for specific purposes in accordance with a task that has implemented. Dynamic load balancing in a distributed system is needed to improve performance and reliability of the system in large scale [14]. Distributed system is usually distributed as resource computing unit which is connected to the network to meet the needs of large-scale and high-performance [15].

Load balancing can be combined with a multi-agent system to automatically perform load sharing in accordance with the specified parameters and configuration. Multi-agent system is a group of an agent that cooperate to perform their task. Definition agent according to Lange is the active agent and can move towards to another node (computer), or exploring the network to collect information, data, or detect changes [16]. Agent has a life cycle in the environment, such as the creation, migration, and disposal. Multi-agent system purpose is to find server computing resource information using a mobile agent. In distributed systems, all computing units will look like single unit groups that cooperate or interact. Distributed system is carried out systematically and regularly to process and distribute the data or information. Load balancing in distributed systems shares or distribute the assigned tasks to different servers to keep workload to a server till given task ends. Load balancing in this research focuses on the algorithm that will be proposed to share the workload equally. Load balancing algorithm is important to determine obtained results. Incoming requests will go through the process of load balancing server until to the server to be processed in accordance with the desired flow algorithm.

In this research, we have designed a load balancing mechanism using reliable algorithm with multi-agent on distributed system. This mobile agent monitors all the activities that occur in the network group. The agents migrate through the nodes in the same group to monitor congestion that may occur. With the integration of this method, the demand for the desired resource can fulfill as possible and full traffic of node can be identified more efficiently [9]. Load balancing system in this research uses an algorithm and java technology called JADE [10]–[13]. JADE programming is like aglets programming that use an agent to find information condition on the server, whether it is large load or still idle (normal). JADE is the successor of mobile agent programming aglets. The information will be sent back to the server and processed. This multi-agent system is cross platform so it can be used anywhere (any OS) based on java. The load balancing algorithm that shares the load is looking at the information on the condition of the server. Thus, there is a central server which controls the load into the server to avoid server overload or idle and monitor agent performance. The research purpose is to develop load balancing algorithm that can share the load equally and reliably on a large scale. Load balancing server can perform load balancing mechanism efficiently to all backend server, no bottleneck, low-risk overload server, and accelerate response time.

Overall, this report is divided into several sections. Section 2 discusses the research methodology. Section 3 discusses the result of this research. Section 4 is the conclusion of this research.

2. RESEARCH METHODOLOGY

3.1. System Overview

System load balancing in distributed environment has been simulated using the multi-agent system with proposed load balancing algorithm. All multi-agent system using JADE run with JDK 1.8 and load balancing software using nginx as a reverse proxy. Overall, the system consists of one physical server machine (server host) and the specification is 6 core processors AMD Phenom II X6 1055T, 16 GB RAM, and 3 T hard drive. The host server is divided into five virtual servers, three virtual servers running on Linux debian 3:16 operating system and 2 virtual servers running on Windows 2012 operating system standard. This virtual system software using proxmox software-based debian 3:16. The host server operating system running on debian Jessie (kernel 3.16).

The entire virtual machine connected over virtual LAN use bridge configuration in proxmox. In this scheme, load balancer server has two NIC, public and private. The public NIC connected directly to the user / users and the private NIC is used to interconnect with the backend servers. Figure 1 is a simple description of implemented topology of the system.

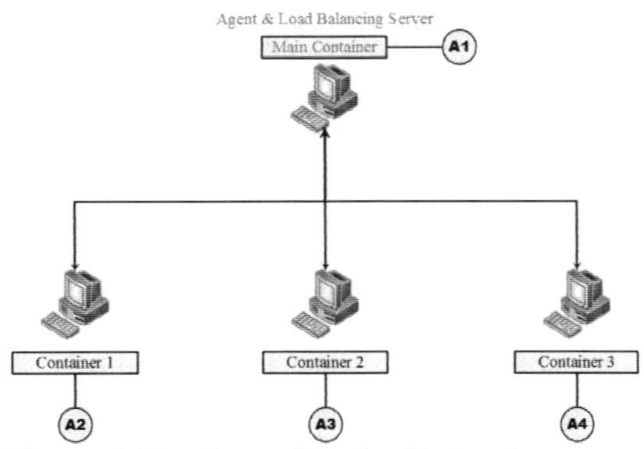

Figure 1. Topology of the load balancing system

Figure 1 shows that the load balancing system is divided into two parts: main container and the regular container. Main container served as a central of information and data from mobile agent, also as a load balancing server. The regular container is the backend server of a service provider. The main container will receive the request from users then be forwarded to the backend server for processing.

The first process from this method is to forward request from *n* request to backend server according to least time first-byte algorithm. Then main server sends a mobile agent to all backend servers. The mobile agent of main server is configured to ask for resource information from each backend server. Once the agent has reached the backend server, backend server will process requests from the main server agent. The resource information will be sent to the main server via mobile agent of the backend server. The mobile agent of the backend server contains information resource backend server and sent to the main server. Afterwards, the main server determines the condition of backend server load that can or cannot perform the service based on provided information by backend server agent. The main server inserts the results into the list of load balancing with the "normal" or "overload". The normal condition is a condition where backend server can perform the service and load server does not exceed the threshold. An overload condition is a condition which the backend server cannot perform the service because workload resource of backend server is high or exceeded the threshold. Proposed threshold in this research is measuring the load of CPU and memory. Threshold values greater than or equal to 90% of CPU load and memory load. The main server sends back mobile agent repeatedly to monitor the condition of backend server resources. Backend server temporary blocked from the list if backend server gets overload condition until back to normal conditions. Furthermore, the main server calculating the request with proposed LFB algorithm. The backend severs processing that request from the user and give back the result to the user through main server. This method of load balancing can share workload equally on all backend server that makes efficient service, keeping large request and has fast response time. Simulation has to performed by comparison of previous research algorithm with LFB algorithm.

3.2. Simulation method

Simulation method is a method to verify a research by the system parameter. Performed simulation method is to obtain quantitative result or information about the system being simulated. The obtained result from this simulation is the average response time, throughput, error number, and a deviation of load balancing system. This simulation method is inspired from previous research by Choi [17] using weight least Connection (WLC) algorithm to perform load balancing across 10 virtual servers with equal server resource, The number of requests simulated in this research was 800 request. WLC algorithm is a combination of weight and least connection algorithm. The least connections algorithm is selects the server with the least number of active connections to ensure that the load of the active requests is balanced on the backend servers. Fewest or least connection would be a priority of this algorithm. Weight algorithm is algorithm to determine backend servers with a higher weight value will receive a larger percentage of connections according to weight values at any one time. This algorithm will repeat back to the first server if a request of the user has reached the last server. The request will be processed equally by WLC algorithm with a weight configuration values that have been determined in according to request per server. Time duration that used to perform this research was 10 seconds. WLC algorithm can maintain the balance between the server and get fast response time.

Based on this method, WLC simulation method has chosen because research environment is similar to previous environment research. Simulation method of this research was using load balancing algorithm described earlier to find the value of response time, throughput, error, and deviation. LFB algorithm was combined with the mobile agent to keep the request from users to the system or server. Agent was set by scheduling to get information resource from each server. This algorithm was compared with WLC algorithm according to simulation method using 800 requests for 10 seconds. The second simulation did a comparison with the WLC algorithm using an 1800 request for 10 seconds. Benchmark application that used in this simulation is Apache JMeter.

In this scheme, each backend server embedded mobile agent ready to provide resource information from each backend server. The main server is the center of gathering information from each backend server. The main server creates a table that contains information resource information from each backend server. This information is combined with load balancing algorithm. The value of resource information is CPU load, memory load and the number of connections. The number of inbound connections is linear with CPU and memory load. Based on that, obtained formula is used to measure the load of a server as follows:

$$load = \frac{\%CPU.\%memory}{N_{request}} \tag{1}$$

The results of that load are then used as a point that determines backend servers are in normal or overload condition. After determining the condition of each backend server, then that result will be combined with the proposed load balancing algorithm. Servers with high load value will be not active or delist from load balancing list until the load value decrease and back to normal.

The response time is a value of the data from this research. Response time value obtained from server response time to the user. The value of the response time can be used as a variable that will be entered into the equation. Average value equation can be determined by using response time variable. The average value equation of response time is as follows

$$Average = \frac{\sum_{i=1}^{n} xi}{n_{request}} \tag{2}$$

Variable n is the number of requests that will be simulated. Variable x is the value of the response time every request. The average value obtained from dividing the amount of response time with a number of requests in accordance with Equation (2). Another equation can be generated by using response time and the number of request variable from user. Median, variance, and standard deviation equations are determined using these formulas:

$$Variance = \frac{n\sum_{i=1}^{n} x_i^2 - \left(\sum_{i=1}^{n} x_i\right)^2}{n\,(n-1)} \tag{3}$$

$$Std.Deviation = \sqrt{variance} \tag{4}$$

$$Throughput = \frac{n}{\Sigma t} \tag{5}$$

The value of variant is used to determine how much distribution value of response time against average. Standard deviation shown in Equation (4) is a value that indicates the level variation of a set of data. Backend server group consists of three servers as described previously. Throughput value of each server in the Equation (5) is obtained from total incoming requests divided by total response time. The value of the resulting error in testing to be performed. There is also have error value to be analysis. This error value is the number of requests that fail or not processed on the server. Information about the server load will remain on the main server as log information. Log information to maintain if there are miss or error in one backend server or agent.

3. Result and discussion

The result of the simulation was compared between the methods of previous research using WLC algorithm [17] with LFB research algorithm. Assessment in this simulation is the value of the response time, average response time, std. deviation, throughput, and an error occurred. The input task was generated by sending 800 requests to the server for 10 seconds with 3 backend server. The values taken from simulation result were compared and analyzed. Results of the response time of load balancing using WLC and LFB are illustrated in Figures 2 and 3.

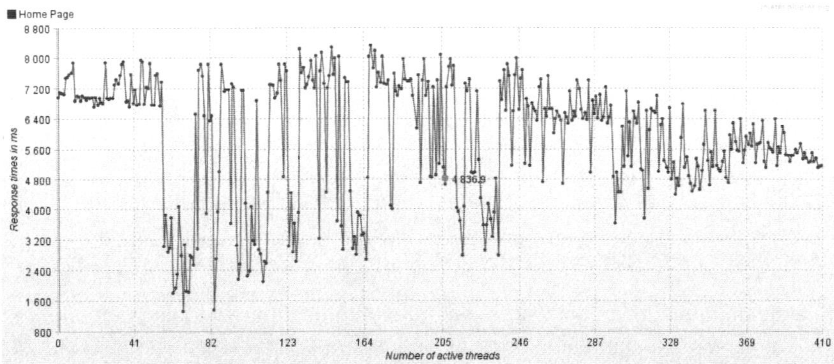

Figure 2. Response time of WLC method for 800 request

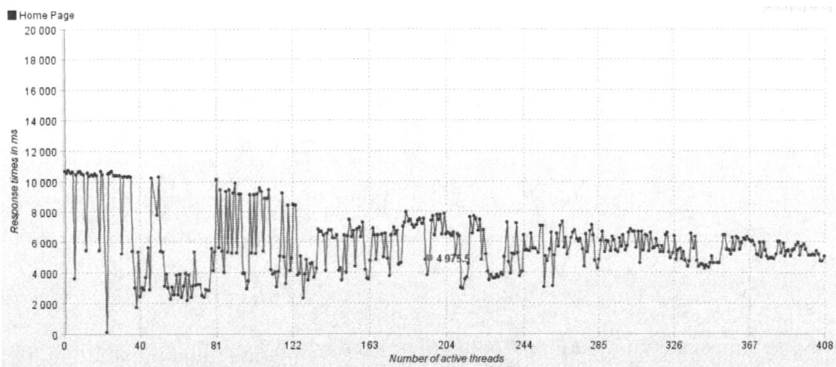

Figure 3. Response time of LFB method for 800 request

Figure 2 describes the response time simulation of WLC with 800 user request for 10 seconds, while Figure 3 describes the response time of LFB with same scheme. WLC method provides faster response time than LFB. WLC response time is less than 8800 ms, 4836 of average response time, 2435.76 of std.deviation, 47.7 of throughput and 0 % error. LFB response time is less than 1100 ms, 4975 of average response time, 3056.35 of std.deviation, 38.7 of throughput and 0 % error. In the first of simulation with 800 requests, WLC is better than the LFB according to the result. The next result is with doing the second simulation. The purpose of second simulation was to see the limits of each server backend and comparison of the limit of each server using WLC and LFB method. The second simulation was divided in two stages: the first stage by giving 600 requests directly to each server and the second stage by sending 1800 request using WLC and LFB method for load balancing. The limit of each server is using 600 requests based on experiments to measure the limit. The error value as the point to be analysis is whether the server has passed the limits and the request become down. Preliminary results of the second simulation for the first stage will be illustrated in Figures 4, 5 and 6.

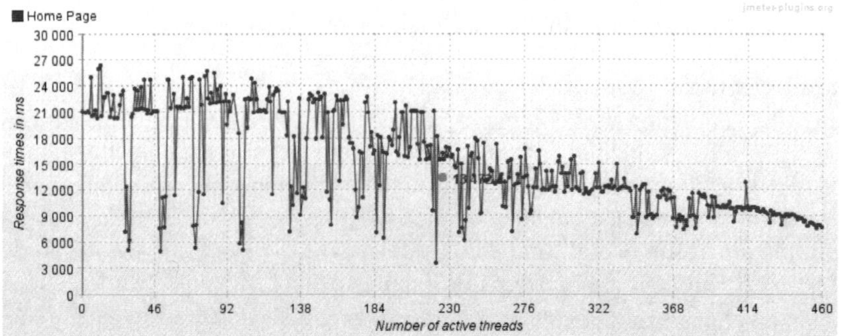

Figure 4. Response time of backend server 1 for 600 request

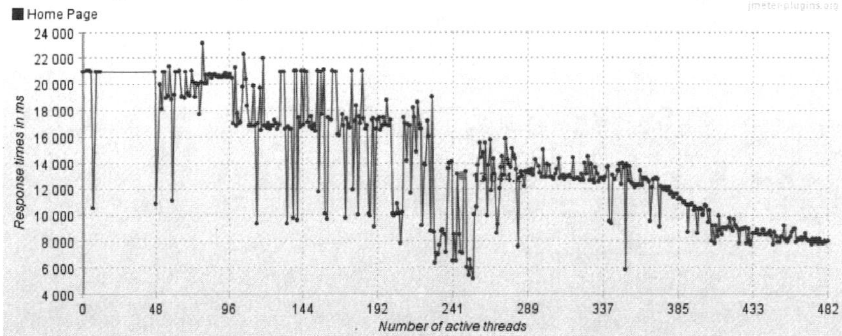

Figure 5. Response time of backend server 2 for 600 request

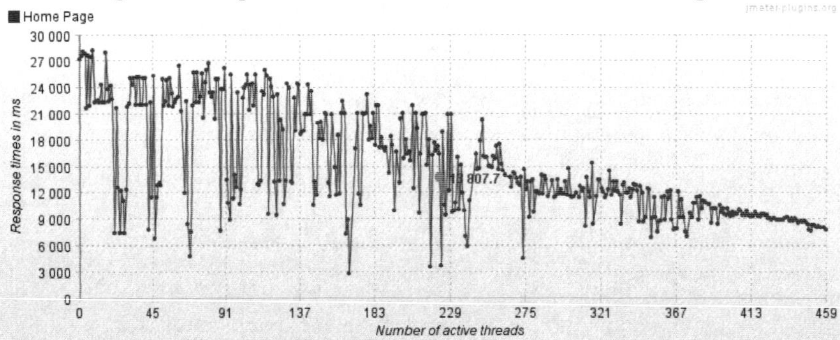

Figure 6. Response time of backend server 3 for 600 request

Figure 4 describes the response time of backend server 1 with 600 requests for 10 seconds. Backend server 1 provides the response time within less than 27,000 ms, average response time of 13,472 ms, std.dev 7165.6 ms, throughput 19.4 and 6.5 % error. Figure 5 describes the response time of backend server 2 with 600 requests for 10 seconds. Backend server 2 provide the response time within less than 24,000 ms, average response time of 13044 ms, std.dev 5703.96 ms, throughput 20.8 and 8.17 % error. Figure 6 describes the response time of backend server 3 with 600 requests for 10 seconds. Backend server 3 provide the response time within less than 30,000 ms, average response time 13807 ms, std.dev 7692.48 ms, throughput of 18.9 and 5.67 % error. Error occurred on all backend servers, identified by the ability limit for backend server on simulation of this research. Backend server 2 has the highest error of 8.17%, followed by backend server 1 at 6.50% and backend server 3 at 5.67%. The total error of the backend server is 20.34%. In second stage of the second simulation comparison between the methods WLC with LFB was performed. The task of simulation was by sending 1800 request simultaneously for 10 seconds. The results of second stage of the second simulation can be described in Figure 7 and 8.

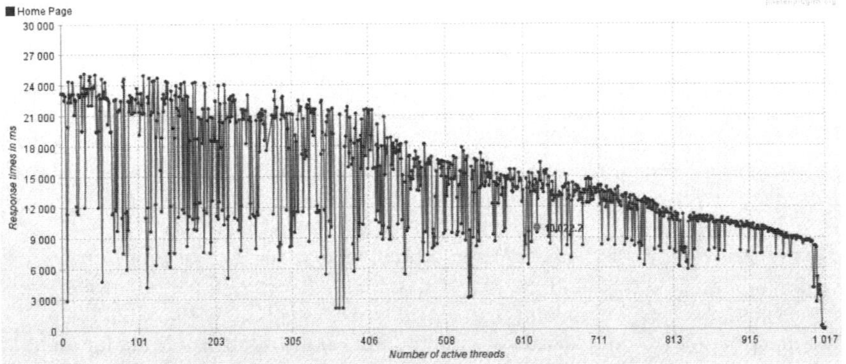

Figure 7. Response time of WLC method for 1800 request

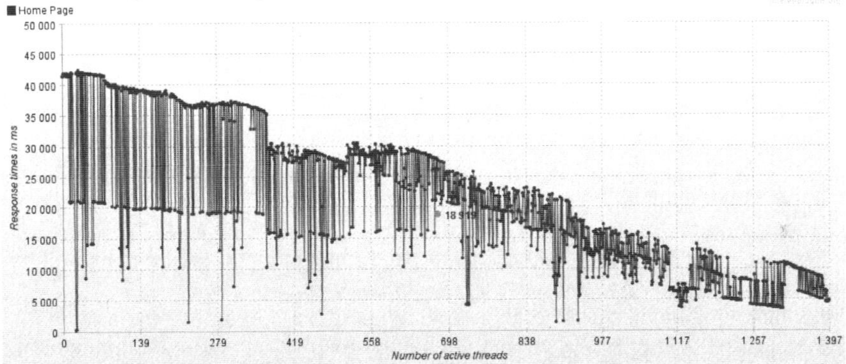

Figure 8. Response time of LFB method for 1800 request

Figure 7 describes the response time of the 1800 request on WLC method for 10 seconds. The result shows response time using methods WLC is less than 27,000 ms, 10,022 ms of average response time, std.dev 8213.64 ms, throughput of 61 and 21.11 % error. Figure 8 describes the response time of the 1800 request on LFB method for 10 seconds. The result shows response time using LFB method is greater than the WLC method, which is less than 45,000 ms, average response time of 18,918 ms, std.dev 13440.64 ms, throughput of 35 and 0 % error. However, the LFB method was more resilient in dealing with a large number of requests. WLC method was only capable of holding within 78.99% from the 1800 request. The LFB method could withstand 100% from the 1800 request although it sacrificed in response time, throughput and others. Based on this second stage of the second simulation showed that the LFB method combined with the mobile agent is more powerful and reliable than the WLC method.

4. Conclusion

System load balancing becomes important in maintaining the reliability of a service. Innovations in making a load-balancing algorithm must be improved. Innovation was created to address the problems that continue to evolve in a distributed system. In this research, load balancing algorithms that can improve reliability and efficient with the mobile agent has been discussed. The proposed algorithms are the least time of first byte (LFB) combined with the mobile agent. LFB algorithm combines the least connection with checks on the time the first byte of data. The mobile agent complements LF algorithm to monitor the state of the resource of each server backend that does not reduce the quality of their service. Based on the results, LFB load balancing method is more reliable and robust compared with WLC method of processing 1800 requests. The WLC method is only capable of processing 78.99% of 1800 while the method of LFB can process 100% from the 1800 request. LFB method is expected to be a reliable load balancing solution in addressing a number of requests on a large scale. However, the response time and system throughput on the results of simulation requires improvement. Improved response time and system throughput would make the system work much better than before.

5. References

[1] J. Cao, Y. Sun, X. Wang, and S. K. Das, "Scalable load balancing on distributed web servers using mobile agents," *J. Parallel Distrib. Comput.*, vol. 63, no. 10, pp. 996–1005, 2003.

[2] Q. Long, J. Lin, and Z. Sun, "Agent scheduling model for adaptive dynamic load balancing in agent-based distributed simulations," *Simul. Model. Pract. Theory*, vol. 19, no. 4, pp. 1021–1034, 2011.

[3] M. A. Metawei, S. A. Ghoneim, S. M. Haggag, and S. M. Nassar, "Load balancing in distributed multi-agent computing systems," *Ain Shams Eng. J.*, vol. 3, no. 3, pp. 237–249, 2012.

[4] A. Yaseen, H. Ji, and Y. Li, "A load-balancing workload distribution scheme for three-body interaction computation on Graphics Processing Units (GPU)," *J. Parallel Distrib. Comput.*, vol. 87, pp. 91–101, 2016.

[5] O. Rihawi, Y. Secq, and P. Mathieu, "Load-Balancing for Large Scale Situated Agent-Based Simulations," *Procedia - Procedia Comput. Sci.*, vol. 51, pp. 90–99, 2015.

[6] A. Singh, D. Junejab, and M. Malhotraa, "Autonomous Agent Based Load Balancing Algorithm in Cloud Computing," *Int. Conf. Adv. Comput. Technol. Appl. (ICACTA- 2015)*, vol. 45, no. Procedia Computer Science, pp. 832–841, 2015.

[7] P. Braun and W. Rossak, *Mobile Agents: Basic Concepts, Mobility Models, and the Tracy Toolkit*. 2004.

[8] V. Baousis, S. Hadjiefthymiades, G. Alyfantis, and L. Merakos, "Autonomous mobile agent routing for efficient server resource allocation," *J. Syst. Softw.*, vol. 82, no. 5, pp. 891–906, 2009.

[9] X.-J. Shen *et al.*, "Achieving dynamic load balancing through mobile agents in small world P2P networks," *Comput. Networks*, vol. 75, pp. 134–148, 2014.

[10] F. Bellifemine, G. Caire, and D. Greenwood, *Developing Multi-Agent Systems with JADE*. 2007.

[11] F. Bellifemine, G. Caire, A. Poggi, and G. Rimassa, "JADE: A software framework for developing multi-agent applications. Lessons learned," *Inf. Softw. Technol.*, vol. 50, no. 1–2, pp. 10–21, 2008.

[12] A. V. Sandita and C. I. Popirlan, "Developing A Multi-Agent System in JADE for Information Management in Educational Competence Domains," *Procedia Econ. Financ.*, vol. 23, no. October 2014, pp. 478–486, 2015.

[13] C. V. Trappey, A. J. C. Trappey, C. J. Huang, and C. C. Ku, "The design of a JADE-based autonomous workflow management system for collaborative SoC design," *Expert Syst. Appl.*, vol. 36, no. 2 PART 2, pp. 2659–2669, 2009.

[14] R. Z. Khan and J. Ali, "Classification of Task Partitioning and Load Balancing Strategies in Distributed Parallel Computing Systems," *Int. J. Comput. Appl.*, vol. 60, no. 17, pp. 48–53, 2012.

[15] Y. Jiang, "A Survey of Task Allocation and Load Balancing in Distributed Systems," *IEEE Trans. Parallel Distrib. Syst.*, vol. 9219, no. c, pp. 1–1, 2015.

[16] D. Lange and M. Oshima, "Mobile agents with Java: the Aglet API," *World Wide Web*, pp. 1–18, 1998.

[17] D. Choi, K. S. Chung, and J. Shon, "An Improvement on the Weighted Least-Connection Scheduling Algorithm for Load Balancing in Web Cluster Systems," *T.-h. Kim al. GDC/CA 2010*, vol. CCIS 121, pp. 127–134, 2010.

IOP Conf. Series: Materials Science and Engineering **190** (2017) 012034 doi:10.1088/1757-899X/190/1/012034

Root Cause Failure Analysis of Stator Winding Insulation failure on 6.2 MW hydropower generator

Agus Adhi Nugroho[1], **Ida Widihastuti**[2], **As Ary**[3]

[1,2]Electrical Engineering Program Faculty of Industrial Technology, Sultan Agung Islamic University, Semarang, Central Java, Indonesia

[3]PT. Indonesia Power UP Mrica

[1]agusadhi@unissula.ac.id, [2] ida_fti@unissula.ac.id

Abstract. Insulation failure on generator winding insulation occurred in the Wonogiri Hydropower plant has caused stator damage since ase was short circuited to ground . The fault has made the generator stop to operate. Wonogiri Hydropower plant is one of the hydroelectric plants run by PT. Indonesia Power UBP Mrica with capacity 2 x 6.2 MW. To prevent damage to occur again on hydropower generators, an analysis is carried out using Root Cause Failure Analysis RCFA is a systematic approach to identify the root cause of the main orbasic root cause of a problem or a condition that is not wanted. There are several aspects to concerned such as: loading pattern and operations, protection systems, generator insulation resistance, vibration, the cleanliness of the air and the ambient air. Insulation damage caused by gradual inhomogeneous cooling at the surface of winding may lead in to partial discharge. In homogeneous cooling may present due to lattice hampered by dust and oil deposits. To avoid repetitive defects and unwanted condition above, it is necessary to perform major maintenance overhaul every 5000-6000 hours of operation.

1. Background

Wonogiri Hydroelectric Power Plant is one of the renewable energy electric power plant managed by PT.Indonesia Power UBP Mrica with 2 x 6.2 MW. Damage in generator winding caused financial losses of Rp. for repair cost Rp. and loss of output Rp. Based on the above problems PT. Indonesia Power does not expect any disruption in operations generating repeated. Therefore the article with the title Root Cause Failure Analysis of Stator Winding Insulation failure on 6.2 MW hydropower generator can be in use as one way to find out the root problem of damage to the generator and found the failure defense task such damage, hence similar damage can be prevented early so that the continuity of power supply for the general population can be met.

RCFA (Root Cause Failure Analysis) is a systematic approach to identify the root cause of the main (basic root cause) of a problem or a condition that is not wanted. RCFA leads to knowledge on the factors contributing to the occurrence of potential failure [1]. By collecting technical data, operation, maintenance, and protection data then converted into a fault tree analysis diagram used to search for the root causes of stator winding insulation failure. Analysis begins loading pattern and operations, protection systems, generator insulation resistance, the cleanliness of the air and the ambient air temperature and vibration generator.

2. Methodology

Manual analysis method performed by fault tree Analysis with collecting parameters such as operating data and maintenance (generator operation, maintenance, system protection test result, winding insulation test and vibration test), then changing the analysis results in to fault tree analysis diagram to facilitate the analysis. Preceded by determining how the top event can be caused by individual event or combination several events that is under them [1]. Analysis is continued by looking into each failure code like electrical analyze include insulation resistance, protection systems, loading pattern, operation pattern, environment analyze includes air temperature and ambient air hygiene, mechanical vibration and cooling system.

3. Discussion

bservations on operating parameters:
From the data obtained that the generator operation is still in the safe operating limits of the maximum limit of 6.17 MW from 6.5 MW and 0.71 MVAR of a maximum limit of 3.7 MVAR and amounted to 6.21 MVA of a maximum limit of 7.75 MVA [4].

Protection system test :
Differential relay on normal condition, there is no interference, because the operating current for diferential relay is followed the Equation (1).

$$i_d = i_1 - i_2 \quad (1)$$

i_d = 3.389 - 3.389 = 0 Ampere,
In the case, there is disruption of the generator phase to ground then it will get id $\neq$ 0 then the differential relay will work [3], how ever, differential relay is still in good condition and it is not trip.

Insulation test :
Using IEEE Standard Std. 43-2000, IEEE Recommended Practice for Testing Insulation Resistance of Rotating calculated by the formula of minimum insulation resistance (UN + 1) MΩ, Un = kV[2]. Testing insulation resistance between the phases U-G is 2.196 GΩ and it is considered as normal, measurement between phases V-G is at 1.35 GΩ (normal) and between phases W-G is 10 KΩ. The last measurement is considered below the standard of 7 MΩ. From the test results, there is indication that the phase W-G has insulation damage.

Insulation polarity index test :
The test is performed by injecting 2500 V dc. The polarity index shows measuring insulation is 10 KΩ and the injection voltage in the test equipment does not come out. Those indications shows potential of short circuit [2], then concluded that the generator at phase-W experienced short circuit to ground.

Measurement of winding resistant (Rdc):
DC resistance measurement indicates in the range 1 μΩ - 5 μΩ values with uncertainty of 1%, in the range of 5 μΩ - 10 μΩ the value uncertainty is 0.5%, in the range 10 μΩ - 200 μΩ the value uncertainty is 0.2%, It is concluded there is no short inter turn on coil according standard NEN-EN-ISO / IEC 17025[3]

Dissipation factor :
To determine the insulation damage caused by the increase in the number of air pockets (voids) in the insulation or increasing resistance in the semiconductor layer, the measurement of dissipation phase U to G with a value of DF 2.83% is normal, phase U to V with a value of DF 6.72% is normal, phase V to G with a value of DF 2.81% is within tolerance, phase V to W with a value of DF 9.67% is beyond the tolerance limit. The phase W to G has an indication of short circuit because the test voltage result is zero. Phase W to the U with a value of DF cannot be tested and it is an indication of short circuit because the voltage test result is zero[2].

Impedance Test :

Impedance Test using AC Current Voltage Injection has the results that Impedance Z for phase U is 1.26Ω, phase – V is 1.25 Ω, and phase W is 1.27 Ω. From this test concluded that the function of the inductance for both three stator winding phase is in normal and reasonable to do with a reinsulation [2].

Clean air and ambient air Temperature:
Winding failure 5 in stator coil insulation caused by the indication of partial discharge on the winding end side or bottom of generator stator. Visually visible damage on the upper side of the coil and bottom coil side of the upper winding phase -W does not appear abnormal. At the bottom of the winding phase –W, dark and white colors indicate short at the stator body as shown in Figure 1.

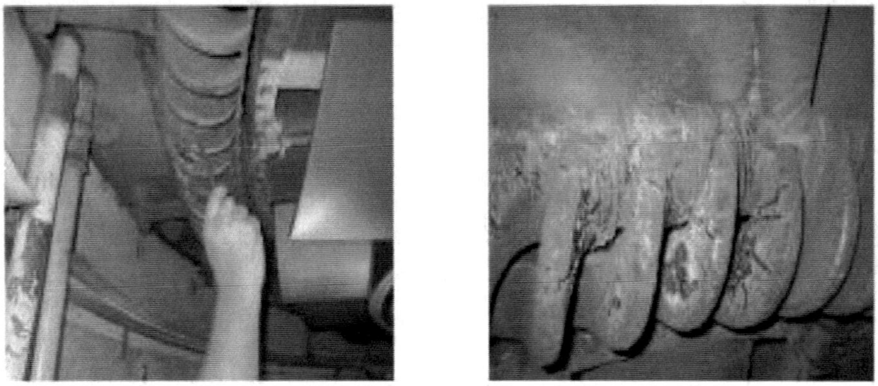

Figure 1. Short circuit W-phase stator generator

Indications of oil and dust deposits on the coil accumulated for a long time in the area shows that the generator stator air quality at generator stator cooling system has poor quality. This evidence is shown by the photo after the pull-out rotor in Figure 2.

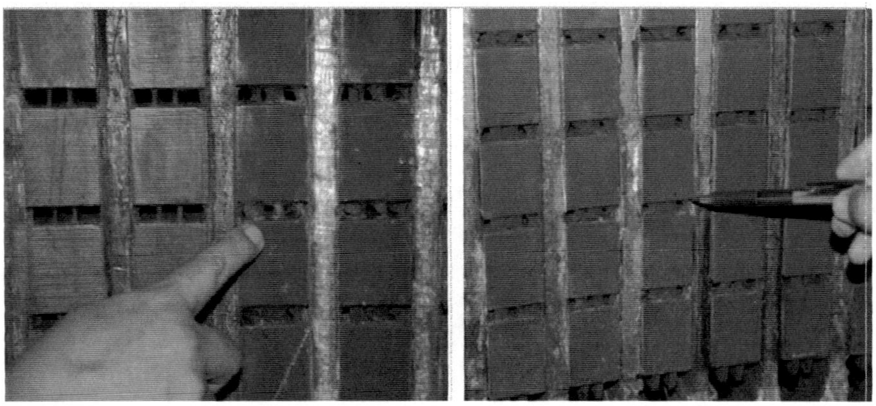

Figure 2. Grid dirty obstruct the flow of cooling air

Figure 2 shows the evidence that by the discharge air temperature over the limit of maximum standards of 55 degree Celsius. Cooling system in hydropower generator stator Wonogiri is not optimal due to the accumulation of dirt, grease, and dust. The accumulation of dirt impedes the flow of air so that surface temperature at different winding is inhomogeneous and it arises a local thermal stress and triggers partial discharge. suspected that the partial discharge already occurs for long duration and cumulates since the last maintenance in 1994 or 18 years ago.

Vibration:
Vibration testing is done with the results of high enough vibration occurs at a pickup point at 3A at 4.04 mm/s and it is relatively normal. From observations of all tests conducted, there is evidence that the root problem of generator stator insulation damage caused by the failure of the generator stator insulation, short circuit in phase W to the Ground, and partial discharge. All these event occurs in sequence and it makes generator stop to operate. inhomogeneous cooling and web condition inside generator which is caused by deposits of dust and grease.

4. Conclusion
From the analysis and discussion of Root Cause Failure Analysis Failure Isolation Generator Stator Winding described above, it can be concluded that:
1. The chosen method to find the root cause damage to the generator stator is Fault tree analysis.
2. Loading pattern and operation pattern generator capability curve is below safe limit 7.75 MVA.
3. Relay differential generator is still in good condition according to 30% settings of and test results.
4. The test results indicate resistance of insulation is breakdown at phase W-G phase with value of 10 KΩ and it is below the allowable minimum standard
5. Air Cooling system in hydropower generator stator Wonogiri does not work optimal due to deposit of oil, dirt and dust blocking the air passage to spread throughout the insulation layer of generator. The condition above causes the temperature of air is over maximum standard limit of 55°C.

5. Suggestion
Generator maintenance is important to be applied for every 50000 to 60000 hours of operation or 5 years in order to anticipate the occurrence of insulation failure at generator stator winding.

References
[1] R. K. Mobley, Root Cause Failure Analysis, vol. 52. 1999.
[2] IEEE, “IEEE Recommended Practice for Testing Insulation Resistance of Rotating Machinery,” IEEE Std 43-2000, pp. 1–28, Mar. 2000.
[3] International Standard Organization, “ISO/IEC 17025 General requirements for the competence of testing and calibration laboratories,” Int. Stand., vol. 2005, pp. 1–36, 2005.
[4] G. Klempner and I. Kerszenbaum, Operation and Maintenance Of Large Turbo Generator. Canada: John Wiley & Sons, Inc., Hoboken, New Jersey., 2004.

The Effects of Cells Temperature Increment and Variations of Irradiation for Monocrystalline Photovoltaic

Faishal Fuad Rahman Soeharto[1], **Hermawan**[2]

[1,2] Department of Electrical Engineering, Diponegoro University, Semarang, Indonesia

[1] faishalfuadr@gmail.com , [2] hermawan.60@gmail.com

Abstract. Photovoltaic cell technology has been developed to meet the target of 17% Renewable Energy in 2025 accordance with Indonesia Government Regulation No. 5 2006. Photovoltaic cells are made of semiconductor materials, namely silicon or germanium (p-n junction). These cells need the light that comes from solar irradiation which brings energy photons to convert light energy into electrical energy. It is different from the solar heater that requires heat energy or thermal of sunlight that is normally used for drying or heating water. Photovoltaic cells require energy photons to perform the energy conversion process, the photon energy can be derived from sunlight. Energy photon is taken from the sun light along with the advent of heat due to black-body radiation, which can lead to temperature increments of photovoltaic cells. Increment of 1°C can decreased photovoltaic cell voltage of up to 2.3 mV per cell. In this research, it will be discussed the analysis of the effect of rising temperatures and variations of irradiation on the type monocrystalline photovoltaic. Those variations are analyzed, simulated and experiment by using a module of experiment. The test results show that increment temperature from 25° C to 80° C at cell of photovoltaic decrease the output voltage of the photovoltaic cell at 4.21 V, and it also affects the power output of the cell which decreases up to 0.7523 Watt. In addition, the bigger the value of irradiation received by cell at amount of 1000 W / m2, produce more output power cells at the same temperature.

1. Introduction

Photovoltaic cell technology has been developed to meet the target of 17% of Renewable Energy based on Indonesian Goverment Regulation No.5/2006[1]. Photovoltaic cells are made of semiconductor materials, namely silicon or germanium (p-n junction). These cells need the light that comes from solar irradiation which brings energy photons to convert light energy into electrical energy. It is different from the solar heater that requires heat energy or thermal of sunlight that is normally used for drying or heating water [2]. For every increment of 1 °C, it can cause the photovoltaic cell voltage of 2.3 mV down to every cell [3]. Sun, as a black object in which a process of fusion of hydrogen atoms into helium, produces heat energy [4]. When the heat continues, it would emit black body radiation spectrum peak heat to give a light color. The color of light depends on the wavelength to be shifted to accommodate the temperature of the object [5], so that the light can be ensured along with the heat. The heat causes the temperature rise of the cell. In addition, the heat makes the environmental hot conditions that raise the temperature of photovoltaic cell. The two-season region at equator brings benefits as well as disadvantages. The benefit is solar energy shines all year round, so it can get more energy. However, the value of the environmental temperature on average is higher than 4-season regions [6], and this condition can raise the temperature of photovoltaic cells.

2. Method

2.1. Simulation

The first step in this simulation is to calculate the value of band gap (the gap between the conduction band and valence band) to the increment in temperature with the equation [7]:

$$E_g(T) = E_g(300K) + \frac{dE_g}{dT}(T - 300K)$$

E_g (300K) = band gap when the temperature of 300 K

T = temperature of photovoltaic cells (K).

After calculating the value of band gap, then the value of I_{sc}, I_o and V_{oc} from photovoltaic cell is calculated by creating MATLAB Simulink simulation [8] by applying the following equations:

$$Isc = \frac{G}{1000}[(Iscr + K_i(Tc - T_r)]$$

$$I_{or} = \frac{Iscr}{[\exp\left(\frac{qV_{ocr}}{nKTr}\right) - 1]}$$

$$Io = Ior\left[\frac{T_c}{Tr}\right]^3 \exp\left[\frac{qE_g}{nK}\left\{\frac{1}{T_r} - \frac{1}{T_c}\right\}\right]$$

K =Boltzman Constant
G =irradiation received by photovoltaic cells (W / m2)
Iscr =short circuit current T cell reference (A)
Ki =temperature coefficient (A / K)
n =factor ideal diode (between 1 and 2)
Tr =reference temperature, 298.15 (°K)
Vocr =open circuit voltage when Tr (V)
Iscr =current Tr short circuit current (A)
Eg =bandgap (eV)
Ior =saturation current time Tr (A)

Table 1. Data result of simulation

T (°C)	Eg (Ev)	Voc (V)	Voc total (V)	Isc (A)	Io (A)	Pm
25	1.100426	0.6028	21.7008	0.59	3.89800E-11	10.07633
27	1.099966	0.599	21.564	0.591	5.28900E-11	10.02978
30	1.099276	0.5933	21.3588	0.5925	8.29300E-11	9.959555
35	1.098126	0.5838	21.0168	0.595	1.72100E-10	9.841432
40	1.096976	0.5743	20.6748	0.5975	3.48500E-10	9.721963
45	1.095826	0.5648	20.3328	0.6	6.90000E-10	9.601148
50	1.094653	0.5552	19.9872	0.6026	1.35400E-09	9.478854
55	1.093526	0.546	19.656	0.605	2.53600E-09	9.35891
60	1.092376	0.5366	19.3176	0.6075	4.71700E-09	9.235793
65	1.091226	0.5272	18.9792	0.61	8.60700E-09	9.111345
70	1.090076	0.5179	18.6444	0.6125	1.54200E-08	8.9873
75	1.088926	0.5085	18.306	0.615	2.71600E-08	8.860196

IOP Conf. Series: Materials Science and Engineering **190** (2017) 012035 doi:10.1088/1757-899X/190/1/012035

It is shown in Table 1 that when the temperatures increase then the value of band gap decreases. When the band gap decreases, it causes the distance between the valence band and the conduction become narrow, furthermore it causes the increment of value of the saturation current diode (Io) as a result of the number of electrons passing narrowed band gap [7]. There is also an increment in the thermal excitation effect on the intrinsic electron charge carriers (ni) [9] [10]. The increment in short circuit current (I_{sc}) is due to the rise of cells temperature (Tc) [8]. Due to the rise of diode saturation current (Io) and the increment in short circuit current (I_{sc}), a decrease in the voltage Voc occurs, because theere is inverse relationship between I_{sc} and Io versus Voc [11]. Voltage Voc decreases in maximum power from photovoltaic cells (Pm) because of the decrease of Voc is greater than the increment of I_{sc} and Io [12].

2.2 Hardware Design

In this study, the experiment module is used for data collection. This module consists of a single photovoltaic cell, and then a series of microcontroller and sensor LM 35 is used to measure temperature. Halogen lamps are used to simulate the sun, because the spectrum of halogen lamp is nearly equal [13] [14]. There are several equipments added such as a direct current cooling fan cooling, resistive load, AC regulator to regulate the level of irradiation of light irradiation, and DC regulator to power the cooling fan. Measurement tools such as multimeter and clamp meter, all equipment components are assembled into an aluminum box as a test site. Figures 1 reveals the flow chart of the simulation based on Matlab, while Figures -4 show the construction of the experiment module.

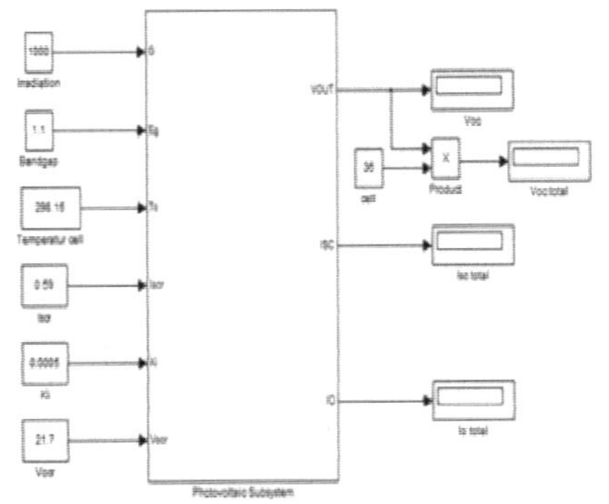

Figure 1. Matlab simulation of photovoltaic

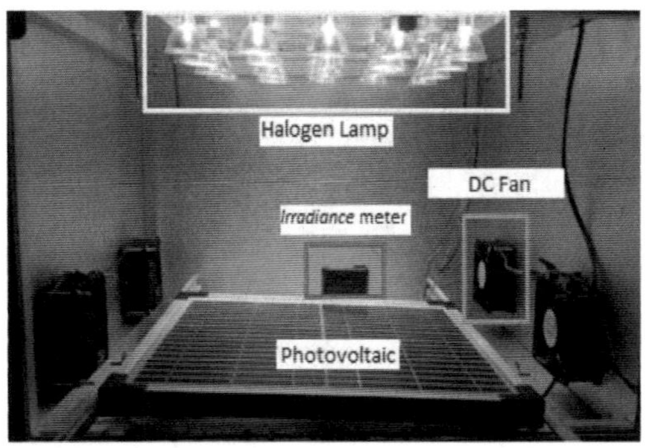

Figure 2. Inside the experiment module

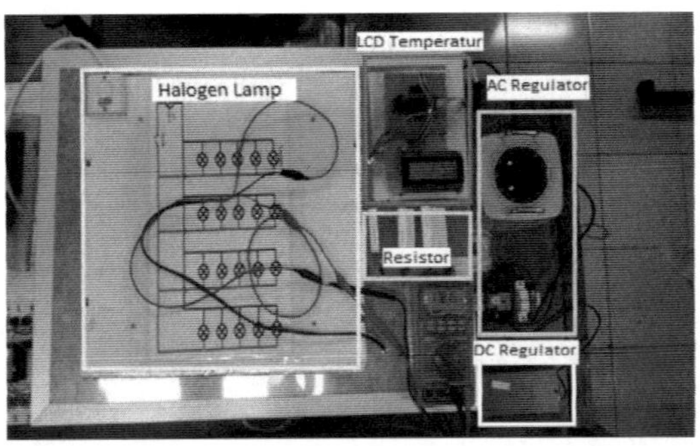

Figure 3. Experiment module from the above

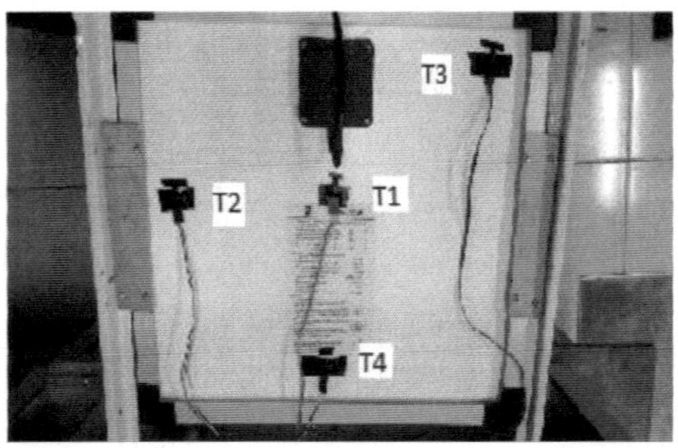

Figure 4. Temperature sensors based on IEC 60891:2009 [15]

With no load condition, the experiment is conducted to measured is the value of open circuit voltage (Voc). On loading condition, several parameters are measured such as short circuit current (Isc), voltage (V), current (A) and the photovoltaic cells temperature (°C). Multiplication of Voc and Isc are called maximum power (Pm). Maximum power is the highest power that can be issued photovoltaic cells. The results of measurement using a multimeter and clamp meter, with variations in the amount of irradiation are 1000 W / m2, 750 W / m2 and 500 W / m^2. Voltage and current data retrieval is done at a temperature value from 27 °C to 80 °C, with a difference of 3 or 5 °C for each increment. Specifications photovoltaic cells are 10 Watt, Voc: 21.7 V, Isc: 0.59 A, Vmp: 18 V, Imp: 0.56 A, in standard test condition (STC) 1000 W / m2, and cell temperature is 25 °C. The tests are conducted at the Laboratory of Electrical Energy Conversion and Power Systems, Diponegoro University. In the first experiment, it is performed in a non- loaded condition with variation of irradiation and temperature.

3. Test Result and Analysis

3.1 Experiment non-loaded and short circuit

In the non-loaded conditions, the open circuit voltage (Voc) reaches a peak value at 21.16 V on irradiation 1000 W / m^2, then 20.81 V on irradiation of 750 W / m^2, and 20.08 V on irradiation of 500 W / m^2. At the highest temperature of 80 °C voltage value drops, i.e. 16.95 V on irradiation 1000 W / m^2, then 16.51 V on irradiation of 750 W / m^2, and 15.84 V on irradiation of 500 W / m2. The best result c occurs when the commencement of variation at a temperature of 27 °C and irradiation 1000 W / m^2. The decline in the value of the voltage it is consistent with the theory, due to the value of the band gap is decreased as a result of rising temperatures and the decrease of the value of band gap. This causes the distance between the valence band (n-type) and conduction (p-type) is narrowed, it allows more electrons to move towards the conduction band and it causes a rise in the diode saturation current (Io). Rising temperature also causes thermal excitation that makes intrinsic charge carriers increments so that the saturation current (Io) rides, when Io rises, it causes the open circuit voltage (Voc) to be lower.

Table 2. Result of open circuit voltage (Voc)

Temperature (°C)	**Voc (V) at 1000W/m²**	**Voc (V) at 750W/m²**	**Voc (V) at 500W/m²**
27	21.16	20.81	20.08
30	20.89	20.52	19.78
35	20.53	20.1	19.41
40	20.14	19.7	18.89
45	19.76	19.33	18.66
50	19.38	18.93	18.28
55	18.96	18.55	17.87
60	18.52	18.12	17.42
65	18.06	17.73	17.07
70	17.73	17.31	16.65
75	17.29	16.89	16.25
80	16.95	16.51	15.84

Table 3. Result of short circuit current (Isc)

Temperature (°C)	Isc (V) at 1000W/m2	Isc (V) at 750W/m2	Isc (V) at 500W/m2
27	0.38	0.26	0.15
30	0.38	0.27	0.15
35	0.38	0.27	0.16
40	0.38	0.27	0.18
45	0.39	0.27	0.18
50	0.4	0.28	0.18
55	0.4	0.29	0.18
60	0.41	0.29	0.19
65	0.41	0.3	0.19
70	0.42	0.31	0.19
75	0.43	0.31	0.19
80	0.43	0.31	0.19

At the lowest temperature of 27 °C, the current 0.38 A is obtained at irradiation of 1000 W / m^2, and 0.25 A at irradiation of 750 W / m^2 and 0,15 A at irradiation of 500 W / m^2. A current value rises with increments temperature value. At the highest temperature of 80 °C, the value of short circuit is variable and follow the irradiance value i.e. 0.43 A on irradiation 1000 W / m^2, and 0.31 A at irradiation of 750 W / m^2, and 0.19 A at the irradiation of 500 W / m^2. The increment in current value it is consistent with the theory posed by a rise of temperature of the cell (Tc). By looking at the equation of equivalent circuit photovoltaic cells [10], it can be seen that the higher the temperature of the cell (Tc), the value of the current short circuit (Isc) also increase.

Table 4. Result of maximum power (Pm)

Temperature (°C)	Pm (W) at 1000W/m2	Pm (W) at 750W/m2	Pm (W) at 500W/m2
27	8.0408	5.4106	3.012
30	7.9382	5.5404	2.967
35	7.8014	5.427	3.1056
40	7.6532	5.319	3.4002
45	7.7064	5.2191	3.3588
50	7.752	5.3004	3.2904
55	7.584	5.3795	3.2166
60	7.5932	5.2548	3.3098
65	7.4046	5.319	3.2433
70	7.4466	5.3661	3.1635
75	7.4347	5.2359	3.0875
80	7.2885	5.1181	3.0096

Maximum power (Pm) is found by multiplying the open circuit voltage (Voc) and short circuit current (Isc). At maximum power, it is found that the higher the temperature, the power value will decrease. The highest point of maximum power produced at a temperature of 27 °C is 8.0408 W on the irradiation of 1000 W / m^2, then 5.4106 W on irradiation of 750 W / m^2, and 3.0120 W on irradiation of 500 W / m^2. At a temperature of 80 °C generated power decreases, i.e. 7.2885 W on the irradiation of 1000 W / m^2, then 5.1181 W on irradiation of 750 W / m^2, and 3.0096 W on irradiation of 500 W / m^2. The power reduction due to the value of open circuit voltage (Voc) is decrease significantly than increment in short circuit current (Isc). The current is not too significant rise because it has a maximum value of 0.05 A, on the contratry the decline voltage is 4.21 V. Achieved maximum power has decreased to 0.7523 Watt at the lowest point T on the maximum irradiation (1000 W / m^2).

4. Conclusion

The output power of the photovoltaic cell type monocrystalline irradiation is affected by several factors. The higher the value of irradiation (1000 W / m^2) then the value of the power (P) generated become bigger at the same temperature. At the trial non-loaded and short circuit condition, open circuit voltage (Voc) will decrease with increment in temperature due to a decrease band gap. Furthermore, the diode saturation current (Io) increases, when the short circuit current (Isc) rises. The result of multiplying the Voc and Isc generates maximum power (Pm) and it decrease to 0.7523 Watt, due to a decrease in Voc at 4.21 V and Isc 0.05 A.

References

[1] Government Regulation No. 5/2006, www.esdm.go.id
[2] Tan, David., Handbook for Solar Photovoltaic (PV) Systems, Energy Market Authority Singapore, 2013.
[3] Makvart, Thomas., Handbook of Photovoltaic, John Wiley and Sons, University of Southampton, London, 2000.
[4] Messenger, Roger., Ventre, Jerry., Photovoltaic System Engineering, Florida, CRC Press, Florida, USA, 1999.
[5] Anggreni, Puti; Makalah Fisika Kuantum, Radiasi Benda Hitam, Jurusan Fisika, Universitas Udayana, 2012.
[6] S, Kurtz., D, Miller., Evaluation of High-Temperature Exposure of Photovoltaic Modules, NREL (National Renewable Energy Laboratory), Philadelphia, 2009.
[7] Yates, Tarn A., E Radziemska, Klugmann., Thermally Affected Parameters of Current-Voltage Characterization of Silicon Photocell, Department of Physics, University of California, Santa Cruz, 2003.
[8] Sutrisno, Elektronika Dasar, Penerbit ITB, 1986.
[9] Kurniawan, Heri., Analisis Pengaruh Temperatur Operasionl dalam Simulasi Karakteristik Arus Tegangan pada Dioda Si Menggunakan FEMLAB, Jurusan Fisika, Universitas Jember, 2013.
[10] Nanjannavar, Viswanath., Gandhi, Palav., LabVIEW Based PV Cell Characterization and MPPT Under Varying Temperatur and Irradiance Conditions, NiuCONE, IEEE, India, 2013.
[11] A. Green, Martin., Solar Cells, Operating Principles, Technology, and System Applications (Prentice-Hall series in solid state physical electronics), Prentice Hall, New Jersey, 1981.
[12] Nelson, Jenny., The Physics of Solar Cells, London Imperial College, London, 2003.
[13] Zeiss Microscopy Online Campus, http://zeiss campus.magnet.fsu.edu/articles/lightsources/tungsten
[14] ASTM E-928,K. A. Emery, Solar Energy Research Institute, Solar Simulators and I-V Measurement Methods, NREL (National Renewable Energy Laboratory), California, USA, 1986.
[15] IEC 60891:2009, https://law.resource.org

IOP Conf. Series: Materials Science and Engineering **190** (2017) 012036 doi:10.1088/1757-899X/190/1/012036

Water Pipeline Monitoring and Leak Detection using Flow Liquid Meter Sensor

R F Rahmat[1], I S Satria[1], B Siregar[1], R Budiarto[2]

[1] Department of Information Technology, Faculty of Computer Science and Information Technology University of Sumatera Utara, Medan, Indonesia
[2] Department of Information System College of Computer Science and Information Technology, Albaha University, Saudi Arabia

romi.fadillah@usu.ac.id, 1114083iss@gmail.com, baihaqi@usu.ac.id, rahmat@bu.edu.sa

Abstract. Water distribution is generally installed through underground pipes. Monitoring the underground water pipelines is more difficult than monitoring the water pipelines located on the ground in open space. This situation will cause a permanent loss if there is a disturbance in the pipeline such as leakage. Leaks in pipes can be caused by several factors, such as the pipe's age, improper installation, and natural disasters. Therefore, a solution is required to detect and to determine the location of the damage when there is a leak. The detection of the leak location will use fluid mechanics and kinematics physics based on harness water flow rate data obtained using flow liquid meter sensor and Arduino UNO as a microcontroller. The results show that the proposed method is able to work stably to determine the location of the leak which has a maximum distance of 2 metres, and it's able to determine the leak location as close as possible with flow rate about 10 litters per minute.

1. Introduction

Water is an essential element for every organism, the needs for providing a good water distribution system is a must. Sometimes, the condition in certain location does not allow us to create a good water distribution system on the ground and the development of constructions causes the current water distribution system to residentials, offices, and industry premises through pipes under the ground.

Flow liquid meter sensor is one of the sensors that is used for this monitoring process. This sensor uses Hall Effect sensor inside it to measure the water flow rate and is placed on a pipe that has a diameter equal to the diameter of the sensor. The sensor will retrieve water flow data by analysing rotation count of the wheel. A microcontroller is required to process this data in order to know how the rate of water flow through [1]. Then the data will be sent by the microcontroller to the end node via internet access.

Clean water is a critical requirement in human life. Commonly, water distribution is done through underground pipes, where the water pipeline will become more difficult to control than the one in the open space. This situation will cause a permanent loss if there is a disturbance in the pipeline such as leakage. Leaks in pipes may be caused by several factors, such as the age of the pipeline, improper installation, and the condition of the environment (natural disaster, etc.). Therefore, an efficient solution is required to detect and to locate the damage in underground pipeline system.

2. Previous Research

Previous researches on water pipeline monitoring and detection in water pipes have been conducted before. Reference [2] conducted a research in monitoring the level of water flow, by utilizing web services and Zigbee as a communication device as well as some sensors such as level sensors, water flow sensors, and temperature sensor. In addition to web monitoring, the owner of the sensor can also get important information about the flow of water via SMS to a personal mobile phone number of the owner.

Reference [3] was conducted to detect leakage of water pipeline. The research investigated the impact of various pipe diameter on pressure of the water flow in the pipes and the temperature changes around the pipe. FSR sensor is used to measure changes in the pipe diameter, and temperature sensors are used to measure the temperature around the pipe. In this research, they used 40 mm diameter PVC pipe with a constant pressure of 3 bar.

The other work in Reference [4] analysed vibration in the pipe wall caused by collisions between the water flows to the pipe wall. Vibration is measured using an MEMS sensor. Leakage is analyzed by comparing the vibration of the normal water flow and the vibration when there is a leak in the pipe. Tests are conducted by varying the pressure from 3 to 10 bar with a constant water flow rate of 300 m^3/hr.

A research that compared the consumption savings of tap water using wireless sensor network has been conducted in Reference [5]. This research uses Rfbee sensors to gather data received from the water flow rate transducer, as a sender and a recipient of the data. Data collected by Rfbee sensors will be transmitted by wireless network to a computer connected directly to the sensor Rfbee. A research to monitor and control the water flow through a web server is carried out in Reference [1]. Monitoring and controlling is done by using Hall Effect Flow Sensors, Arduino, Raspberry PI, and Solenoid Electro Valve. Hall Effect Flow Sensor with Arduino will measure fluid flow, while Raspberry PI will control solenoid electro valve, which is used to close or to open the flow of fluid through the pipe.

3. Methodology

General architecture of the experiment set up in this research shown in figure 1.

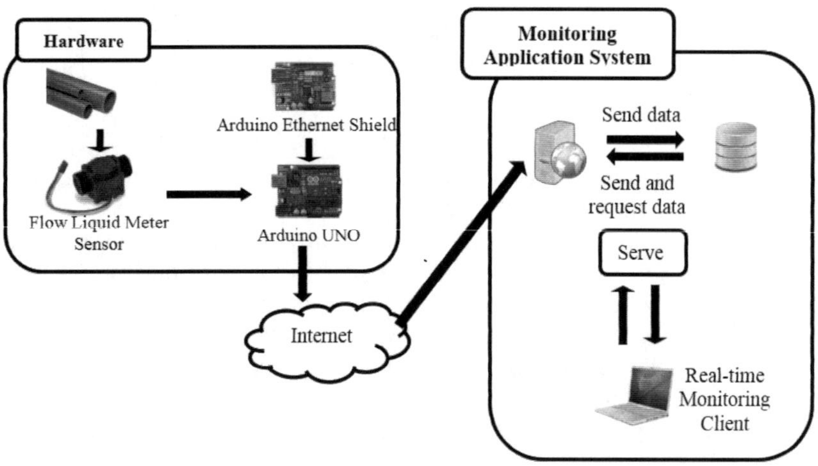

Figure 1. General System Architecture

Hardware

This section shows data retrieved by the sensor and then transmitted by Arduino to the monitoring application system using the Arduino Ethernet shield. The process starts from water that flows in the pipeline and go through the liquid flow meter sensor placed at the end of the pipe. The sensors then will collect data from the water which flows through it, and will be sent to Arduino. Arduino has some pins that serve as a place for data processing and power input. Flow Liquid Meter Sensor will transmit data to Arduino via digital pin 2, while GND pin and 5V pin will be connected to the sensor as the power supply. Then, Arduino calculates the water flow per second based on this data. The obtained data will be stored in a server on the PC's localhost first, then it will be immediately sent to the real-time monitoring application. Data from Arduino is stored to localhost server on the PC via an Ethernet shield, an additional module used to establish connection to the PC by accessing the PC's IP address [6]. When the connection is established, Arduino will access the PC and send the data using the POST method. Stackable Ethernet shield will be directly connected to the PC using the RJ-45.

A program to detect whether there is a leak in the pipe also will be created in Arduino. Leakage will be detected by Arduino when the normal initial flow of water continues to decrease slowly until the water flow stops at a certain speed, and until it is neither continues to decrease nor return to normal anymore. Arduino will process the data in order to obtain the location of the leak and send it to the monitoring application system. The data is measured from wheel rotation count inside the sensor caused by flow of water through sensor. The wheel rotation count is further processed in order to get the data of water flow passing through the sensor every second using equation (1), where n = number of wheel rotation, Q = liquid flow rate (m^3/sec), and c = calibration factor. Calibration Factor has its own constant value of 7.5, depends on the type of sensor used.

$$Q = \frac{n}{c} \tag{1}$$

Every water flow rate data received per second will be compared by Arduino. If there is a significant reduction in normal water flow rate, then the data of it will be stored temporarily until the flow is stable, stability is measured by comparing the normal and potential leakage data. Once the data is stable, the normal and the potential leakage data will be processed to get the velocity of the normal water and the flow after leakage by using equation (2) [7], where v is water velocity (m/s), Q is water flow rate (m^3/s), and A is area of the pipe (m^2).

$$v = \frac{Q}{A} \tag{2}$$

The data of water velocity and time during the reduction will be processed to know how much acceleration does the flow generated by using equation (3) [7]. And based on the acceleration and time data, the location of the leakage on the pipe can be obtained by using the equation (4) [7], where x is distance (m), v is final velocity (m/s), u is initial velocity (m/s), a is acceleration (m/s^2), and t is time (s).

$$v = u + at \tag{3}$$

$$x = \frac{v^2 - u^2}{2a} \tag{4}$$

The pseudocode of the process to detect location of the leak using Arduino can be described as in figure 2.

```
Keep the number of wheel rotation sensor ();
Calculate water flow rate ();
If (water flow rate < normal water flow rate)
        If (water flow rate == leak water flow rate and
         the sending of leak detection data != 1)
                Calculate normal velocity () ;
                Calculate leak velocity () ;
                Calculate water acceleration ();
                Calculate location of leak ();
                Send leak detection data = 1 ;
        Else
                Leak water flow rate = water flow rate ;
                Increase time ;
Else if (water flow rate > normal water flow rate)
        normal water flow rate = water flow rate ;
        Time = 0 ;
        The sending of leak detection data = 0 ;
Else
        Time = 0 ;
        The sending of leak detection data = 0 ;
Send data to server ();
```

Figure 2. Pseudocode to detect the leakage location.

IOP Conf. Series: Materials Science and Engineering **190** (2017) 012036 doi:10.1088/1757-899X/190/1/012036

The location of the leak detection process in accordance with the pseudocode using sample data of the wheel rotation number on sensors that are sent sequentially to Arduino (79, 66.75, 66.75) can be illustrated as follows.

1. Keep the number of wheel rotation sensor (*pulseCount* = 79)
2. Change the number of wheel rotation sensor to get water flow rate (*flowrate* = 10.53 L/min)
3. Check whether the water flow rate data is smaller, larger, or equal normal flow rate data.
4. If value of water flow rate that first received is greater than the normal water flow rate data which initially is 0 and further if the water flow rate data is larger than normal water flow rate data, data will be saved as a normal water flow rate (*flowrateNormal* = 10.53 L / min). Time and leak detection data variable set to 0 to state that the data of leakage have not been sent.
5. If the water flow rate is not smaller and larger than normal, the data is considered as stable and does not have any problem. Time and leak detection data variable is set to 0 to state that the data about leakage have not been sent yet.
6. Water flow rate data has been checked and will be sent to the server.
7. Keep the next number of wheel rotation sensor (*pulsecount* =_66.75).
8. Repeat steps 2 to obtain water discharge (*flowrate* = 8.9 L / min) and step 3.
9. If the water flow rate data is smaller than the normal water flow rate, leakage is detected, check whether the data is the same as the water flow rate after leakage, check the data have been sent or not. If water flow rate is not the same as the water flow rate after leak, store it into the water flow rate after leak data (*flowrateLeak* = 8.9 L / min) and the time will increase. The time will continue to increase until water flow rate declared stable. Then repeat step 6.
10. Keep the next number of wheel rotation sensor (*pulsecount* = 66.75).
11. Repeat steps 2 to obtain water discharge (*flowrate* = 8.9 L/min) and step 3
12. Repeat steps 9. If water flow rate is equal to the water flow rate after leak, indicates that water flow rate as stable and check if the data of leakage has not been sent yet, then the Arduino will calculate the velocity of normal water flow rate (*velocityNormal* = 0.55 m/s), the speed of the water flow rate after leak (*velocityLeak* = 0.46 m/s), the water acceleration (acceleration = -0.09 m/s^2) and the location of the leak (*leakDistance* = 0.50 m) and then stores the value (*datasent* = 1) to declare that the data leakage has been sent to server. Then repeat step 6
13. The leak location detection will be done continuously until the sensor stops transmitting the data of the wheel rotation.

The location of leakage in this pipe will be sent by Arduino to localhost server on PC to be stored and will immediately transmit this data to the web server to then is represented back to the client in the form of notification and information that there had been a leak. This notification contains information about the date, time, position and distance of the leak location in pipe from the sensor.

Web Server

The Web server will act as the server and perform data processing within Arduino, databases, and client. Web server will receive the water flow rate data sent by the PC. When the Web server receives data of the leakage from the Arduino, the web server will store the information of the time when the leakage was detected, in accordance with the time on the database server side. Information about the time when a leak is detected and the location of the leak in the pipe will be sent by a web server to the client in a notification.

Real-time Monitoring Client

The client will be accessing the web page on the web server to monitor and only specific client can get permission to access the page. The page will contain the chart of the water flow through the pipe, and each chart is automatically updated every second and displayed in a certain time interval. In addition, the information about used sensor will be displayed on the web page. The client can also see a notification if there is a leak in the pipe and see the data of the previous water flow on the page.

IOP Conf. Series: Materials Science and Engineering **190** (2017) 012036 doi:10.1088/1757-899X/190/1/012036

4. Result and Discussion

System performance testing is conducted to determine the performance in monitoring and detecting the location of the leak in an underground pipe water flow. A prototype is designed to test the system performance by using PVC pipe ½" and with 5 meters' length. This pipe has five holes with a distance of each other 0.77 meters, 1.55 meters, 2.08 meters, 2.58 meters and 3.1 meters from the sensor. Every hole on this pipe will be tested with three different variations of water flow rate which is 5 litters per minute, 10 litters per minute and 15 litters per minute. The design of the prototype is shown in Figure 3.

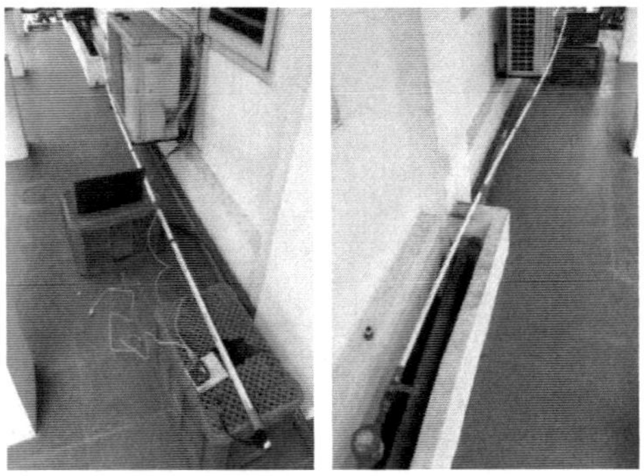

Figure 3. Prototype design

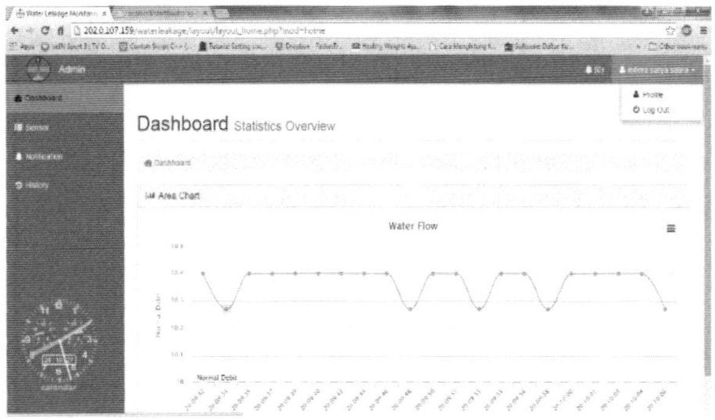

Figure 4. Results of water flow rate monitoring

At the first time of testing, normal water flows through the pipe and the sensor, with the condition of each holes are closed. A hole will be opened and the Arduino will detect that a leak is occurring. This detection result will appear on the page notification.

Testing the performance of the system at the time of water flow rate monitoring shows that the graph automatically moves every two seconds, which is supposed to be one. This caused by the delay time when data is sent from the PC to the web server and it is normal. The results of the monitoring system performance in the water flow is displayed in figure 4.

The system performance tests in detecting the location of the leak is performed 10 times for each variation of the water flow rate on every hole. Each of the test data is converted into the form of 3D graphics, as can be seen, there is a comparison of this data from every hole. The graph in the hole of 0.77 meters, 1.55 meters and 2.08 meters more stable when compared to the graph on the hole of 2.58 meters and 3.1 meters. Stable in this case is the results of the leak detection not changing yet. 3D graphics of the data produced by system testing is depicted in figure 5.

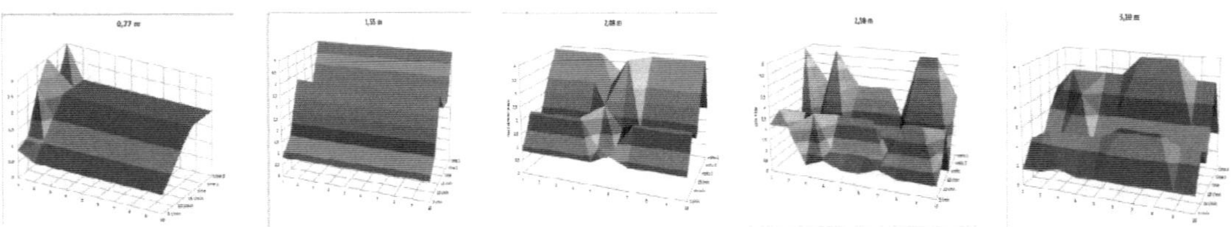

(a) (b) (c) (d) (e)

Figure 5. 3D graphics testing results (a) Distance hole 0.77 meters; (b) Distance hole 1.55 meters; (c) Distance hole 2.08 meters; (d) Distance hole 2.58 meters; (e) Distance hole 3.10 meter

The average results of the leakage location are compared with the actual location of the leak, to see how much difference it made. Based on the difference of average results for the leakage location, it can be seen that at a distance of 1.55 meters, 2.08 meters and 3.10 meters, smallest difference of average results belongs to water flow rate of 10 litters per minute. At a distance of 0.77 meters, smallest difference of average results belongs to water flow rate of 5 litters per minute and at a distance of 2.58 meters, smallest difference of average results belongs to water flow rate of 15 litters per minute. The difference of average results is summarized in table 1.

Table 1. Difference of average results

Actual Location (m)	Flow Rate (l/min)	Difference of Detection Result (m)
0,77	5	0,19
	10	0,284
	15	0,871
1,55	5	0,728
	10	0,025
	15	2.209
2,08	5	1,18
	10	0,22
	15	0,464
2,58	5	1,64
	10	0,92
	15	0,432
3,10	5	2,227
	10	0,514
	15	1,541

Based on figure 5, and the difference of average results in table 1, it can be observed that the system is able to work stably to determine the location of the leak that comes closest to the actual location of the leak in a maximum distance of 2 meters and using water flow rate of 10 litters per minute. This is due to the fact in determining the processing time of water flow rate reduction from the normal until the stable water flow rate after leak is more accurate if the water flow rate is 10 L/min.

5. Conclusions and Future Work

Computation of science, especially fluid mechanics and kinematics physics is implemented on microcontroller. Arduino chip-based ATmega328P can be used to detect the location of leaks in pipes by using a data rate of water flow and the system is able to work stably to determining the location of the leak with a maximum distance of 2 meters and it can determine the location of the leak closest to the actual location of the leak with an average flow rate of 10 litters per minute.

One plan in the future work is to extend the detection ability to more than 2 meters, which later on can be developed by considering the time duration of the reduction in water flow rate when there is a leak so the accuracy of the leakage location is improved. Another future work is to implement the detection system to branched pipelines as well as using various water flow rates.

References

[1]. Suresh N, Balaji E, Anto K J and Jenith J 2014 Raspberry PI based liquid flow monitoring and control *Int. J. Research in Engineering and Technology (IJRET)* **3** 122-125

[2]. Rani M U, Kamalesh S, Preethi S, Shri C K C and Sungaya C 2013 Web based service to monitor water flow level in various applications using sensors *Int. J. Biological, Ecological and Environmental Sciences (IJBEES)* **2** 119-122

[3]. Sadeghioon A M, Metje N, Chapman D N and Anthony C J 2014 SmartPipes: Smart wireless sensor networks for leak detection in water pipelines *J. Sensor and Actuator Network (JSAN)* **3** 64-78

[4]. Rizwan M and Paul I D 2015 Leak detection in pipeline system based on flow induced vibration methodology in pipeline *Int. J. Science and Research (IJSR)* **4** 3326-3330

[5]. Yano I H, Oliveira V C, Araujo E V Campagnuci A G, Fabiano B and Demanboro A C 2014 Wireless sensor networks for measuring the consumption of save water taps *American J. Applied Sciences* **11** 899-905

[6]. Chaudari R P and Chakraborty S 2014 Ethernet based, industrial furnace control and data acquisition *Int. J. Emerging Technology and Advanced Engineering (IJETAE)* **2** 124-128

[7]. Verma H C 1999 *Concepts of Physics* (Bharati Bhawan: Kanpur)

IOP Conf. Series: Materials Science and Engineering **190** (2017) 012037 doi:10.1088/1757-899X/190/1/012037

Segmentation and classification models validation area mapping of peat lands as initial value of Fuzzy Kohonen Clustering Network

Erwin , Saparudin , Muhammad Fachrurrozi

Informatics Engineering, Computer Science Faculty, Universitas Sriwijaya, Indonesia

erwin@unsri.ac.id

Abstract. Ogan Komering Ilir (OKI) is located at the eastern of South Sumatra Province, $2^0$30'-$4^0$15' latitude and $104^0$20'-$106^0$00' longitude. Digital image of land was captured from Landsat 8 satellite path 124/row 062. Landsat 8 is new generation satellite which has two sensors, Operation Land Manager (OLI) and Thermal Infra-Red Sensor (TIRS). In pre-processing step, there are a geometric correction, radiometric correction, and cropping of the digital images which resulting coordinated geography. Classification uses maximum likelihood estimator algorithm. In segmentation process and classification, grey value spread is into evenly after applying histogram technique. The results of entropy value are7.42 which is the highest of result image classification, then the smallest entropy value in the result of correction mapping are 6.39. The three of them prove that they have enough high entropy value. Then the result of peatlands classification is given overall accuracy value = = 94.0012% and *overall kappa value* = 0.9230 so the result of classification can be considered to be right.

1. Introduction

Image segmentation is one of the complicated problems in image processing. The target which has to be achieved from image processing is partitioning the image into multiple areas which mean according to some features, so there are some features in one area and others area are not same. Image segmentation is a feature based on pixel classification procedure, then the clustering analysis is a viable way to apply the image segmentation [1]. Image segmentation techniques automatically classify adjacent pixels to contiguous region according to similarity criteria of property pixels. The object can be better than a pixel, in terms of knowing their neighbors along with spatial connection and spectral between pixels.

Some of the features affect the segmentation result depending on several things: parameters scale, shape, Smoothness, and harmony. Parameters scale is a measure that specifies the maximum value of allowed heterogeneity in generating image objects where for heterogeneous data, the objects that are generated will be smaller than more homogeneous data and by modifying parameters scale value can be made the diverse size of image objects. An indirect form can specify color criteria, stating what percentage of spectral values in the layer image that will contribute to the overall criteria of their homogeneity.

Image segmentation includes part of image processing [2]. Image segmentation aims to separate the region object with the background region so that the object in the image is easy to be analyzed in order to identify objects [3].

The utilization of neuro fuzzy systems is for automatic image segmentation and detect the edge is one of new breakthrough that will certainly add to the advanced digital world particularly image [4].

The whole procedure of image classification of remote sensing aims to group all the pixels in the image into thematic cover class and land utilization. Conventional classification technique uses the unsupervised technique as well as supervised technique while taking decision method can be used minimum distance method, parallelepiped and maximum likelihood. However, because of the clustering process, the partition is very sensitive in the steps of clustering center initialization so that causes tend to stick on local optimization, therefore it needs validation on the model.

Several methods can be used to find a model segmentation optimization and classification such as Fuzzy Kohonen Clustering Network for Hazard zonation with Algorithm Kringing for Interpolation Data, identifying flood area using Classifier which is known as Fuzzy Kohonen Local Information C-Means (FKLICM) with the assistance or the comparison using HKFCM-r Classifier [5], comparing learning methods in clustering on pattern recognition , extraction, VQ, Image Segmentation, and data mining [6], Analysis Entropy using spatial information by adopting general characteristics Hopfield Neural Network (HNN) and multi-synapse neural network (MSNN) [7], Comparing a Segmentation of different color images [8]. Test the effectiveness of the maximum iteration t and initialization maximum value 0 m on the behavior of FKCN[9]. Obtaining and Analyzing information about object or earth, region or symptom[8]. The concept of fuzzy gives the set to handling a vague and inaccurate data [10]. It can be divided into two main categories : Supervised and unsupervised [11]. The algorithm used is a modified version of the Kohonen Algorithm which found in network nerve. So, this algorithm can also refer as Fuzzy Clustering Kohonen Network (FKCN) [12].

2. Statistic Characteristic

One of the Characteristic of digital image textures which can be calculated from statistics is the intensity value of the gray scale in the image that is mean, standard deviation, histograms, and entropy value. In this research, all those three parameters are calculated and used to validate the model. Mean indicates the average value of the matrix which formed from the value of the pixels of an image while the deviation standard indicates the spread of data. Mathematically the calculation of mean and standard deviation using equations (1)

$$\bar{x} = \frac{1}{n}\sum_{i=1}^{n} x_i$$

$$std = \sqrt{\frac{1}{n}\left(\sum_{i=1}^{n} x_i - \bar{x}\right)^2} \tag{1}$$

2.1. Histogram

The histogram is a graph that shows the spread of the values of the pixels of the image, then it can be known the value of relative frequency. In addition, the histogram can indicate the level of brightness and contrast of an image. Histogram calculation mathematically can be calculated using equation (2)

$$I_t = \frac{f_t}{f}, \quad t = 0,1,\dots,K-1 \tag{2}$$

where I_t is histogram value, f is sum of all pixel

The distribution of I_t and f_t provide information about the appearance of the image. The image histogram is distributed evenly on the whole level of grayscale which means having a good level of contrast when histogram accumulates in the dark area means having Dim image and histograms which accumulatein the light area (high intensity) means showing a bright image.

2.2. Histogram
Image Entropy is statistic to measure the randomness of distribution value pixels. The image is called as perfect if have zero entropy. Entropy can be used as a measure of statistics on the characterization of the texture of the greyscale image input. Entropy formulated by equation (3).

$$E = -\sum_t I_t log_2 I_t \quad (3)$$

3. Materials and Method
The subject of the research is an area of peatlands distribution in the province of Ogan Komering Ilir (OKI) South Sumatera which is geographically situated between 104°20 ' and 106°00 ' East longitude and 2°30 ' to 4°15 ' South latitude. The data used in this study i.e. Landsat 8 image downloaded from the USGS website with path/row 124/062 with photo shoot on 17-07-2015 Landsat7 Image and ETM orthorectification in 2010 which will be used for geometric correction of reference. Steps Image data processing on Landsat 8 which done in this research are described in section 3.1 until 3.3.

3.1. Radiometric and Geometric Correction
Image data of the Landsat 8 is done using the image of Landsat 7 ETM orthorectification (the image has been processed and corrected geometric). While radiometric corrections were done to fix the value of the pixels to match with the histogram adjustment technique. While the geometric correction is done for corrected between the coordinates of the pixel digital image field controls and topographic map coordinates.

3.2. Cropping
The image is done to get the research areas with a view to being able to do image data processing that is more focused, detailed and optimal. With expectation generates the image of a representative and continuous. The cutting image has value to other utilities, namely reducing the areas that will be examined in accordance with the area of interest. The image cuts can be carried out in accordance with the desired polygon shape such as restrictions on the territory of the County, district or village. So, image cropping (cutting images) can be useful to facilitate the performance of a person while being observing the image, especially in restricting certain regions.

3.3. Classification of land cover
Land cover classification is used the guidance classification by using the maximum likelihood algorithm and based on land cover types on the map. In this algorithm the pixels classified as a particular object according to the shape, size and orientation of the samples in the feature space. While the accuracy of the algorithm can be computed using the confusion matrix with the given tolerance limit i.e. ≥ 80%.

Maximum Likelihood Algorithm

1. Supposed random variable $x_1, x_2, \dots, x_n$ has joint probability distribution $f(x_1, x_2, \dots, x_n; \theta)$
2. Form of Likelihood function as $L(\theta) = f(x_1, x_2, \dots, x_n; \theta)$
3. Maximum Likelihood function (Max $L(\theta)$) with method:
$\frac{dL(\theta)}{d\theta} = 0$ or $\frac{dLog\ L(\theta)}{d\theta} = 0$

4. If more than one parameter:

$$L(\theta_1, \theta_2, \dots, \theta_n) = \prod_i^n f(x_i; \theta_1, \theta_2, \dots, \theta_n)$$

$$\frac{dL(\theta_1, \theta_2, \dots, \theta_n \theta)}{d\theta_1} = 0, ---, \frac{dL(\theta_1, \theta_2, \dots, \theta_n \theta)}{d\theta_n} = 0.$$

4. Result and Discussions

Histogram for 3 (three) mapping image is the map of correction result (radiometric and geometric) and map of classification result are counted using Matlab 2009a. Those are presented in figure 1 as :

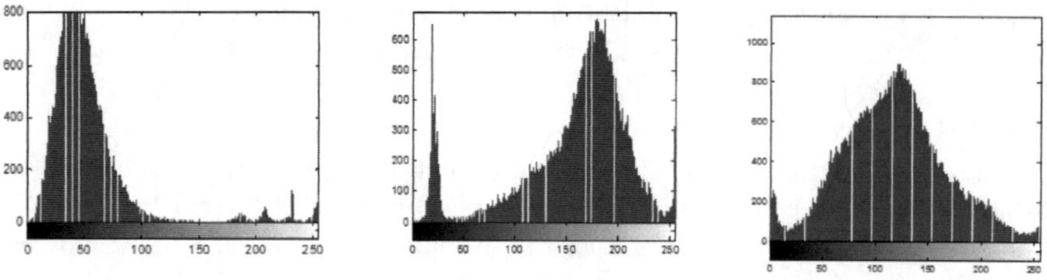

a b c

Figure 1. Histogram a. Map of correction result, b. Map of segmentation result and c. Map of classification result

From histogram result which can be seen in figure 1. Map image of correction result originally has heap histogram in a low area (0-100). The heap histogram in this area is caused the map is too dark. After the segmentation process has done, the histogram is seen to start spreading and in the classification process, the histogram is seen spreading in all grey area (0-255). Spread histogram evenly means the image result into better quality and better level brightness. Calculation value of statistic value RGB (red, green and blue) is in the form of entropy, mean, and deviation standard for three type of image is seen in table 1 below:

Table 1. The Statistic of Correction Geometry, Segmentation and Classification Results

Statistics	Corrected Map	Segmentation	Classification
Matrix Dimension	193x223x3	227x226x3	309x304x3
Entropy of Image	6.39	7.69	7.42
Red			
Entropy	6.50	7.36	7.60
Mean	53.32	157.71	115.51
Standard Deviation	39.04	54.34	51.07
Green			
Entropy	6.30	6.80	7.34
Mean	43.69	134.50	108.89
Standard Deviation	19.98	47.42	47.10
Blue			
Entropy	6.28	6.87	5.48
Mean	43.80	76.49	18.95
Standard Deviation	19.99	32.28	17.55

Based on table 1, it is shown that the highest entropy value in classification image result is 7.42, while the smallest entropy value in the map of correction result is 6.39.Statistic of blue color shows the smallest entropy value for corrected map and map of classification result are 6.28 and 5.48, while in the map of segmentation result shows green color has the smallest entropy value is 6.89. The three of image prove that those have enough high entropy value. It means that the image is still far from excellence. The image can be said excellence if it has zero in entropy value. Zero entropy value can be earned if image histogram evenly in all of the parts.

In this research, the image of red color shows peatlands classification result and generate overall accuracy value = 94.0012% and overall kappa value = 0.9230. With that result, classification can be considered to be right. Contour plot for the three map show in figure 2, as:

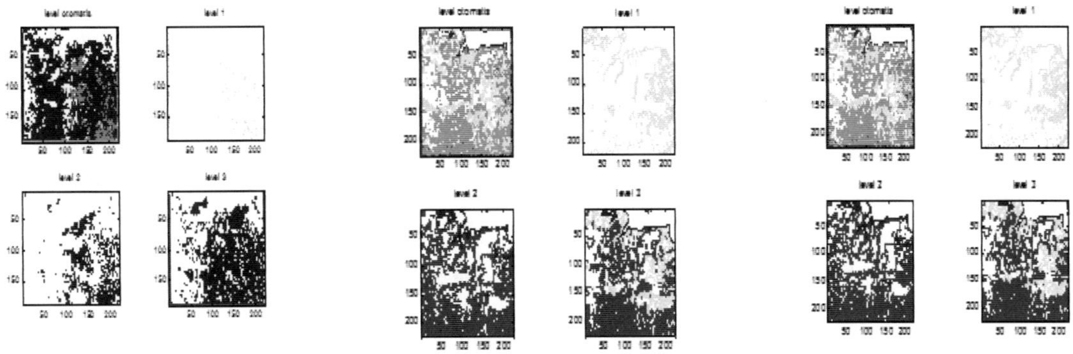

Figure 2. Contour Plot with level automatically, 1, 2 and 3, a. corrected map b. map of segmentation result and c. map of the classification result

5. Conclusion

Utilization histogram technique can be applied in segmentation process and classification in the image of peatlands mapping which causes the spread of grey value evenly. The highest entropy value in classification image result is 7.42, while the smallest entropy value in correction mapping result is 6.39. The three of image prove that those have enough high entropy value. The result of peatlands classification is given overall accuracy value = 94.0012% and overall kappa value = 0.9230 so the classification result is considered correct.

Acknowledgements

Thank you so much to reviewers for all review result which has completed this journal. This journal is powered by grants research Unggulan Kompetitif Universitas Sriwijaya scheme.

References

[1]. L. Bosheng, W. Yuke, and L. Jiangping, "A noise-resistant fuzzy kohonen clustering network algorithm for color image segmentation," *Proc. 2009 4th Int. Conf. Comput. Sci. Educ. ICCSE 2009*, pp. 44–48, 2009.

[2]. N. I. Jabbar and M. Mehrotra, "Application of Fuzzy Neural Network for Image Tumor Description," *Proc. World Acad. Sci.*, vol. 1, no. 8, pp. 575–577, 2008.

[3]. A. Z. Arifin and A. Asano, "Review Paper Image segmentation by histogram thresholding using hierarchical cluster analysis," vol. 1, pp. 1–4, 2006.

[4]. Shweta Lawanya Rao and C.L.Chandrakar, "An Adaptive Neuro-Fuzzy System for Automatic Image Segmentation and Edge Detection," *Int. J. Data Warehous.*, vol. 4, no. 1, pp. 33–40, 2012.

[5]. K. K. Singh and A. Singh, "Identification of flooded area from satellite images using Hybrid Kohonen Fuzzy C-Means sigma classifier," *Egypt. J. Remote Sens. Sp. Sci.*, 2016.

[6]. K. L. Du, "Clustering: A neural network approach," *Neural Networks*, vol. 23, no. 1, pp. 89–107, 2010.

[7]. B. Huo and F. Yin, "Image Segmentation using Neural Network and Modified Entropy," vol. 8, no. 3, pp. 251–260, 2015.

[8]. N. I. Jabbar and S. I. Ahson, "Modified fuzzy Kohonen clustering network for image segmentation," *2010 Int. Conf. Financ. Theory Eng. ICFTE 2010*, vol. 3, pp. 176–179, 2010.

[9]. N. I. Jabbar, "Estimation Clusters Based Preprocessing in Fuzzy Kohonen Clustering Network for Image Segmentation," *Int. Conf. Artif. Intell. Image Process.*, pp. 207–210, 2012.

[10]. A. Hamdy, "Fuzzy Logic for Enhancing the Sensitivity of COCOMO Cost Model," vol. 3, no. 9,

pp. 1292–1297, 2012.
[11]. V. Jecheva and E. Nikolova, "Some clustering-based methodology applications to anomaly intrusion detection systems," *Int. J. Secur. its Appl.*, vol. 10, no. 1, pp. 215–228, 2016.
[12]. D. V. Fernandes, T. Thomas, N. Joseph, and J. Joseph, "Image Segmentation using Fuzzy - Kohonen Algorithm," no. Icmlc, pp. 176–179, 2011.

IOP Conf. Series: Materials Science and Engineering **190** (2017) 012038 doi:10.1088/1757-899X/190/1/012038

Implementation of 128 bits Camellia Algorithm for Cryptography in Digital Image

Bagus Satrio Waluyo Poetro

Department of Informatics, Faculty of Industrial Technology, Sultan Agung Islamic University, Semarang, Indonesia

Bagusswp@unissula.ac.id

Abstract. Nowadays information technology requires stronger cryptographic algorithms. Camellia algorithm is also known for its suitability in terms of the implementation of both software and hardware as well as a high level of safety. The digital image is an image f (x, y) which having the spatial coordinates, and brightness levels are discrete. Unlike text messages, the image data has special features such as high redundancy and a high correlation between pixels. This research conducted a cryptographic process of the digital image using the Camellia algorithm. Comparisons were made on three digital image format .bmp, .jpg, .png with 128 bits key block Camellia algorithm. Results shows that Camellia cryptographic algorithms in digital image can successfully produce encrypted images. In addition, the same algorithm can also reproduce the image when decryption process.

1. Introduction

Cryptography is the study of science and art to keep a message or data information to be safe. Nowadays information technology requires stronger cryptographic algorithms [1]. One of important criterion for a good algorithm is that it should be practically efficient for legitimate users [2].

Camellia algorithm was developed by NTT (Nippon Telegraph and Telephone Corporation) and Mitsubishi Electric Corporation in 2000. Camellia algorithm support 128 bits block key which is same with AES (Advanced Encryption Standard) algorithm [3]. AES and Camellia algorithm can be concluded that the length of character input and key length is proportional to the process time of encryption and decryption [4]. Also Camellia algorithm is also known for its suitability in terms of the implementation of both software and hardware as well as a high level of safety [5].

The digital image is an image f (x, y) which having the spatial coordinates, and brightness levels are discrete [6]. Unlike text messages, the image data has special features such as high redundancy and a high correlation between pixels [7]. Digital image can be quickly moved and copied without any loss of information or image quality [8].

This paper proposes implementation of Camelia Algorithm to understand the performance of the algorithm.

2. Methods

Research method in this study can be explained as follows:

2.1. Camellia Algorithm

Camellia algorithm is a symmetric key cryptography algorithm that works on a block size of 128 bits with a key length of 128 bits, 192 bits or 256 bits [5]. Phases of Camellia Algorithm in this research are described as follows:

2.1.1. Encryption Process.

a) Separating the original image into 3 layers (Red, Green, Blue).
b) Insert 128 bits key (16 characters) and do the key scheduling process until produce subkeys which will be used in encryption process.
c) Initializes long rows & columns from original image every 128 bits (16 pixels, 4x4 size).
d) Change the shape of block data becoming array.
e) Encrypting block data from original image every 128 bits (16 pixels) until maximum size from the original image with the 128 bits key from the key scheduling process.
f) Change the shape of array from encryption process becoming block data again.
g) Repeat step 3 until 6 at each of image layer (Red, Green, Blue).
h) Merge 3 layers (Red, Green, Blue) which will be knwon as encrypted image.

2.1.2. Decryption Process.

a) Separating the encrypted image into 3 layers (Red, Green, Blue).
b) Insert 128 bits key (16 characters) and do the key scheduling process until produce subkeys which is opposite in encryption process.
c) Initializes long rows & columns from encrypted image every 128 bits (16 pixels, 4x4 size).
d) Change the shape of block data becoming array.
e) Decrypting block data from encrypted image every 128 bits (16 pixels) until maximum size from the encrypted image with the 128 bits key from the key scheduling process.
f) Change the shape of array from encryption process becoming block data again.
g) Repeat step 3 until 6 at each of image layer (Red, Green, Blue).
h) Merge 3 layers (Red, Green, Blue) which will be known as decrypted image (it should be look like original image).

2.2. Key used

The key which was used in this research have 16 characters or 128 bits long. The key selected was '123456789abcdefg'.

2.3. Data used

Digital image which is intended in this research was 3 images with a still image formats. bmp, png and. jpg. The size of the image used is square which mean that is the length and width of the image should be the same. The maximum size used in this research was 512 x 512 pixels.

The imagery used is high-frequency image such as the image of Barbara (a) [9], image that contain high-quality and low-cost as image of Bridge (b) [10], and complex image such as the image of Mandrill (c) [11]. Figure 1 show three digital images which was used in this research.

Figure 1. Images used for testing

2.4. *System Design*
General overview of the encryption or decryption process used in this research as in Figure 2 and Figure 3. Figure 2 explained that Camellia algorithm encrypted the original image simultaneously with the help of 128 bits encryption key which was mentioned before. When finished, the image of the encryption process called the encrypted image. While Figure 3 explained that the encrypted image can be decrypted with 128 bits key which was exactly the same when the encryption process. After the decryption process completed the encrypted image turned back to original image.

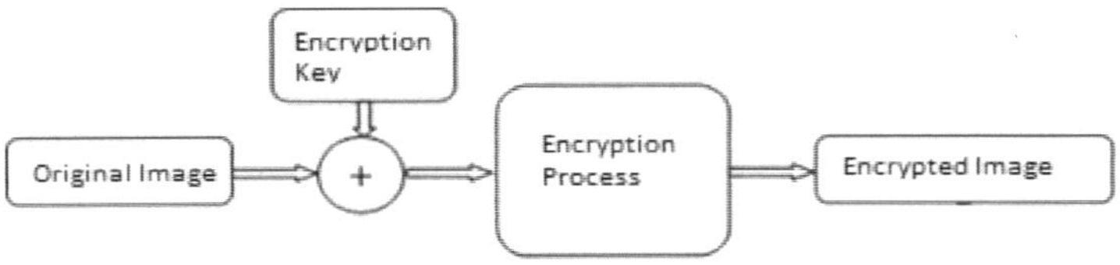

Figure 2. Encryption Process

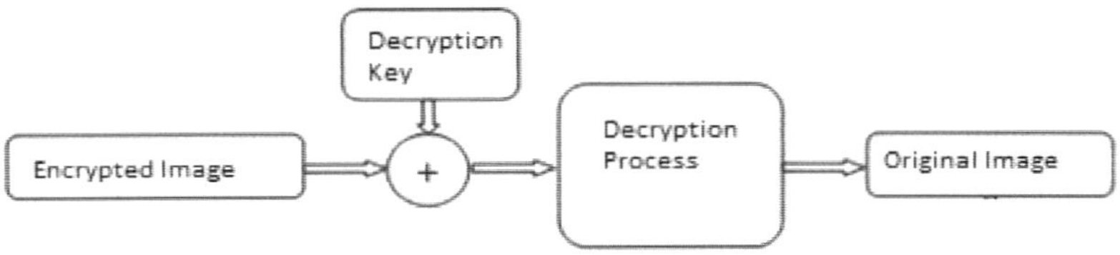

Figure 3. Decryption Process

2.5. *Image similarity measurement*
To show how much exactly the same original image was, this research used euclidean distance measurement method. This method was chosen because image similarity measurement can be done partly by using euclidean distance calculation [12].

3. Results and Discussion

3.1. *Shape of encrypted image*
The image results are shown in the encryption process when testing was similar in every format (a,b,c). In Figure 4 shows the results from the encryption process of Camellia algorithm against the format. bmp,. jpg,. png. While in Figure 5 shows the results from the decryption process of Camellia algorithm with exactly the same 128 bits key at encryption process against the format .bmp, .jpg, .png.

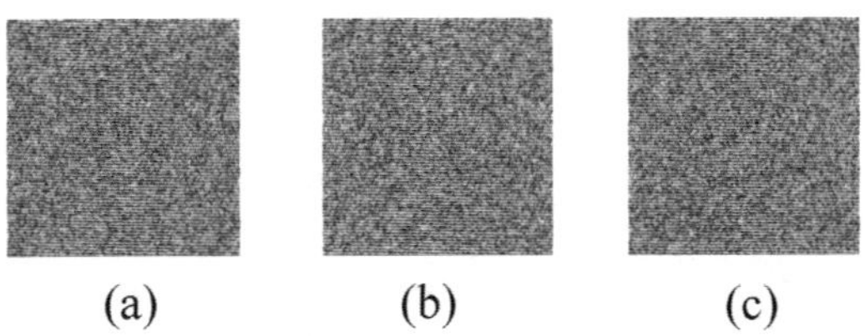

Figure 4. Encryption process results for .bmp , jpg and png format

(a) (b) (c)

Figure 5. Decryption process results for .bmp , jpg and png format

Figure 5 shows that .bmp, .jpg and .png format which have been encrypted was turned back exactly the same with original images without any loss or faulty condition. Based on Table 1 all images were calculated using the euclidean distance method and the results reveal that every single image has distance close to 0. This means that getting close to 0 implies the more similar the images are.

Table 1. Results of euclidean distance calculation

No	Image Name	Format		
		BMP	JPG	PNG
1	Barbara	0,534	0,921	0,620
2	Bridge	0,511	0,901	0,609
3	Mandrill	0,523	0,914	0,632

4. Conclusion

Based on the research and discussion that has been done, it is concluded that:

a) Camellia cryptographic algorithms in digital image can successfully produce encrypted images.
b) Every encrypted image was back to original image when decryption process.
c) BMP format was more getting close to 0 than PNG and JPG which mean more similar than others.

References

[1] B.S.W. Poetro, et al, "Kriptografi Citra Digital dengan Algoritma Rijndael dan Transformasi Wavelet Diskrit Haar," S.Kom. undergraduate thesis, Dept., Informatics., Diponegoro Univ., Semarang, Central Java, Indonesia, 2010.

[2] W. Mao, Modern Cryptography : Theory and Practice, Prentice Hall PTR, New Jersey, USA, 2003.

[3] Specification of Camellia – a 128-bit Block Cipher, NTT and Mitsubishi Electric Corporation, 2000.

[4] Sabriyanto, "Analisis Perbandingan Performansi Algoritma Camellia dan AES Pada Blok Cipher," ST. undergraduate thesis, Dept., Elect and Telco, Telkom Institute, Bandung, West Java, Indonesia, 2009.

[5] M. Matsui et al., "A Description of the Camellia Encryption Algorithm, Request for Comments : 3713", 2004.

[6] S. Nugroho, "Sistem Pendeteksi Wajah Manusia pada Citra Digital," M.Cs thesis, Dept., Computer Science, Gadjah Mada University, Yogyakarta, Indonesia, 2004.

[7] E.I.F. El-Ashry, "Digital Image Encryption," M.Sc thesis, Dept., Electronic Engineering, El Monofeya, Menofia University, Egypt, 2010.

[8] S.D. Anderson, "Digital Image Analysis : Analytical Framework for Authenticating Digital Images," M.S thesis, Dept., Engineering and Applied Science, University of Colorado Denver, Denver, USA, 2001.

[9] H. Elkamchouchi and M.A. Makar, "Measuring encryption quality of bitmap images encrypted with Rijndael and KAMKAR block ciphers,", Proceedings Twenty second National Radio Science Conference (NRSC 2005), pp. C11, Cairo, Egypt, 2005.

[10] E.C. Larson and D.M. Chandler, "Most Apparent Distortion : Full Reference Image Quality Assesment and the Role of Strategy," Journal of Electronic Imaging, Vol 19, Issue 1, USA,

2010.

[11] L.P. Dung, "Perception Based Quality Measure for Image Compression and Transmission," M.IT thesis, Dept. Computer Science and Engineering, Monash University, Australia, 1998.

[12] B.S.W. Poetro and A. Labellapansa, "Pengelompokan Gejala Penderita Kolesterol melalui Pola Iris Mata menggunakan Moment Invarian dan Euclidean Distance," Prosiding Seminar Nasional Ilmu Komputer Universitas Diponegoro 2012, pp 117, Publisher : Graha Ilmu, 2012.

IOP Conf. Series: Materials Science and Engineering **190** (2017) 012039 doi:10.1088/1757-899X/190/1/012039

The Automation System Censor Speech for the Indonesian Rude Swear Words Based on Support Vector Machine and Pitch Analysis

S N Endah[1], D M K Nugraheni[1], S Adhy[1] and Sutikno[1]

[1] Department of Informatics, Universitas Diponegoro, Jl. Prof Soedharto, Kampus UNDIP Tembalang, Semarang, Indonesia

sukma_ne@yahoo.co.id

Abstract. According to Law No. 32 of 2002 and the Indonesian Broadcasting Commission Regulation No. 02/P/KPI/12/2009 & No. 03/P/KPI/12/2009, stated that broadcast programs should not scold with harsh words, not harass, insult or demean minorities and marginalized groups. However, there are no suitable tools to censor those words automatically. Therefore, researches to develop a system of intelligent software to censor the words automatically are needed. To conduct censor, the system must be able to recognize the words in question. This research proposes the classification of speech divide into two classes using Support Vector Machine (SVM), first class is set of rude words and the second class is set of properly words. The speech pitch values as an input in SVM, it used for the development of the system for the Indonesian rude swear word. The results of the experiment show that SVM is good for this system.

1. Introduction

One of the powerful tools for pattern recognition that uses a discriminative approach is a Support Vector Machine (SVM) [1]. SVMs use linear and nonlinear separating hyper-planes for data classification [2][3]. It is not only for speech recognition [4], but it can be applied for another application like image processing [5][6] and bioinformatics [7]. From this research, the classification using support vector machine has a high degree of accuracy. In speech recognition, SVM is also can collaborate with other method, such as Hidden Markov Model (HMM) [8][9].

In Indonesia, the research related to Indonesian speech recognition relatively limited. Reliability of SVM can be used to classify whether the speech word is a rude word or properly word. Therefore, we propose a system of an intelligent software to automatically censor the speech words for rude swear word in Indonesian using SVM. Input in SVM is the pitch value of speech.

The proposed system can be used for broadcasting. Based on Law No. 32 of 2002 concerning the broadcasting, it is noted that broadcasting as mass communication activities have a function as a medium of information, education, wholesome entertainment, control and social glue [10]. To fulfill this purpose, the content broadcast was arranged in Law No. 32 of 2002 Article 35 and Article 36 and reinforced the Indonesian Broadcasting Commission Regulation No 02/P/KPI/12/2009 and No. 03/P/KPI/12/2009. In that rules, stated that a broadcast programs cannot curse with curse words, not harass, humiliate or degrade minorities and marginal groups such as community groups with specific jobs (domestic workers, security guard), a group that has a deviation (hermaphrodites), group size and physical form outside the normal (tongos teeth, fat, midget, eyes squinting), clusters have physical disabilities (deaf, blind, mute), which have a defective batch or mental retardation (autism, idiot) , and group people with specific diseases (HIV / AIDS, leprosy, epilepsy, Alzheimer's, alarmed) [11,12].

Recently, for recorded or live show broadcasting programs censored by giving voice beep at the words that cuss words are rude, abusive or insulting. However, it is hard to be broadcast with a censor for live show. This is a concern for the Indonesian Broadcasting Commission (KPI) that presented by Nina Mutmainah, Member of the Central KPI Coordinator cum Releases Fill in KPI Discussion with Trans Corp Group at Trans Corp Building, on Monday, February 13, 2012 [13]. For that, it takes intelligent software capable of censoring the words by automatically and real time. In order to fulfill this concern, the proposed system must be able to know in advance the unspoken words including slang words or not. One of the ways to recognize the unspoken words can be done in classifying words, what is included as rude swear words or properly words by using SVM.

2. Design System

Design of the proposed system for this application can be view in Figure 1. It has two processes, which are training process and testing process. This application input was in the form of words or a phrase that consisting of two words. In the training process, either rude or properly words extracted features by taking the pitch of each word. Those values then carried out by using SVM training. Result of the training process is used for classification in the testing process by SVM method. If it is classified as a spoken rude word, then the word will be replaced with a beep sound.

3. Experiments

Experiments carried out by recording 27 people which aged of 19 to 21 years old, consisting of 19 men and 8 women. Recording is done with an average sampling frequency of 44100 Hz, mono channel, and 16-bit resolution. Everyone saying 31 rude swear words which consisting of 25 words and 6 phrases, and 31 properly words which consisting of 25 words and 6 phrases. Each speech data sought his pitch values as input in SVM. Examples of spoken words can be seen in Table 1.

Table 1. Example an Indonesian spoken words and phrases

No	Rude Swear Word or Phrase	Properly Word or Phrase
1	Bajingan	Kemeja
2	Bangsat	Pulang
3	Keparat	Minum
4	Tolol	Manis
5	Brengsek	Hallo
6	Setan Alas	Sampai Jumpa
7	Otak Udang	Sepeda Motor
8	Dasar Sinting	Selamat Pagi

Experiments were conducted to distinguish male and female voice. Each sound category also distinguished between cuss words which consisting of one word and the phrase. Each experiment has different words of data. The data used as training data is also different from testing data. A positive class in the form of words or phrases is categorized as rude or curse word, while a negative class are categorized as properly word.

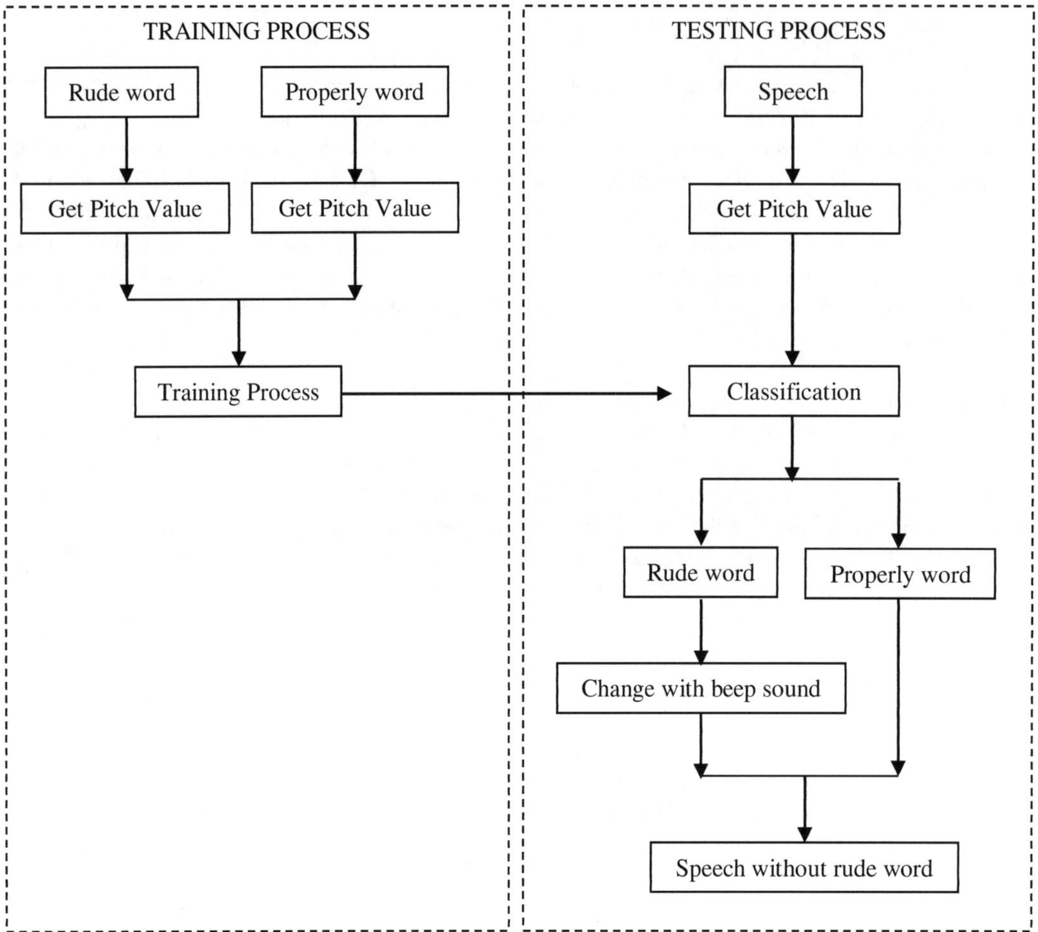

Figure 1. Design system

Tests are conducted to have several scenarios, which are described in section 3.1 until 3.3.

3.1 First Scenario
Testing in the first scenario is done by using the same amount of training data and the testing data. There are 25 training data which consisting of 12 positive classes and 13 negative classes and testing data which consisting of 13 positive classes and 12 negative classes. For each gender, the data carried out for 5 times experiments. The data used in each experiment is different. Table 2 shows the results of first scenario.

3.2 Second Scenario
The second scenario used the same amount of training data and the testing data, which is six phrase data with the amount of data of each class are varies, ranging between 2-4 good phrases. Experiments for each gender data also done for five times. Table 3 shows the results of second scenario.

3.3 Third Scenario
The third scenario used the amount of training data that is equal to a quarter of the testing data and vice versa. Experiments done only for two times for each gender. The third scenario experiment results can be seen in Table 4.

Table 2. First scenario experiment result

Voice	Experiment	Sum of Error Data	Degree of Accuracy (%)
Woman	1	6	76
	2	3	88
	3	5	80
	4	7	72
	5	4	84
	Average		80
Man	1	5	80
	2	1	96
	3	6	76
	4	7	72
	5	8	68
	Average		78,4

Table 3. Second scenario experiment result

Voice	Experiment	Training Data			Testing Data			Sum of Error Data	Degree of Accuracy (%)
		K +	K -	Σ	K +	K -	Σ		
Woman	1	3	3	6	3	3	6	1	83,3
	2	4	2	6	2	4	6	0	100
	3	2	4	6	4	2	6	2	66,7
	4	3	3	6	3	3	6	2	66,7
	5	2	4	6	4	2	6	2	66,7
		Average							76,68
Man	1	3	3	6	3	3	6	0	100
	2	4	2	6	2	4	6	0	100
	3	2	4	6	4	2	6	1	83,3
	4	3	3	6	3	3	6	0	100
	5	4	2	6	2	4	6	2	66,7
		Average							90

Table 4. Third scenario experiment result

Voice	Experiment	Training Data			Testing Data			Sum of Error Data	Degree of Accuracy (%)
		K +	K -	Σ	K +	K -	Σ		
Woman	1	5	5	10	20	20	40	11	72,5
	2	20	20	40	5	5	10	2	80
Man	1	5	5	10	20	20	40	3	92,5
	2	20	20	40	5	5	10	2	80

4. Analysis Results

Each experiment has different words of data. Example in Table 2, for women voice in experiment 1 and experiment 3, although the number of training data and testing data is equal, but the error data is different. Table 5 and Table 6 show the words of data that used for training data and testing data.

Table 5. The training data for experiment 1 and experiment 3

No	K^+ / Rude word ($\sum$ =12)		K^- / Properly word ($\sum$ =13)	
	Experiment1	Experiment 3	Experiment1	Experiment 3
1	Bajingan	Sialan	Hallo	Kemeja
2	Idiot	Laknat	Pagi	Celana
3	Bangsat	Jahanam	Bagus	Baju
4	Kunyuk	Iblis	Baik	Mau
5	Perek	Gembel	Cantik	Obat
6	Pecun	Setan	Pintar	Minum
7	Jablay	Brengsek	Saya	Pulang
8	Bencong	Keparat	Aku	Manis
9	Bego	Budek	Kamu	Sayang
10	Goblok	Bolot	Kita	Ganteng
11	Geblek	Sarap	Makan	Siapa
12	Bodho	Tolol	Ayo	Dimana
13	-	-	Mau	Kemana

Table 6. The testing data for experiment 1 and experiment 3

No	K^+ / Rude word ($\sum$ =13)		K^- / Properly word ($\sum$ =12)	
	Experiment1	Experiment 3	Experiment1	Experiment 3
1	Sialan	Bajingan	Kemeja	Hallo
2	Laknat	Idiot	Celana	Pagi
3	Jahanam	Bangsat	Baju	Bagus
4	Iblis	Kunyuk	Kemana	Baik
5	Gembel	Perek	Obat	Cantik
6	Setan	Pecun	Minum	Pintar
7	Brengsek	Jablay	Pulang	Saya
8	Keparat	Bencong	Manis	Aku
9	Budek	Sinting	Sayang	Kamu
10	Bolot	Goblok	Ganteng	Kita
11	Sarap	Geblek	Siapa	Makan
12	Tolol	Bodho	Dimana	Ayo
13	Sinting	Bego	-	-

Pitch value of rude word is greater than the pitch value of properly word. It is happened either male or female voice. It can be occurred because if we utter a rude word usually more loader than if we utter a properly word.

From Table 2 shows, that the average for male voices produces accuracy levels of 78.4% and for women 80% accuracy. From Table 3 shows that the average for male voices produce accuracy levels of 90% and for female 76.68% accuracy. An error in the classification is likely caused by the pitch value data of value of zero. A value of zero may occur due to any of the deduction at the time of the recording process. In this case, the cut is more advanced. Therefore, the initial frequency is the frequency that obtained before someone speaks. In addition, to sample data word phrases are relatively small (only 6) so that the training process hyper plane formed cannot cover the data for the data that being tested.

The different amount of training data and testing data produce different level of accuracy. Smaller training data cause lower level of accuracy and vice versa. This occurs because the more training data form perfectly SVM hyper plane. The recent condition is contrast to male voice data. This occurs because the attributes of male voices are more than female.

5. Conclusion

Based on the discussion it can be concluded that the SVM method can be used to develop an automation systems censor speech for the Indonesian rude swear word by classify speech into rude swear and properly words with high accuracy, in the ranges of 72% up to 95%. High pitch tone or a greeting word value can be used as a feature or an attributes for classification process.

References

[1]. Moore R 1994 Twenty things we still don't know about speech (Proc.CRIM/ FORWISS Workshop on Progress and Prospects of speech Research and Technology) p 9
[2]. Cristianini N and Taylor J S 2000 An Introduction to Support Vector Machines and Other Kernel-Based Learning Methods (Cambridge Press University)
[3]. Hsu C, Chang C, and Lin C 2008 A Practical Guide to Support Vector Classification. Department of Computer Science (Natural Taiwan University Taiwan)
[4]. Guo G and Li S 2003 Content Based Audio Classification and Retrieval by SVMs IEEE Trans. Neural Networks 14 209-15
[5]. Irtaza A, Jaffar M and Mahmood T 2014 Semantic Image Retrieval in a Grid Computing Environment Using Support Vector Machines The Computer Journal 57 205-16
[6]. Widyanto M, Endah S and Hirota K 2010 Human Behavior Classification Using Thinning Algorithm and Support Vector Machine Journal of Advanced Computational Intelligence and Intelligent Informatics 14 28-34
[7]. Nugroho A, Witarto A and Handoko D 2003 Application of Support Vector Machine in Bioinformatics (Proceeding of Indonesian Scientific Meeting in Central Japan, December 20, Gifu-Japan)
[8]. Sloin A et al. 2008 Support Vector Machine Training for improved Hidden Markov Modeling, IEEE Transactions on Signal Processing 56 172-88
[9]. Kocsor A et al. 2004 Kernel Based Feature Extraction with a Speech Technology Application, IEEE Transactions on Signal Processing 52 2250-63
[10]. Soekarnoputri M (President of Indonesia) 2002 Law No 32 of 2002 about Broadcasting. Reglement of the Republic of Indonesia in 2002 No 139 (Jakarta)
[11]. Sendjaja S 2009 The Indonesian Broadcasting Commission Regulation No 02/P/KPI/12/2009 about Broadcasting Code of Conduct (Jakarta)
[12]. Sendjaja S 2009 The Indonesian Broadcasting Commission Regulation No 03/P/KPI/12/2009 about Standard Broadcasting Program (Jakarta)
[13]. RG The Indonesian Broadcasting Commission -Trans Corp Discussion: Avoid the Sensitive Broadcasting, http://www.kpi.go.id/index.php?option=com_content&view=article&id=30378%3Akpi-hindari-tayangan-sensitif&catid=14%3Adalam-negeri-umum&lang=id, 7 March 2012

IOP Conf. Series: Materials Science and Engineering **190** (2017) 012040 doi:10.1088/1757-899X/190/1/012040

Noise Reduction in Breath Sound Files Using Wavelet Transform Based Filter

M F Syahputra[1], S I G Situmeang[2], R F Rahmat[1], R Budiarto[3]

[1] Department of Information Technology, Faculty of Computer Science and Information Technology University of Sumatera Utara, Medan, Indonesia
[2] College of Engineering, National Chung Cheng University, Taiwan
[3] Department of Information System College of Computer Science and Information Technology, Albaha University, Saudi Arabia

E-mail: nca.fadly@usu.ac.id, samuel103m@cs.ccu.edu.tw, romi.fadillah@usu.ac.id, rahmat@bu.edu.sa

Abstract. The development of science and technology in the field of healthcare increasingly provides convenience in diagnosing respiratory system problem. Recording the breath sounds is one example of these developments. Breath sounds are recorded using a digital stethoscope, and then stored in a file with sound format. This breath sounds will be analyzed by health practitioners to diagnose the symptoms of disease or illness. However, the breath sounds is not free from interference signals. Therefore, noise filter or signal interference reduction system is required so that breath sounds component which contains information signal can be clarified. In this study, we designed a filter called a wavelet transform based filter. The filter that is designed in this study is using Daubechies wavelet with four wavelet transform coefficients. Based on the testing of the ten types of breath sounds data, the data is obtained in the largest SNRdB bronchial for 74.3685 decibels.

1. Introduction

To make diagnosis of a human respiratory system, health practitioner or doctor uses a device called a stethoscope. With it, medical practitioners can hear the sound of the lungs to diagnose a person's physical condition and health. Breath sounds heard through the stethoscope, is still mixed with other noises like the sound of the heart, skin friction with the stethoscope, and chest movements. It is difficult for health practitioners to diagnose a person's physical condition and health. In signal processing, other voices or extraneous noise is commonly called signal interference or noise.

The development of science and technology in the field of healthcare increasingly provide convenience in diagnosing respiratory system problem. Breath sounds recording is one example of these developments. Breath Sounds are recorded using a digital stethoscope, and then stored in a file with sound format. As with conventional methods, respiratory sounds are not free from interference signals. Therefore, filter noise or signal interference reduction is required so that breath sounds components that contain information signal can be clarified. Research on reducing signal interference had been done since the last few years, but has not been able to give a satisfying result.

Previous studies related to respiratory noise filter, including the separation of heart and lung sounds from the breath sounds with modified spectro-temporal representation [1], and reduction of heart sounds from lung sounds recordings using adaptive filter [2] have been carried out.

Furthermore, researches on the wavelet transform including separation of discontinuous adventitious sounds from vesicular sounds using wavelet transform stationary-non-stationary filters [3], the separation splits the non-stationary input signals (heart sounds) from the stationary part (lung sounds) using a stationary-non-stationary wavelet transform filters [4], and the reduction of heart sounds from lung sounds have been recorded at low and medium flow rate using wavelet transform based filter [5] have also being done.

In this research, we applied wavelet transform based filter to reduce interference signals and white noise in the breath sounds recording, so we get a clearer signal without losing important information contained in the signal. With breath sounds recording filtered, health practitioner or doctor will get more detailed information for a better diagnosing method.The difficulty of diagnosing the disease through the breath sounds recording is that it is mixed with other noises which makes it difficult to diagnose a person's physical condition and health. A wavelet transform based filter is used with wavelet daubechies and global threshold in the breath sounds recording to produce a better breath sounds signal for diagnosis of diseases in the respiratory. In short, this work attempts to reduce noise so that the breath sounds recording can be used to provide a better diagnosis.

2. Methodology

2.1 Respiratory sounds

Respiratory sounds are all sounds related to respiration including breath sounds, adventitious sounds, cough sounds, snoring sounds, sneezing sounds, and sounds from the respiratory muscles. Voiced sounds during breathing are not included in respiratory sounds. [6].

Respiratory sounds can be classified into two groups: breath sounds and adventitious sound (abnormal) [7]. Breath sounds produced from healthy subject's chest called normal breath sounds. Normal breath sounds include both the inspiration and expiration. Both occur when the air moves in and out during regular breathing cycle. Adventitious sounds are additional respiratory sounds in the breath sounds. This sound occurs unexpectedly during regular breathing cycle.

There are several types of abnormal breath sounds. The four most common types are rales, rhonchi, wheezing, and stridor [8]. Rales are small clicking, bubbling, or rattling sounds in the lungs. They are heard when a person breathes in (inhales). And are believed to occur when air opens in closed air spaces. Rales can be further described as moist, dry, fine, and course. Rhonchi are sounds that resemble snoring. They occur when air is blocked or air flow becomes rough through the large airways. Wheezing are high-pitched sounds produced by narrowed airways. They are most often heard when a person breathes out (exhales). Wheezing and other abnormal sounds can sometimes be heard without a stethoscope. Stridor is wheeze-like sound heard when a person breathes. Usually it is due to a blockage of airflow in the windpipe (trachea) or in the back of the throat. Fort the characteristics of Lung Sound and Noise, the peak sound of normal lung is usually found at frequencies below 100 Hz, where lung sound energy decreases sharply between 100-200 Hz, but can still be detected at or above 800 Hz with a sensitive tool [2]. According to Earis & Cheetham [9], noises such as the sound of the respiratory muscles, chest motion sounds, heart sounds, and other low-frequency noise, have a frequency between 50 to 150 Hz [9].

2.2 Wavelet transform

Fourier method specifies only the spectral content of a signal in the frequency domain. The disappearance of time information during Fourier transformation for preservation during the incident is not considered. This condition can be ignored if the signal is stationary. However, for stationary signals such as speech, time and frequency data is important to avoid significant loss of information in the signal. Wavelet analysis can be used as an alternative method to solve the problem in Fourier method [10]. By using the concept of multi resolution wavelet analysis (for example, the representation of time and frequency scaling) to produce precise decomposition of the signal to obtain an accurate representation, detailed characteristics such as small discontinuities, similarity, and even higher order derivation hidden by conventional Fourier analysis can be revealed.

Wavelet is a family of functions $\psi_{a,b}(t)$ derived from a base wavelet $\psi(t)$, called the "mother wavelet", by dilation and translation [11], as shown in (1) as an example.

$$\psi_{a,b}(t) = \frac{1}{\sqrt{a}}\psi\left(\frac{t-b}{a}\right), a > 0, b \in \Re \quad (1)$$

Wavelet analysis is basically shifting and scaling a limited form of energy called the *mother wavelet* $\psi(t)$ of the desired signal. So, that the discrete wavelet transform can be written as in (2).

$$\psi_{j,k}(t) = 2^{j/2}\psi(2^j t - k) \quad (2)$$

2.3 Signal-to-noise ratio

Signal-to-noise ratio can be generally defined as the dimensionless ratio of signal power to the noise power contained in a record (3) [12].

$$SNR = \frac{P_{signal}}{P_{noise}} = \left(\frac{A_{signal}}{A_{noise}}\right)^2 \quad (3)$$

where

P_{signal} = mean of signal power
P_{noise} = mean of noise power
A_{signal}= *root mean square* (RMS) of signal amplitude
A_{noise} = *root mean square* (RMS) of noise amplitude

2.4 General architecture of the proposed method

Explanation of the components contained in the general architecture as shown in figure 2 is as follows:

a. Respiratory voice recording file is used as input. Respiratory sound is a combination of sound lungs and signal interference (noise).
b. Sounds can be played and printed out in the form of a signal.
c. Then respiratory sound is read by system and decomposed into an array of type byte and stored in the data [].
d. Array Data [] is converted into the form of an array of type double.
e. Array Data [] repeatedly decomposed according to the specified level of decomposition, produces two arrays, each array having half of the length of the data array []. The first array called a low pass filter and a second array called a high pass filter.
f. In each array, apply wavelet transform to the coefficients.
g. Both arrays are reassembled in the data array [] with a low pass filter placed in the first half and the high pass filter is placed at the end of the half.
h. Array Data [] is passed through a threshold, resulting in two arrays, array and the array of respiratory sound signal noise.
i. Perform repeatedly reconstruction as many as the level of reconstruction that has been assigned to each array.
j. Change the order in the data array [] of the previous half and half low pass filter high pass filter, be alternating low pass filter high-pass filter to each array.
k. Reapply the wavelet transform coefficients of each array.
l. Array Data [] and is then converted from an array of type double into an array of type byte. Audio formats and file name that has been set, is applied to the data [].
m. An array of data [signal] breathing sounds is rebuilt into a breathing sound file and the data array [] noise rebuilt into a beam of noise.

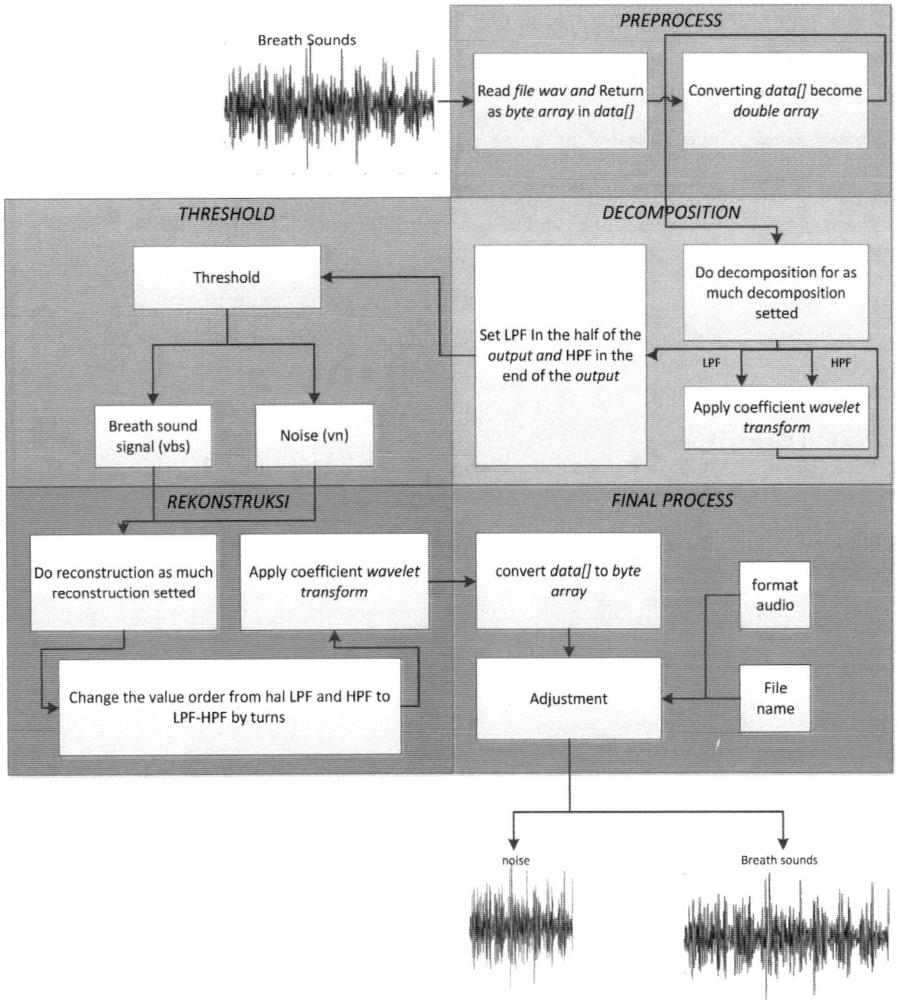

Figure 1. *General Architecture*

3 Experiments, Results and Discussion

3.1 Data

The data used in this research is breath sounds records, available from Littmann company. The existing data can be viewed as 10 different sounds with 10 different type of sounds. The data is described as: (*type, source*) = (bronchial, left lower lobe), (coarse crackles, right lower lobe), (fine crackles with deciduous bronchial, right middle lobe), (fine crackles, lung basis), (inspiratory stridor, tracea), (normal tracheal, tracea interscapular), (normal vesicular, right and left lower lobe), (pleural friction, right middle lobe), (rhonchus, right lower lobe), (wheezing, left lower lobe).

3.2 System analysis

Systems analysis aims to identify the system's development. Analysis is needed as a basis for system design. In this study, there is a five preprocess stage which is decomposition of the signal, the threshold stage, reconstruction stage, and the final stage process.

To test the designed system, a scenario is prepared to measure the sound of respiratory filter and the impact of the system with the objective and subjective criteria. Objective criteria is the signal-to-noise ratio (SNR) and signal display. While subjective criteria is the human auditory signal and the noise which is being observed.

A breathing sound recording of data, ie, bronchial signal used in this work is shown in Figure 2.

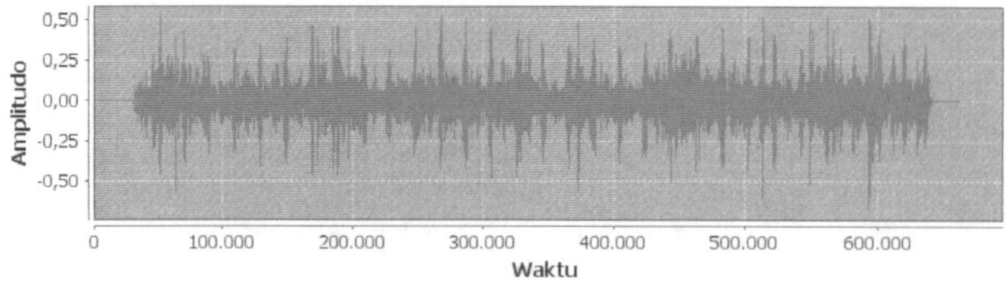

Figure 2. Bronchial sound signal before reduced

Measuring processes of noise reduction using wavelet transform based filter to the data in Figure 2 are as follow.

1. Change the data into a byte array type. After the array of type byte obtained, change again into the array of type double.
2. Perform second level decomposition of the data.
3. Perform the threshold stage using the following equation:

$$t = \sigma.\sqrt{2logN} \quad (4)$$

4. Perform level 2 reconstruction to the data.
5. Change the data back into the form of a byte array type.

Figure 3 and Figure 4 exhibits the filtered Bronchial signal and the noise of Bronchial signal of Figure 2, respectively.

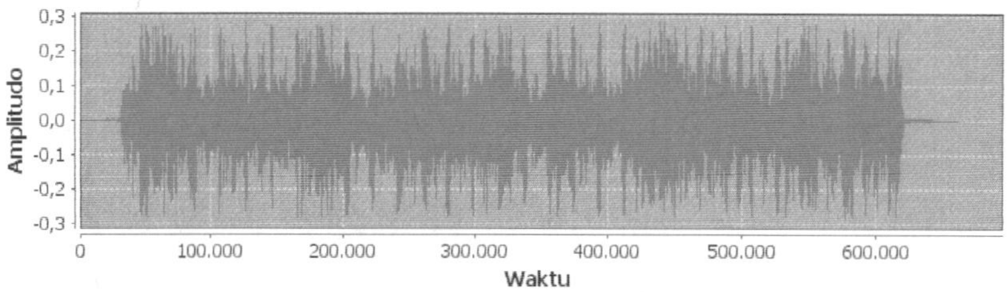

Figure 3. Bronchial signal after through filter

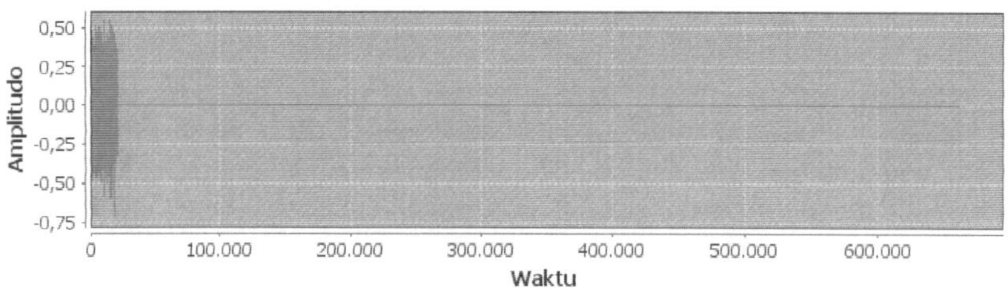

Figure 4. Noise signal after through filter

3.3 Results

Experiment is done to investigate the signal-to-noise ratio (SNR) of each data. The results are shown in Table 1.

Table 1. Noise Reduction Testing Results of Breath Sounds

No.	Type	SNR	SNR_{dB}
1.	Bronchial	2.7343069662486665E7	74.36847268923987
2.	Coarse Crackles	3703.999776899587	35.68670951851344
3.	Fine Crackles with Deciduous Bronchial Sound	1136888.1450540074	60.557177378843406
4.	Fine Crackles	333675.91665527434	55.23324862210189
5.	Inspiratory Stridor	223332.40934887607	53.48951751195278
6.	Normal Tracheal Sound	551447.7650258812	57.41504380900588
7.	Normal Vesicular Sound	15976.861637033	42.03491474128431
8.	Pleural Friction	4389520.347856363	66.42417066561016
9.	Rhonchus	294575.37479824974	54.69196438921298
10.	Wheezing	686282.2578168823	58.365027713927475

4 Conclusion

The experimental results showed that the biggest SNR_{dB} value is 74.36847268923987 decibel which is obtained from the reduction of bronchial. If the threshold value is too large, the noise cannot be reduced from the breath sounds signal. Conversely, if the threshold value is too small, breath sounds signal will loose a lot of important information.

As for our future work, we plan to utilize machine learning methods in order to increase the accuracy of the breath sounds analysis. In addition, we are not going to use the filters that have the level of decomposition and reconstruction greater than 2.

References

[1] Falk T H and Chan W 2008 Modulation filtering for heart and lung sound separation from breath sound recordings *Proc. the 30th Annual Int. Conf. the IEEE Engineering in Medicine and Biology Society*, p 1859-1862

[2] Sukresno F, Rizal A and Iwut I 2009 Reduksi suara jantung dari rekaman suara paru-paru menggunakan filter adaptif dengan algoritma recursive least square *Proc. SENTIA 2009*, p A1-A7

[3] Hadjileontiadis L J and Panas S M 1997 Separation of discontinuous adventitious sounds from vesicular sounds using a wavelet-based filter *IEEE Transaction on Biomedical Engineering* **44** 1269-1281.

[4] Hadjileontiadis L J and Panas S M 1998 A wavelet-based reduction of heart sound noise from lung sounds *Int. Journal of Med. Informatics* **52** 183-190

[5] Hossain I and Moussavi Z 2003 An overview of heart-noise reduction of lung sound using wavelet transform based filter *Proc. the 25th Annual Int. Conf. the IEEE EMBS* p 458-461

[6] Sovijärvi A R A, Dalmasso F, Vanderschool J, Malmberg L P, Righini G and Stoneman S A T 2000 Definition of terms for applications of respiratory sounds *European Respiratory Review 2000* **10** 597-610

[7] Baydar K S, Ertuzun A and Kahya Y P 2003 Analysis and Classification of Respiratory Sounds by Signal Coherence Method. *Proc. the 25th Annual Int. Conf. the IEEE EMBS* p 2950-2953

[8] Schriber A 2011 Breath sounds. http://www.nlm.nih.gov/medlineplus/ency/article/007535.htm

[9] Earis J E and Cheetham B M G 2000 Current methods used for computerized respiratory sound analysis *European Respiratory Review* **10** 586-590

[10] Walker S L & Foo S Y 2003 Optimal wavelets for speech signal representations *Systemics, Cybernetics and Informatics* **1** 44-46

[11] Cohen A and Kovačevič J 1996 Wavelets: The mathematical background. *Proc. IEEE* p 514-522.

[12] Johnson D H 2006 Signal-to-Noise Ratio. http://www.scholarpedia.org/article/Signal-to-noise-ratio

IOP Conf. Series: Materials Science and Engineering **190** (2017) 012041 doi:10.1088/1757-899X/190/1/012041

Tongue Segmentation Using Active Contour Model

Saparudin[1], Erwin[2], Muhammad Fachrurrozi[3]

[1,3]Informatics Engineering, Computer Science Faculty, University of Sriwijaya.
[2]Computer Engineering, Computer Science Faculty, University of Sriwijaya.

Email: saparudin@unsri.ac.id; erwin@unsri.ac.id; mfachrz@unsri.ac.id

Abstract. Tongue is an organ of the human body for tasting sense. Healthy conditions can be known from observation of the surface tongue by an expert. Before analyzing the tongue, feature extraction process is needed to segment the tongue from image, so it is possible to develop an application that can segment the tongue image from opened mouth image. This research uses Canny Edge Detection and Active Contour method. Canny Edge Detection is used to find the edges of tongue. This method has four steps: Smoothing Gaussian Filter, Finding Gradients, Non-maximum Suppression, and Hysteresis Thresholding. After finding the tongue edge, Active Contour Model will be generating energy that can pull into edges curve that is already defined and cropping that to produce tongue image. Testing result of this research yield an accuracy rate of 75%, by which from all 40 tongue images, 30 are successfully segmented.

1. Introduction

Detection image of human organs is widely developed in the computer technology, especially the technology of image processing software. Several studies of organ detection system have been widely developed, such as iris detection, ear detection, face detection, and tongue detection. The tongue is an organ in the human body which has very important role in human survival. Tongue has a very detailed pathological of its surface and has a different contours of each individual. Figure 1 shows the image of a human tongue.

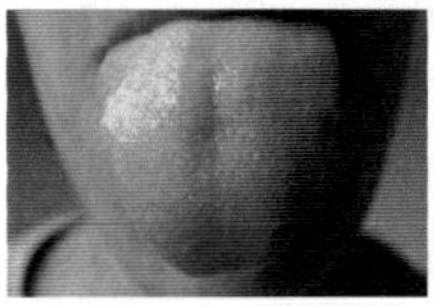

Figure 1. Image of Human Tongue

It is difficult to segment an object from the image which is reveal central part of face (mouth area). Because not only tongue but also lips, teeth, mouth itself, and others object are the objects of mouth area. The problem of this image segmentation is how to get the tongue among other objects in that mouth area.

One of the model usually used for the image segmentation is Active Contour Model [1][2]. Active contour model, also called snakes, is a curve that is used to trace the boundary object of image. This model is a contour detection model that able to locate accurately. This model can be used to segment an image by adjusting the contours and the edges, which is separate the object with others around it.

One of the previous study about tongue segmentation is discussed by [3], a study applying Active Contour Model that able to segment the tongue. In this study, the color image of middle face converted into grayscale image, the separating the tongue object from others object around it using thresholding. Active Contour model is applied on last process to obtain a complete image of the tongue [4]. Other research which focuses on tongue image as the object is a [5] research, adaboost method is applied to find tongue location, then color cancellation is used to clear the area around mouth and tongue. Active Contour Model is used in the final stage to produce a complete image of the tongue. This study used 300 test data of mouth area image, 282 successfully segmented image or has a success rate of over 90%.

2. Data Collecting

This research data is taken from Biometric Research Center website, Hongkong. The data of tongue image is provided into a sample database image of the tongue [5]. There are 12 samples of tongue image which is provided by the Biometric Research Center website. That sample images are given show the mouth with tongue clean condition.

3. Methods

The development of active contour models for tongue segmentation through this following steps: (1) conversion of color images to grayscale, (2) Canny edge detection, and (3) the implementation of active contour models at the end of the process. The first process will be applied in this study is grayscale because it can help to do binary image process in canny edge detection as the second process. The first thing to do is take the input color image then convert it into grayscale image. Every pixel on color images are summed and then divided by three as shown in Equation (1), in order to obtain an intensity values for grayscale images. This result will be used for the next step, canny edge detection [6].

$$S = \frac{R+G+B}{3} \tag{1}$$

Where,

S = value of grayscale;
R = value of red;
G = value of green;
B = value of blue;

Next, apply the canny method to detect the image edges. Canny method has four steps, consist of smoothing (Gaussian filter), finding gradient, non-maximum suppression and thresholding. It is possible if all data images have some noise on it. Noise can make the system doing something wrong when do edge detection. Therefore, the first step should be smoothed by using a Gaussian filter. The essence of the Gaussian filter is the standard deviation with the value of σ = 1.4 is shown in equation (2).

Equation (2) is a 5 x 5 convolution matrix Gaussian filter, with σ=1.4 :

$$g(x,y) = \frac{1}{159}\begin{bmatrix} 2 & 4 & 5 & 4 & 2 \\ 4 & 9 & 12 & 9 & 4 \\ 5 & 12 & 15 & 12 & 5 \\ 4 & 9 & 12 & 9 & 4 \\ 2 & 4 & 5 & 4 & 2 \end{bmatrix} \tag{2}$$

Basically, Canny algorithm for finding point of edge on grayscale image area with strong intensity contrasts, this area founded by using image gradient. The gradient of each pixel image has been applied by Sobel operator. The second step is to estimate the gradient in the x and y directions. This is shown in Equation (3) and (4).

$$K_{GX} = \begin{bmatrix} -1 & 0 & 1 \\ -2 & 0 & 2 \\ -1 & 0 & 1 \end{bmatrix} \tag{3}$$

$$K_{GY} = \begin{bmatrix} 1 & 2 & 1 \\ 0 & 0 & 0 \\ -1 & -2 & -1 \end{bmatrix} \tag{4}$$

Where,
KGX : sobel x filter
KGY : sobel y filter.

The edge can be obtained as distance using Pythagoras in equation (5). Gradient for x axis was symbolized as Gx while gradient for y axis as Gy.

$$|G| = \sqrt{{G_x}^2 + {G_y}^2} \tag{5}$$

Where,
G : the value of image gradient
Gx : the value of gradient for x axis
Gy : the value of gradient for y axis.

The image that has large gradient value often show quite clearly edges, but that condition may cause it is can not show exactly where the edges are. To determine the actual edges, the edges must be determined and stored as shown in Equation (6).

$$\theta = \arctan\frac{|G_y|}{|G_x|} \tag{6}$$

Where θ shows the value of the angle for edge position.

In the third step conversion of the blurred edges will be done, then sharper edges will be obtained. Basically this is done by keeping all local maxima in the gradient image and delete everything else.

By implementing energy-minimizing function, starting with the placement of deformable objects curve, snake energy find out the point image which has the most minimum energy with equation (7), it will make a curve locate to that point. In the end of the process, deformable curve will occupy the boundary of the defined object in that image, so information about the points of deformable curve can be determined.

Snake can be implemented as these following formula:

$$E_{snake} = \int_0^1 E_{snake}\big(V(s)\big)ds = \int_0^1 E_{int}\big(V(s)\big) + E_{image}\big(V(s)\big) + E_{con}\big(V(s)\big)ds \tag{7}$$

Where,
$E_{int}(V(s))$: internal energy
$E_{image}(V(s))$: image forces
$E_{con}(V(s))$: external energy

Snake energy is formulated from three types of energy, in the form of internal energy, external energy, and the image force. The function α (s) and β (s) in equation (8) affects the snake performance, such as membranes and thin container.
Internal energy is defined in this function:

$$Eint = (\alpha(s) \mid vs(s) \mid^2 + (\beta(s) \mid vss(s) \mid^2)/2 \quad (8)$$

Where,
α(s) : stretching function
β (s) : bending function

In the image force, there will use three types of functions, namely line, edge and terms are defined in equation (9).

$$Eimage = Wline + WedgeEedge + WtermEterm \quad (9)$$

Where,
Eline : line functional energy
Eedge : edge functional energy
Eterm : termination functional energy

Image force can be considered as a style on the image that will attract deformable curve to the desired feature. Line functional is the simplest kind of image force. Functional line was centered on the intensity of the image itself. Line functional defined in equation (10).

$$Eline = I(x, y) \quad (10)$$

Line functional will determine where are deformable curve would be located, if the value of the coefficient is positive then the deformable curve will locate toward the dark line, otherwise deformable curve will be locate to the bright line. Edge functional used to pull deformable curve toward a corner and is defined in equation (11).

$$Eedge = -\mid \Delta I(x, y) \mid^2 \quad (11)$$

Some of the external energy to be used in equation (12) and (13). Spring energy can be formulated as follows:

$$Ex = -k(l - l_0) \quad (12)$$

where k is a constant value of boundary, l is the distance of contraction or extension spring, and lo is the distance of spring relaxation.
Repulse energy can be defined as:

$$Ex = 1/d^2 \quad (13)$$

where d is the distance to a particular location.

4. Experiment Result and Analysis

Percentage (%) as the experiment result can be defined through equation (14):

$$\text{percentage (\%)} = \frac{Number\ of\ image\ category}{number\ of\ all\ testing\ image}\ x\ 100\ \% \quad (14)$$

From the percentage calculation using equation (14), 75% of percentage is obtained from all the tongue segmentation image, 70% of percentage is obtained from primary data, and 80% from the secondary data.

This research use 40 test images for the experiment and produce the result as follows:

a. 30 of the test images are successfully segmented;
b. 14 of the primary test images are successfully segmented;
c. 16 of the secondary test images are successfully segmented;
d. 10 of the test images are unsuccessfully segmented;

Based on the experiment results, the software can be successfully run as the test planning that already define before. From the scenario experiment, a conclusion that show the test images (noisy image and non-noisy image) which is can accurately segmented or not will be obtained. The test images are used must be clearly images, it means the image does not have other objects besides that tongue itself because its can disturb when entered the edge detection process. The distance when test images are taken also determine the success of segmented or not, if the distance exceeds from the mouth area, other objects such as a mustache or beard would be captured and it may affect when entered the edge detection process, it means the edge tongue will not be detected. Configuration of Double thresholding also affect the final result of segmentation, the configuration of upper 30 and lower 10 do not produce too much and thick edge so the result is quite accurate. Otherwise the configuration of upper 15 and lower 5 produces very large and thick edge that it disturbs the segmentation process, it may affect the end result (inaccurate result).

5. Conclusions

1. Tongue Segmentation Software that implements Active Contour Model using mouth area image (as the center part of face) has been successfully developed.
2. This developed software is able to segment human tongue. The accuracy rate shows how successful the experiment to be done and it is obtained from 40 test images, they are: 75% for all the tongue segmentation image, 70% from primary data, and 80% from the secondary data.
3. The test images which have noise or other objects (besides tongue) will affect the edge detection process, it may make the image cannot segmented.

Acknowledgments

Thank you very much to the all reviewers for reviewing to complete this article. The article is supported by Unggulan Perguruan Tinggi Kementerian Riset Teknologi dan Pendidikan Tinggi Tahun 2016 research scheme and Sriwijaya University.

6. References

[1] X. Wang, C. Ballangan, and D. Feng, "Elastic & Efficient Three-Dimensional Registration for Abdominal Images," 2006.

[2] D. Zhang, "Combination of Polar Edge Detection and Active Contour Model for Automated Tongue Segmentation," *Third Int. Conf. Image Graph.*, vol. 1, pp. 270–273, 2004.

[3] Y. Yang and Z. Hong, "Medical segmentation using sobolev optical flow," *Proc. 4th Int. Conf. Image Graph. ICIG 2007*, pp. 432–436, 2007.

[4] J. Ning, D. Zhang, C. Wu, and F. Yue, "Automatic tongue image segmentation based on gradient vector flow and region merging," *Neural Comput. Appl.*, vol. 21, no. 8, pp. 1819–1826, 2012.

[5] Z. Fu, W. Li, X. Li, F. Li, and Y. Wang, "Automatic tongue location and segmentation," *ICALIP 2008 - 2008 Int. Conf. Audio, Lang. Image Process. Proc.*, pp. 1050–1055, 2008.

[6] Z. Xu, K. Yuan, and W. He, "An implementation method of Canny edge detection algorithm on FPGA," *2011 Int. Conf. Electr. Inf. Control Eng. ICEICE 2011 - Proc.*, no. 60875051, pp. 3958–3962, 2011.

IOP Conf. Series: Materials Science and Engineering **190** (2017) 012042 doi:10.1088/1757-899X/190/1/012042

A Framework for Remote Monitoring System

Cahyo Crysdian

Jurusan Teknik Informatika, Universitas Islam Negeri Maulana Malik Ibrahim Malang

cahyo@ti.uin –malang.ac.id

Abstract. Remote monitoring system becomes an important facility to support observation activities for various natural disasters. In many incidents of natural disaster such as volcano eruptions, the available monitoring systems installed closely to disaster area were damaged due to extreme condition raised by the event. The temperature of disaster site could suddenly increase to hundred degrees of Celsius, drowned in a water flood or even trapped in a toxic heating gas. Therefore, it is important to have observation facility that is installed far away from disaster area. This research is an exploratory study to develop the framework for remote monitoring system. It includes hardware requirement and algorithm definition that cover system lenses and a set of image processing algorithm. The framework delivers a promising preliminary result towards the effort for remote monitoring system development.

1. Introduction

The last decade becomes the witness of natural disasters hitting Indonesia continuously. Floods, landslides, hurricanes, earthquakes, volcanic eruptions, forest fires, even tsunami wave came and went. Various sources noted that the total losses caused by natural disasters has reached the value of trillions of rupiah and thousands of deaths. In addition to being influenced by natural factors, the damage also influenced by the unpreparedness of anticipating and managing natural disasters. It is closely related to the lack of early detection and observation facilities that has been installed and operated. Many cases demonstrate the absence of early detection facilities and amenities observation led to substantial material losses and claimed many casualties, like the tsunami in Aceh at 2004 and in Cilacap at 2006.

For facilities that have available observations, some cases even show damage to the equipment installed in the disaster area, such as when the eruption of Mount Kelud in 2007. Damage to the tool is primarily due to the extreme conditions of the disaster area such as the emergence of natural gas heat as well as the surface temperature of the earth. Therefore, the facility of early detection and observation that installed at a considerable distance from the disaster site yet delivers a good observation result become indispensable.

Meanwhile, the need for remote facility of monitoring system is increasing by nowadays. This is triggered by the magnitude of the benefits that can be generated by this facility, primarily as a tool for observing catastrophic incident or natural phenomena. This system could be used to reveal a variety of natural phenomena that need timely and long-range observation [1,2]. Such natural phenomena are often found in Indonesia, for instances are the migration of groups of birds crossing the oceans and the development of a variety of wildlife, or even the happening of various natural disasters which have claimed many casualties as described above.

2. Motivation

The Indonesian archipelago, located between the edge three plates of the earth namely the Pacific, Eurasia, and Australia plate. This country has also about 60 volcanoes that are still active, 17,000 islands, vast forests, and inhabited by diverse flora and fauna. The geographic conditions make the region of Indonesia has a rich collection of flora and fauna, has a nice view, but is susceptible to natural disasters such as volcanic eruptions, forest fires, hurricanes, tsunamis, earthquakes, and others. Therefore, in relation to the exploration of natural resources and the efforts to minimize vary losses which can be caused by these natural disasters, the facility that can be used to monitor a variety of natural phenomena in Indonesia is urgently required. The system must be able to be used to observe terrestrial objects over long distances considering many natural phenomena that cannot be monitored from close range.

Nowadays various attempts have been made by researchers to build digital monitoring system such as afforded by Yao et al. [3,4] and Shirvaikar [8]. However, the existing monitoring system is capabe to only monitor terrestrial object in the order of hundreds of meters with blurred output quality and has narrow view angle. When it is compared with the distance required to monitor natural disaster such as volcano eruption, the existing condition is still far below the expectation. Although some systems have been built to reach the object in longer distance in the field of astronomy, such as the Hubble and several large observatories [9], such systems is not intended to observe terrestrial object in detail. Therefore, research activities to establish remote monitoring system with a capability to reach terrestrial objects in distant with good detail quality are indispensable.

3. Related Work

Remote monitoring system is an application of imaging that aims to retrieve data from a distant event in the form of pictures or visual data. Some factors that affect the performance of these systems have widely been known by researchers such as the focus, exposure, and enlargement of images [10,11]. To monitor an object in close range, focus, lighting, and image magnification can be handled with ease. So, to produce good quality of output for any objects of interest that are located closely to the monitoring facility is trivial. But not for distant objects, since objects that are located in distant would have small size when it is viewed from a monitoring point. Besides, the amount of intensities of the constituent components from a distant object is also relatively small compared to the objects that are closely located [5]. This leads to the difficulties to manage focus, exposure and image magnification for distant objects. Therefore, it is not surprising that a lot of effort to build a remotely digital monitoring system facing problems for recording visual data [4,8,11].

Various additional methods have also been proposed by researchers to solve the above problems, such as digital enlargement [6,7], parameter calibration image recording [8,11], and image restoration [4]. Nevertheless, the result of these methods still contains many weaknesses such as less sharpness of the output, increasing noise in visual data, and last but not least increasing complexity of the system. Meanwhile, Lintu and Magnor [5] has a unique approach towards solving this problem. The monitoring results that are unfavorable to distant objects is replaced by a set of images that are stored and have been available in the database. This method seems to become a breakthrough in the field of remote monitoring system, especially when the approach is associated with learning distant objects, for instance in the field of astronomy to study the structure of the galaxies. However, this approach is not appropriate to be implemented for terrestrial objects due to limitation of the system that is unable to display the output in real time, particularly when the object to be monitored constantly changing in timely basis. Few examples of these events are the activities to monitor natural events such as the migration of wild life animals or the happening of natural disasters.

4. Framework Development

Development of a framework for remote monitoring system has been conducted by considering the following aspects:

4.1. Data Acquisition

The parameter used to acquire visual data from real objects is the focus, magnification (zoom), lighting, and sensor size. The focus is used to collect all the light from a real object to the observation point. Focus is governed by a system of lenses. Enlargement is used to increase the size of captured images. Enlargement is affected by the longest focal distance divided by the shortest focal distance that can be reached by the lens system. The greater the result of this division, the stronger the magnification can be achieved. Lighting unit is used to ensure that captured objects emit enough light to be shaped into a digital image. In practice lighting unit is regulated by two mechanisms, namely the aperture size and the shutter speed. Aperture is the size of the hole that passes light to the lens system. The larger the hole, lighter intensities would be passed to a set of light sensor to shape the image. Meanwhile the shutter speed is the speed of opening and closing the curtain on the aperture. The faster a shutter is operated would generate dimmer digital image due to less light passed the lenses system. Sensor is a device used to capture and convert light into digital data in the form of images. The size of sensor affects the smoothness of image. Larger sensor would produce more delicate and higher resolution image.

4.2. Image Quality

To measure the quality of images quantitatively can be done by forming a histogram and see the spread of pixel values contained in an image. Values measured using this method is the value of color distribution, where a good digital image will show a clear difference in the distribution of colors for the main object and the background. In addition, a digital image quality can also be measured using a light intensity parameter, namely the amount of incoming light energy at the time the object is recorded. This is apparent from the degree of brightness of the image. To perform the measurement of light intensity of a digital image is done by changing the condition of the image into a single color, usually done by changing the color to gray level or referred to the grayscale image. Another way is by measuring the spatial resolution of the image, which measures the number of pixels making up the digital image. More number of pixels contained in digital image would show more detailed condition of real objects.

4.3. Image Zooming

Efforts to increase the size of digital image have been made by researchers such as research conducted by Morse and Schwartzwald [7], Yao et. al [3,4] and Sajjad et al [6]. Morse and Schwartzwald perform digital zoom using interpolation method by applying the level set to return the pixel value into its original condition. This method has only been tested on a low magnification, which is two to four times zooming and therefore it does not meet the need to enlarge the remote object which is at a distance of hundreds to thousands meters. Digital image magnification for long distances up to hundreds of meters is made by Yao et al [3,4]. This is done by arranging the optic device in hardware. However, the system displays blurred and trembling output. Other effort as performed by Sajjad et al [6] is only capable of displaying a mechanism with small magnification levels, i.e. four times magnification and has a shortage for full-color image enlargement.

4.4. Framework Design

Design framework that is done in this activity are shown in Figure 1, which consists of two stages as follows: The first stage is the data acquisition phase, which begins with preparing the platform for producing digital images mainly taken from a distance of hundreds to thousands of meters. Then the results of data acquisition would shape the characteristics of the developed remote monitoring system by comparing with the digital image taken from a short distance. The second stage is the main part of the study, namely by processing and enhancement of digital image from a distance using a variety of digital image processing algorithms for the purpose of enlarging the size of the image, so that objects acquired from hundreds to thousands of meters away from he observations point can be displayed properly. The first step that must be done is to determine the distribution of the color using the histogram to determine the difference between the existences of the main object with the background. This process if followed by the sharpening process of the main object by increasing the contrast value of the digital image. This is done by the interpolation process to enlarge the dimensions of the main object and the filtering process to clarify the edges of the main object. Result of the developed framework is depicted in Figure 1.

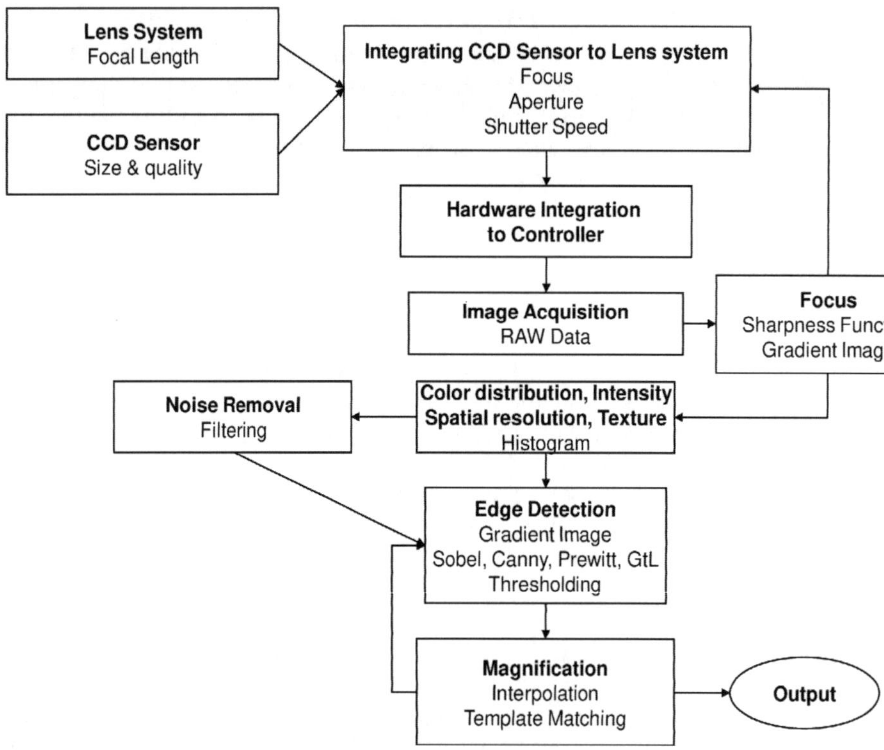

Figure 1. Developed framework for remote monitoring system.

5. Experiment & Discussion

Few experiments have been conducted to a partial component of the framework proposed above. The intention is to reveal any obstacle restricting the framework from reality. By employing a half top part of the framework for capturing distance objects compared to its close counterpart, some facts can be disclosed such as illustrated in Figure 2. Experimental results depicted in Figure 2 show the difference presentation of object captured from a long range and from a common distance of pocket camera. The differences include histogram and frequency domain presentations, in which some components are missing from the picture obtained from a distance. These facts disclose the cause why distance object lost its detail compared to its short-range counterpart.

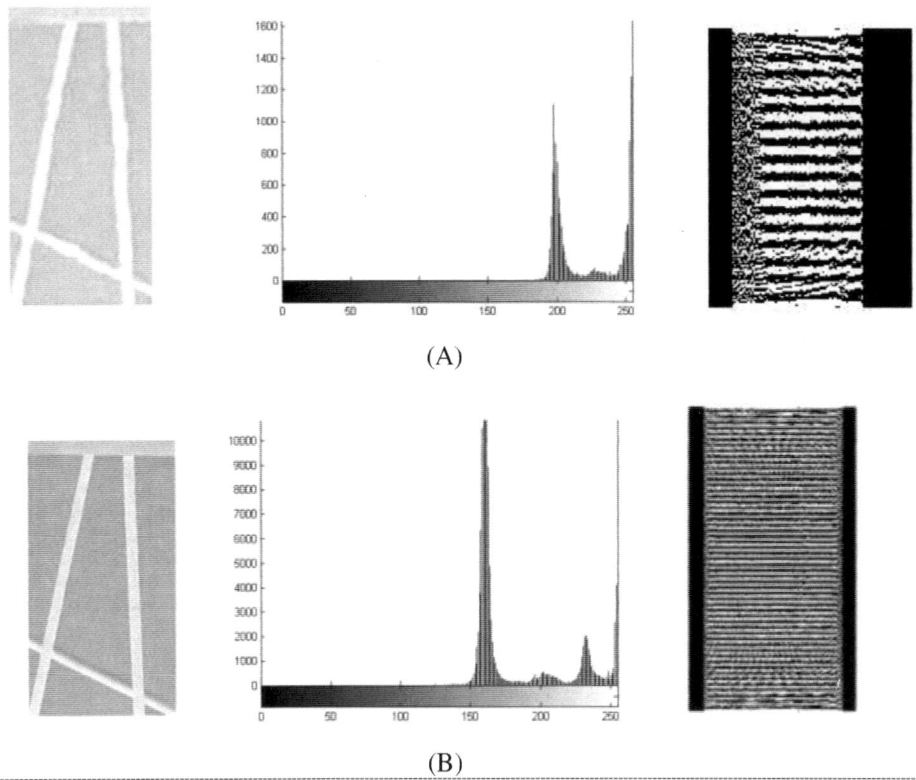

Figure 2. Experimental results. (A) Distance object (B) Short range object

6. Conclusion

The framework to build remote monitoring system has been developed in this study. It consist of two major stages namely lenses system and image processing. The first stage is to acquire digital image captured from real objects using three well-known parameters i.e. focus, aperture and shutter speed that are produced from integrating the focal length of lenses system and the size as well as the quality of image sensor. This stage is done in an image capturing device that equipped with lenses system. Result of captured image is then supplied to image processing to precede image with a set of algorithm which include focus sharpening, noise removal using a set of digital filters, detection of object boundary based on edge detections and image magnification algorithm based on interpolation. The algorithms also include histogram equalization and detection of spatial resolution in order to support the optimal focus obtained from distant object. The framework produced by this study become a promising preliminary result to further developed remote monitoring system based on visual data processing.

References

[1] BridLife. (2006). Monitoring of Wild Birds. BirdLife International, www.birdlife.org, 11 April 2006.

[2] Colozza, A; Doke, J.L. (2005). High-Altitude, Long-Endurance Airships for Coastal Monitoring. National Aeronautics and Space Administration, NASA Glenn Research Center, NASA/TM-2005-213427, gltrs.grc.nasa.gov, February 2005.

IOP Conf. Series: Materials Science and Engineering **190** (2017) 012042 doi:10.1088/1757-899X/190/1/012042

[3] Yao, Y; Abidi, B.R.; Abidi, M.A. Auto Focussing in Extreme Zoom Monitoring: A System Approach with Application to Faces. 2nd International Symposium on Visual Computing, Lake Tahoe, NV, November 2006.
[4] Yao, Y; Abidi, B.R.; Abidi, M.A. Extreme Zoom Monitoring: System Design and Image Restoration. Journal of Multimedia, Vol. 2, No. 1, February 2007.
[5] Lintu, A.; and Magnor, M. An Augmented Reality System for Astronomical Observations. Proceedings of the IEEE conference on Virtual Reality 2006.
[6] Sajjad, M.; Khattak, N.; Jafri, N. Image Magnification Using Adaptive Interpolation by Pixel Level Data-Dependent Geometrical Shapes. World Academy of Science, Engineering and Technology, 2007.
[7] Morse, B.S.; Schwartzwald, D. Image Magnification using Level-Set Reconstruction. Proceedings of the 2001 IEEE Computer Society Conference on Computer Vision and Pattern Recognition, 2001.
[8] Shirvaikar MV. An optimal measure for camera focus and exposure. Proceedings of the IEEE SSST 2004.
[9] NASA. Great Observatories – Beyond Einstein. NASA, http://universe.nasa.gov/program/observatories, 2007.
[10] Li, M.; Lavest, J.M. Some Aspects of Zoom Lens Camera Calibration. Technical Report ISRN KTH/NA/P—95/03—SE, Dept of Numerical Analysys and Computing Science, KTH (Royal Institute of Technology, February 1995.
[11] Vatani, N.N.; Roberts, J. Automatic Camera Exposure Control. Online Repository of en.scientificcommons.org, 2007.

Utilization of Digital Image Processing In Process of Quality Control of The Primary Packaging of Drug Using Color Normalization Method

Danang Erwanto[1], Sri Arttini Dwi Prasetyowati[2], Eka Nuryanto Budi Susila[2]

[1]Magister Teknik Elektro, Fakultas Teknologi Industri, Universitas Islam Sultan Agung – Semarang
[2]Fakultas Teknologi Industri Universitas Islam Sultan Agung – Semarang

Email: danang.erwanto@gmail.com

Abstract. In the process of quality control, accuracy is required so that the improper drug packaging is not included into the next production process. The automatic inspection system using digital image processing can be applied to replace the manual inspection system done by humans. The image captured from the vision sensor is RGB image which is then converted into grayscale. The process of converting RGB image into grayscale image is performed using the color normalization method to spread the data of RGB colors at each pixel. From the software of image processing using the color normalization method that have been created, it shows grayscale images on the drug object which have degrees of gray higher than the grayscale image section of the background when the degree of the R, G or B color of drug is higher than the degree of the R, G, B color on the background of packaging. The determination of threshold value indicates that the binary image of the drug is white and a binary image of the background of drug packaging is black.

Keywords: Color Normalization, Object of Drug, Thresholding, Object Counting, Outside Corners and Inside Corners.

1. Introduction

In the process of quality control, accuracy is required so that the drugs with improper or failed packaging are not included to in the next production process, e.g. the process of secondary packaging. However, the process of quality control which is done manually by humans has a weakness because it cannot be done continuously due to physical fatigue, added with problems of accuracy that allows some failure in primary packages that escapes the quality control process. Therefore, automatic inspection system can be applied to replace the manual inspection system done by humans.

Computer vision is mostly applied in many industries especially those requiring automatic inspection of the goods produced at each stage of the production line [1]. Computer vision is an automatic process integrating a large number of processes for visual perception, such as image acquisition, image processing, classification, recognition, and making a decision [2]. This research discusses about the use of digital image processing using color normalization method use to perform a visual inspection of primary packaging of drug blister shape using webcam as a sensor. Using the model of normalization, an object with a particular color can be detected and free from the influence of changes in light intensity from the outside [3].

2. Method

The steps implemented in this research are: environmental preparation, image capturing, grayscale, thresholding, objects counting, objects separation as shown in the flowchart in figure 1.

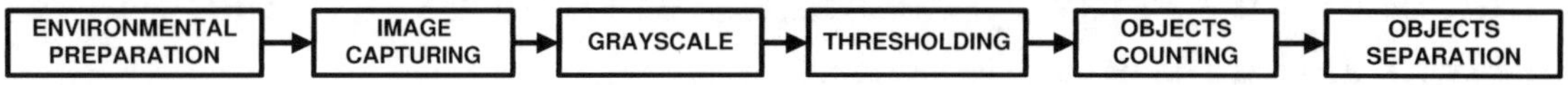

Figure 1. Research Steps

3. Results and Discussion

The environment of image capturing of the object needs to be conditioned so that the image of the object gained achieve ideal conditions. The environmental setting includes: a conveyor belt as background object is colored in contrast to the object, the lighting is set in order to have sufficient illumination and there is no influence of outside light that can disturb the object or the camera when the image capture process, reducing interference around the object. The velocity of the conveyor is set so that the object can be captured as well by a camera. It is set on 0.03 m / sec with PWM duty cycle of 230. Here the speed of the processing capture as well as the image processing per frame are affected by the processing speed of a processor.

In the stage of picture taking, image with a size of 320 x 240 pixels is taken using a webcam placed on the top of the object. Before carrying out to image acquisition, camera height position is set in such a way so that all areas of package can by a camera properly.

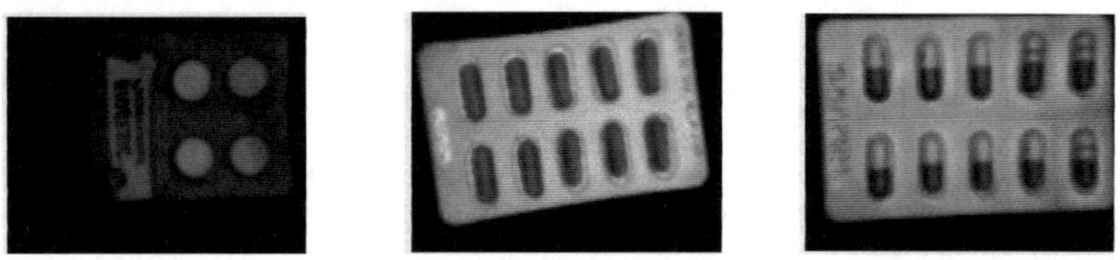

(a) (b) (c)

Figure 2. (a) RGB image of the first test samples, (b) RGB image of the second test samples, (c) RGB image of the third test samples

Images taken from image capture process generate an image in the form of an RGB image. To simplify the model of image, the RGB image to be converted into a grayscale image using the colors normalization method to parse the RGB data at each pixel. Colors Normalization in one method to determine the intensity of certain basic color of an image grayscale obtained from the comparison between certain base color intensity with a total intensity in each color channel by using the following equation:

$$r = \frac{R}{R+G+B}, = \frac{G}{R+G+B}, = \frac{B}{R+G+B} \quad (1)$$

Where R = component of red color of the RGB image, G = component of green color of the RGB image, B = component of blue color of the RGB image, r = red color after normalization, g = green color after normalization, b = blue color after normalization. Result from normalization causes the difference of illumination of each pixel which does not depend on the lighting when the image taking for their sum of $r+g+b=1$.

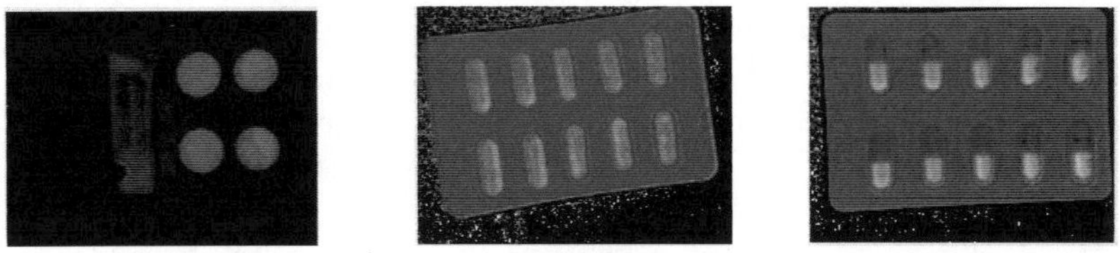

(a) (b) (c)

Figure 3. (a) Image grayscale result of the color normalization to red color on the first test sample, (b) Image grayscale result of the color normalization to green color on the second test sample, (c) Image grayscale result of the color normalization to blue color on the third test sample

The conversion of gray image into binary image (threshold) is done by determining the certain threshold value in order to obtain an error as small as possible. From this threshold value, it can be determined whether a pixel will be converted into black or white colors.

Threshold process is used to define the object and background information. In the threshold process, a pixel containing information object and having degree of gray value higher than the threshold value specified is then rated as 255 while pixels of the background and has a degree of gray value lower than the threshold value specified is rated as 0. Therefore, object of drug will be colored white and all the background of image will be colored black. The determination of giving value 0 and 255 in the binary image is done by using of the following equation,

$$g(x,y) = \begin{cases} 0 \; if \; f(x,y) < K_{Th} \\ 255 \; if \; f(x,y) \geq K_{Th} \end{cases} \quad (2)$$

Where g(x,y) = binary image results of the threshold process at the point (x,y). F(x,y) = the grayscale image was resulted from the colors normalization of at the point (x,y), KTh = threshold value. From the threshold test, the data is shown in Table 1.

Table 1. The result of threshold test to 3 sample

Number t est sample	*Grayscale* image result from c olor normalization	Image result from threshold process	Threshold va lue obtained
I	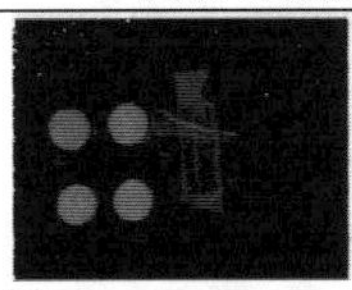 **Note :** Color normalization = red		0,3
II	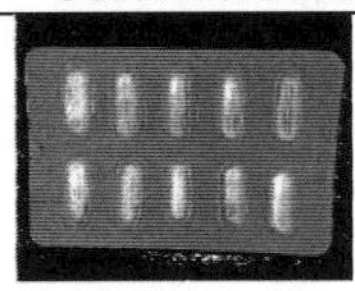 **Note :** Color normalization = green		0,4
III	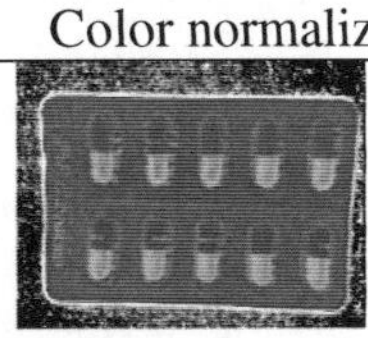 **Note :** Color normalization = blue		0,4

From the binary image generated at the stage of threshold process, number of objects can be calculated by using the total of possible outside and inside corners at a pattern of object by using the following equation of 4.

$$T = \text{abs}\left(\frac{a-b}{4}\right) \qquad (3)$$

Where T = number of objects, a = number of possible outside corners, b = the number of possible inside corner. From Equation 4, number of objects from the calculation results by the system (based on the calculation of outside and inside corner) should be rounded so that number of objects produced is not in a fraction.

Table 2 . The result of calculation from the first type of sample using digital image processing

Image Test Sample	**Binary Image**	**Number of Corner**		**Number of Drug**		**Note**
		Outside	**Inside**	**Before Rounded**	**After Rounding**	
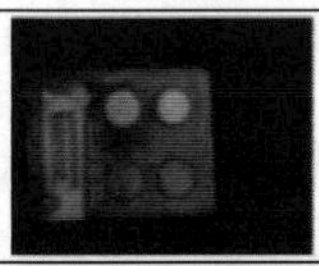		51	59	2	2	Packaging fail
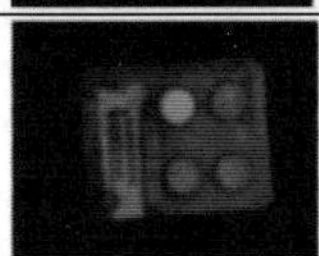		23	27	1	1	Packaging fail
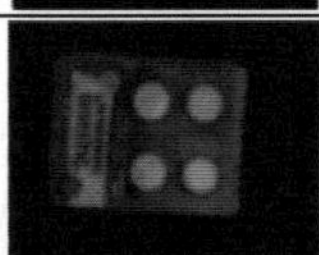		98	115	4,25	4	Packaging pass
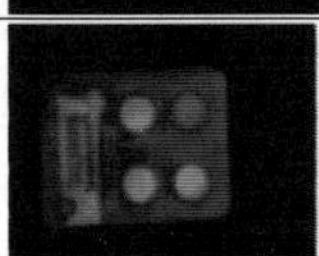	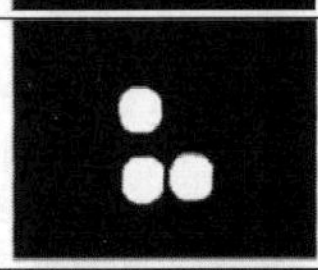	77	89	3	3	Packaging fail
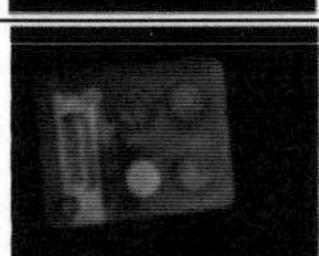		23	28	1,25	1	Packaging fail
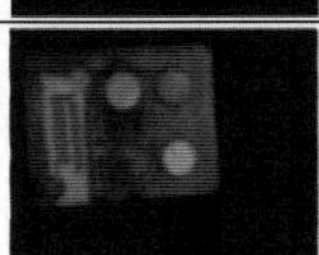	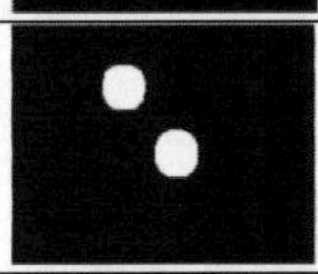	50	58	2	2	Packaging fail
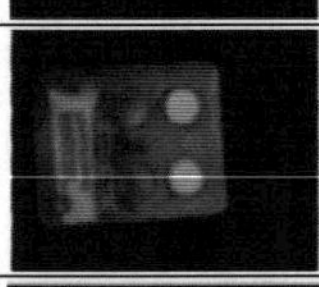	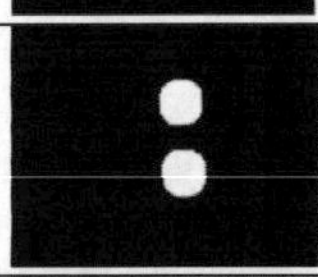	48	56	2	2	Packaging fail
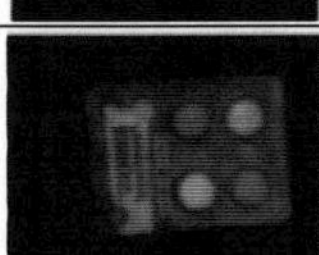	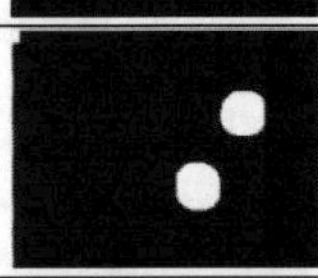	52	61	2,25	2	Packaging fail
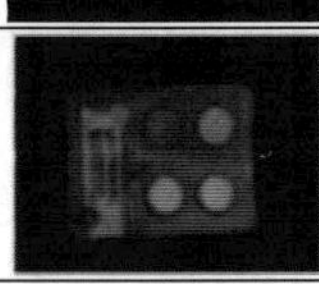	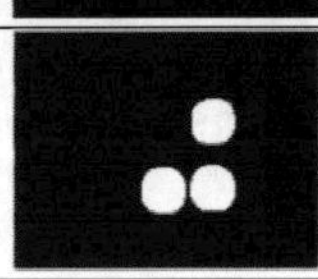	77	89	3	3	Packaging fail

IOP Conf. Series: Materials Science and Engineering **190** (2017) 012043 doi:10.1088/1757-899X/190/1/012043

Image Test Sam	Binary Image	Number of Corner		Number of Drug		Note
	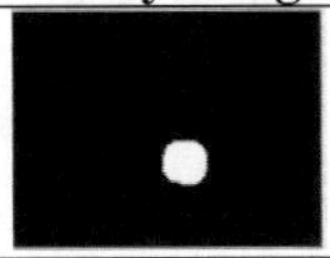	24	28	1	1	Packaging fail

From the test results of counting the number objects of drugs using digital image processing on the 10 samples of the first type of drug package, the results indicates that the system can calculate and detect defects drug properly (no failure).

4. Conclusion

Based on test results, it can be concluded that control system for the quality control of drug packaging that has been made can differentiate blister package of drug was defective and without disabilities at a speed of 0.03 m / sec. From the software of image processing using the color normalization method that have been created showing grayscale images on the object drug have degrees of gray is higher than the grayscale image section the background when the drug has the degree of the R, G or B color is higher than the degree of the R, G, B color on the background of packaging. The determination of threshold value indicates that the binary image of the drug is white and a binary image of the background of drug packaging is black.

Reference

[1] Dawson, K., & Howe. 2014. *A practical Introduction to Computer Vision with Open CV*. United Kingdom : John Wiley & Sons Ltd.

[2] Munir, R., 2004. *Pengolahan Citra Digital Dengan Pendekatan Algoritmik*. Bandung : Informatika

[3] Kusumanto, R.D., & Tompunu, A.N., 2011. *Pengolahan Citra Digital Untuk Mendeteksi Obyek Menggunakan Pengolahan Warna Model Normalisasi RGB*. Prosiding Seminar Nasional Teknologi Informasi & Komunikasi Terapan (SEMANTIK). Semarang : Universitas Dian Nuswantoro.

The Analysis of Alpha Beta Pruning and MTD(f) Algorithm to Determine the Best Algorithm to be Implemented at Connect Four Prototype

Lukas Tommy[1)] Mardi Hardjianto[2)] Nazori Agani[3)]

[1)2)3)]Graduate Program of Master of Science in Computer Science, Budi Luhur University
Jl. Ciledug Raya, Petukangan Utara, Kebayoran Lama, South Jakarta 12260

lukastommy92@gmail.com[1)] mardi.hardjianto@budiluhur.ac.id[2)]
nazori.agani@gmail.com[3)]

Abstract. Connect Four is a two-player game which the players take turns dropping discs into a grid to connect 4 of one's own discs next to each other vertically, horizontally, or diagonally. At Connect Four, Computer requires artificial intelligence (AI) in order to play properly like human. There are many AI algorithms that can be implemented to Connect Four, but the suitable algorithms are unknown. The suitable algorithm means optimal in choosing move and its execution time is not slow at search depth which is deep enough. In this research, analysis and comparison between standard alpha beta (AB) Pruning and MTD(f) will be carried out at the prototype of Connect Four in terms of optimality (win percentage) and speed (execution time and the number of leaf nodes). Experiments are carried out by running computer versus computer mode with 12 different conditions, i.e. varied search depth (5 through 10) and who moves first. The percentage achieved by MTD(f) based on experiments is win 45,83%, lose 37,5% and draw 16,67%. In the experiments with search depth 8, MTD(f) execution time is 35, 19% faster and evaluate 56,27% fewer leaf nodes than AB Pruning. The results of this research are MTD(f) is as optimal as AB Pruning at Connect Four prototype, but MTD(f) on average is faster and evaluates fewer leaf nodes than AB Pruning. The execution time of MTD(f) is not slow and much faster than AB Pruning at search depth which is deep enough.

Keywords : Board Games, Connect Four, Artificial Intelligence, Alpha Beta Pruning, MTD(*f*)

1. Introduction

Connect Four is a two-player game which the first player chooses disc colour. Players then take turns dropping coloured discs into a vertically suspended grid. The objective of Connect Four game is to connect four of one's own discs next to each other (vertically, horizontally, or diagonally)[1].

In its implementation, usually Connect Four is played by humans against other humans or computer. If against computer, an artificial intelligence (AI) is required by the computer in order to play properly like human.

There are many AI algorithms that can be implemented to board games like Connect Four, i.e. random, brute force, greedy and minimax as well as its variants such as negamax, alpha beta (AB) Pruning and negamax AB Pruning.

Research on the implementation and comparison of AI algorithms at Connect Four and other board games had been done before. Research[2] implements minimax and AB Pruning algorithm, while research [3] implements random and greedy algorithm at Connect Four. Research[4] implements negamax and AB Pruning, while research[5] implements brute force and negamax AB Pruning algorithm at Othello.

There are problems in the existing AI algorithms, random, brute force and greedy algorithm execution time are fast, but return less optimal move than minimax variants[3][5]. Minimax variants, return optimal move, but their execution time are long if the search depth (d) is deep enough[5].

To solve above problems, it is necessary to perform algorithms analysis and comparison to determine the suitable ones to be implemented at Connect Four. The suitable algorithms means optimal in choosing move and its execution time is not slow at search depth which is deep enough. The selected algorithms are AB Pruning and MTD, which later analysed and compared in terms of win percentage, execution time and the number of evaluated leaf nodes (LN). AB Pruning is selected because optimal like minimax but faster[4][2]. Negamax AB Pruning is not selected because it evaluate same LN as AB Pruning even though its implementation is simpler [6]. MTD(f) is selected because it is claimed to be simpler, more efficient and on average is better than the previous algorithms [7][8][9].

The research on AB Pruning and MTD (f) algorithm analysis and comparison has some limitations. The used AB Pruning is standard AB Pruning. The used Connect Four board is standard size, i.e. 7 columns and 6 rows. The tested parameters are evaluation value of each move, win percentage, the number of evaluated LN and the execution time of the algorithm. The used depth in the experiments are 5 through 10. The research only includes computer versus computer mode, where both computers use same depth.

With the analysis and comparison of both algorithms it is expected to determine the best algorithm to be implemented at Connect Four with deep enough depth.

2. Theoretical basis

2.1. Connect Four

There are some rules in playing Connect Four according to[10], namely:

a. Played on a grid of 7 columns 6 rows size.
b. Players take turns occupying a hole that only can be occupied once.
c. Only bottom of the empty hole in a column that can be occupied.
d. The first player who occupies four holes consecutively (orthogonally or diagonally) is the winner.

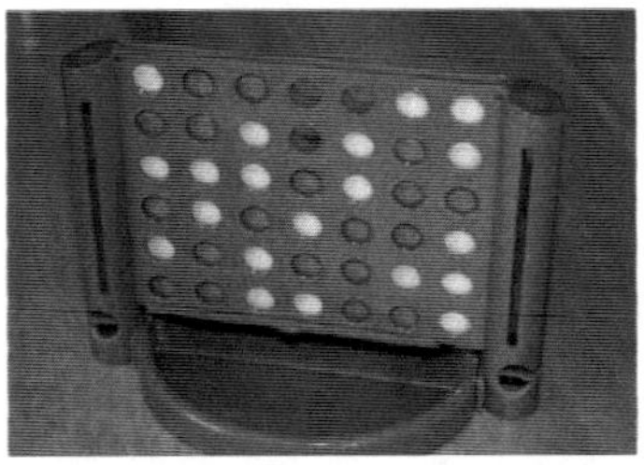

Figure 1. Connect four board [10]

2.2. Alpha Beta (AB) Pruning

AB Pruning computes the same optimal move as minimax, but more efficient by eliminates sub trees which can be proved irrelevant. AB pruning can be applied to trees of any depth and it is often possible to prune not only leaves, but the entire sub trees[11]. Example of AB Pruning searching process can be seen in Figure 2.

There are two parameters in AB Pruning that describe bounds on the backed-up values that appear anywhere along the path. Alpha (α) is the value of the best choice which is found so far at any choice point along the path for MAX. Beta (β) is the value of the best choice which is found so far at any choice point along the path for MIN.

IOP Conf. Series: Materials Science and Engineering **190** (2017) 012044 doi:10.1088/1757-899X/190/1/012044

AB Pruning updates the values of α and β as it goes along and prunes the remaining branches at a node. Pruning is done as soon as the value of the current node is known to be worse than the current α value for MAX or the current β value for MIN. The pseudocode of AB Pruning is shown Figure 2.

```
function ab(node, depth, α, β,MAXPlayer)
  if depth = 0 or node is a terminal node
      return the heuristic value of node
  if MAXPlayer
      v ← -∞
      for each child of node
          v ← max(v, ab(child, depth-1,
α, β, FALSE))
          α ← max(α, v)
          if β ≤ α
              break (* β cut-off *)
      return v
  else
      v ← ∞
      for each child of node
          v ← min(v, ab(child, depth-1,
α, β, TRUE))
          β ← min(β, v)
          if β ≤ α
              break (* α cut-off *)
      return v
```

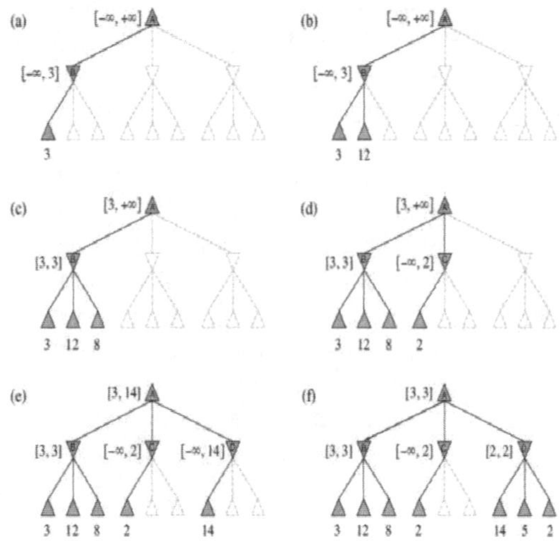

Figure 2. Searching process of AB pruning [11]

2.3. MTD(f)

MTD(f) or Memory-enhanced Test Driver with node n and value f works by calling AlphaBeta (AB) with memory a number of times with a search window of zero size. Each AB call returns a bound on the minimax value which is stored in upperbound and lowerbound, forming an interval around the true minimax value for that search depth[8]. The pseudocode of MTD(f) algorithm is as following:

```
function MTDF(root,f,d)
  g ← f
  upperbound ← +INFINITY
  lowerbound ← -INFINITY
  repeat
      if g = lowerbound then beta ← g +
1 else beta ← g;
      g ← ABMemory(root,beta-1, beta, d)
      if g < beta then upperbound ← g
else lowerbound ← g;
  until lowerbound >= upperbound;
  return g
```

MTD(f) uses a window of zero size, so that on each call AB, will either fail high or fail low. Fail high returning lowerbound, while fail low returning upper bound on the minimax value. Zero window calls cause more cutoffs, but return less information (i.e. a bound on the minimax value). MTD(f) has to call AB a number of times to find that information, nevertheless. The overhead of re-exploring parts of the search tree in repeated calls to AB disappeared. This is because MTD(f) uses AB that stores and retrieves the nodes it sees in memory. An entire sub tree in MTD(f) is aborted when a cutoff happens, furthermore, the recursive search code does not spawn re-searches anymore. The AB in MTD(f) can be simplified to use a single input bound, instead of both alpha and beta, since alpha is always one less than beta($\alpha = \beta - 1$) [8].

MTD(f) calls an AB that stores nodes in memory as it has determined their value. That nodes value then retrieved when a re-search happens, making MTD(f) become efficient. MTD(f) would re-explore most of those nodes each time a pass happens, if nodes storing are not done[8]. The Pseudocode of AB with memory is as following:

```
function ABMemory(n,alpha,beta,d)
  if retrieve(n) = OK then
    if n.lowerbound >= beta then return
n.lowerbound;
    if n.upperbound <= alpha then return
n.upperbound;
    alpha ← max(alpha, n.lowerbound)
    beta ← min(beta, n.upperbound)
  if d = 0 then g ← evaluate(n)
  else if n = MAXNODE then
    g ← -INFINITY; a ← alpha;
    c ← firstchild(n);
    while (g < beta) and (c != NOCHILD)
do
        g ←max(g,ABMemory(c,a,beta,d-1))
        a ← max(a, g);c ←
nextbrother(c);
  else
    g ← +INFINITY; b ← beta;
    c ← firstchild(n);
    while (g > alpha) and (c !=
NOCHILD)do
        g←min(g,ABMemory(c,alpha,b,d-1))
        b ← min(b, g);c ←
nextbrother(c);
  if g <= alpha then n.upperbound ← g;
store n.upperbound;
  if g >= beta then n.lowerbound ← g;
store n.lowerbound;
  return g
```

The access of transposition table takes place in the retrieve and store calls. The lines around retrieve make sure that if a value is present in the table, it is used instead of continuing the search. The store function is needed to make sure that the table is filled with values as they become available. In a real program, the best move can also be stored in the transposition table, and upon retrieving, that best move is searched first [8].

In MTD(f) searching process, if the value of the first evaluated leafnode (LN) is less than beta, upperbound(f +) will be returned and the minimax value is determined by MAX solution tree. Meanwhile, if the value of the first evaluated LN is greater than or equal to beta, lowerbound (f -) will be returned and the minimax value is determined by MIN solution tree [12].

At MAX solution tree, all child of MAX node are expanded, while only the first child of MIN node is expanded. At MIN solution tree, all child of MIN node are expanded, while only the first child of MAX node is expanded[13].

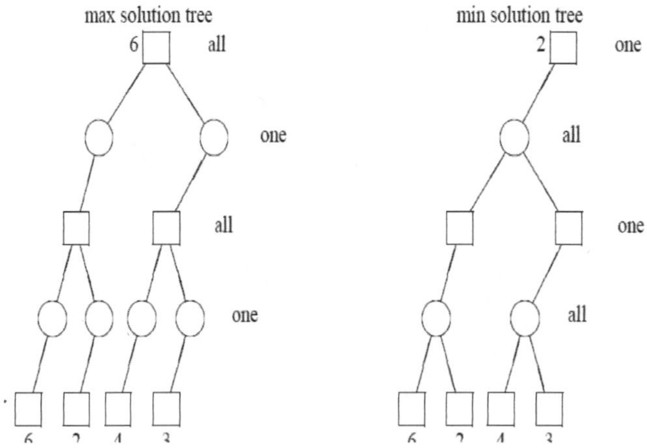

Figure 3. Solution tree [12]

2.4. Related Work

Lestari and Winata[2] implement minimax and AB Pruning algorithm at Connect Four. Experiments are performed on the time difference in taking the best move between minimax and AB Pruning computer. At depth 2 and 4, the time difference between the minimax and AB Pruning computer is not different enough. Minimax computer at depth 6 requires a long time to take the best move, unlike AB Pruning. The time required to take the best move by AB Pruning computer is less than minimax.

Sarhan et al. [3] implement random and greedy algorithm at Connect Four. The implemented algorithms are good enough at Connect Four which is played in real-time. Greedy algorithm may not returns optimal move, but it has met the system requirements without too much overhead.

Handayani et al. [4] implement negamax and AB pruning at Othello. Negamax algorithm is not efficient when used alone, because the search space is way too big. AB Pruning cutoff and reduce the search space, so that the search and evaluation process can be done faster.

Siswanto et al. [5] implement brute force and negamax AB Pruning algorithm at Othello. Brute force computer as fast as possible puts discs in any boxs as long as the move is valid, regardless of the box value of the taken position. Meanwhile, negamax AB Pruning computer apply the principle of zero-sum and heuristic value in the search tree in choosing box. This causes the taken box will be more optimal than brute force, but its execution time is long enough if the search depth is deep enough.

Gunawan et al. [7] implement negascout and MTD(f) algorithm at Othello. Based on 12 experiments versus downloaded Othello cyclog computer, negascout won 10 times, while the MTD (f) won 11 times. MTD(f) algorithm on average is more efficient than negascout, in terms of the number of searched nodes and average execution time at depth 6.

Shibahara et al. [9] compare MTD(f), negascout and AB Pruning algorithm to 500 randomly generated SHOGI game trees with fixed depth. The game trees are made as close as possible to the actual SHOGI game. MTD(f) has higher performance than the other two algorithms, even at depth 6, MTD(f) execution time is 1,8 times faster than negascout. MTD(f) performance will be more superior if the costs for evaluating the position can not be ignored like the actual game.

Plaat et al. [14] compare aspiration negascout (AspNS), AB Pruning, MT-Dual*, MT-SSS* and MTD(f) algorithm at Checkers, Othello and Chess. The compared factors are the number of evaluated leaf nodes and the number of nodes. Based on the experiments of 3 games, MTD (f) is the best, followed by AspNS, AB Pruning, MT-Dual* and lastly MT-SSS*. The number of MTD(f) leaf nodes on average is 5-10% fewer than AspNS (depend on the game), even MTD(f) also outperformed AspNS in terms of the number of nodes. The execution time of MTD(f) at Checkers and Othello is 5% faster than AspNS, while in Chess MTD(f) is 9-16% faster.

2.5. Research Object Review
Research objects which become the focus of this research are the optimality and the speed of AB Pruning and MTD(f) algorithm. Algorithm optimality is the optimality (evaluation value) of holes chosen by computer to achieves win. Algorithm speed is the execution time required by computer and the number of evaluated leaf nodes in choosing those holes.

2.6. Conceptual Framework
The conceptual framework that will be used in solving the problem formulations in this research is as shown in Figure 4

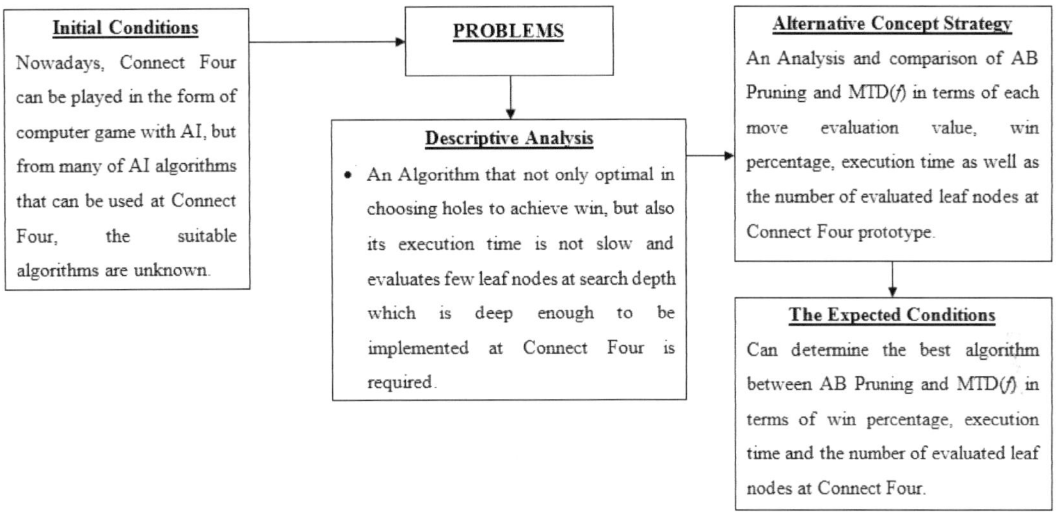

Figure 4. Conceptual framework of research

2.7. Hyphothesis
Based on the conceptual framework that has been described, the provisional guess to the research results that will be done are MTD (f) algorithm is guessed to be as optimal as AB Pruning with equal win percentage, however its execution time is faster and evaluates fewer leaf nodes at Connect Four prototype. The execution time of MTD(f) is guessed to be not slow at depth which is deep enough.

3. System and application design
Literature study method is the first step in this research, where the theoretical basis which related to the analysis of AB Pruning and MTD(f) algorithm at Connect Four on some literature and other references are studied.

The second step in this research is observation. Observation method is carried out by playing Connect Four game which is available online and offline, then analyse the used algorithms.

The analysis and comparison of AB pruning and MTD(f) algorithm will be conducted at Connect Four prototype. Both algorithms are used by the computer in choosing holes. The analysed and compared parameters are the optimality (evaluation value) of chosen holes and the number of wins, as well as the execution time and the number of evaluated leaf nodes.

The environment used in the experiments is a laptop with a dual-core processor @ 2.4GHz, 4GB memory and Windows 7 64-bit operating system.

Experiments are carried out by running Connect Four prototype with computer X (AB pruning) versus computer O (MTD(f)) mode with a same depth for both algorithms. The tested depths are 5 through 10 and at each depth, experiments are carried out with the first turn is computer X and the first turn is computer O.

Each condition is tested 10 times and the selected execution time for each move to be compared is the average of mentioned 10 test. Based on experiments which have been carried out, information such as the average execution time, the number of evaluated leaf nodes and the win percentage in each condition are acquired. These information then analysed to determine which is the best between both algorithms.

The results of this research are an algorithm which is the fastest, most optimal and evaluate the least leaf nodes at Connect Four can be determined. The suitable algorithm may also be determined to be applied at Connect Four with deep enough depth

3.1. Algorithm Design

The pseudocode of algorithm which is designed for computer versus computer mode is as following:

```
Select who moves first
Select search depth
turns ← 1
LN_X ← 0; LN_O ← 0; Pass_O←0;
Order ← [3,4,2,5,1,6,0]
Show empty Connect Four board
WHILE gameover=False DO
  IF turn="X" THEN
    Count start time
    Do AB Pruning function with depth n
    Update LN_X
    Choose hole and print evaluation value
    execution_time← current time-start time
    turn←"O"
  ELSE
    Count start time
    Do MTD(f) function with depth n
    Update LN_O; Update Pass_O
    Choose hole and print evaluation value
    execution_time← current time-start time
    turn←"X"
  END IF
  Update Connect Four board
  Print execution_time
  Check H,V,D segments at board
    IF connected X coin >=4 THEN
      winner←"X"
      gameover←True
    ELSE IF connected O coin >=4 THEN
      winner←"O"
      gameover←True
    END IF

  turns+=1
  IF turns > 42 THEN
    winner←"Draw"
    gameover←True
  END IF

END WHILE

Print turns, winner, LN_X,LN_O,Pass_O;
```

3.2. Evaluation Function Design

An AI algorithm requires evaluation function so that the computer can gives value to the current state of the game board. This is done so the computer can chooses the best hole from the existing holes to win the game.

At standard Connect Four board, there are 69 win segments at board, which is 24 horizontal segment (H), 21 vertical segments (V) and 24 diagonal segments (D). Each segment consists of four holes and each hole is weighted by the number of segments that can be created from that hole[15], as shown in Figure 5. The computer will consider to search the hole with highest weight first.

3	4	5	7	5	4	3
4	6	8	10	8	6	4
5	8	11	13	11	8	5
5	8	11	13	11	8	5
4	6	8	10	8	6	4
3	4	5	7	5	4	3

Figure 5. Weighted holes of connect four

There are some rules in evaluating segment, i.e. [16]:

a. 0 for empty holes or there are different kind of coins
b. 1 for a coin of one kind
c. 10 for two coins of one kind
d. 100 for three coins of one kind

e. Infinity (∞) for four coins of one kind (win)

The value of the segment will become positive if the evaluated segment is the coin of MAX player(X) or negative if the evaluated segment is MIN player (O).

The value returned from the designed evaluation function is the sum of evaluation value of all player's segments minus the sum of evaluation value of all opponent's segments.

3.3. Experimental Design

Experiments are carried out by running computer versus computer mode at Connect Four prototype with various different conditions. The number of game conditions that will be tested are 12 with details as following:

a. X moves first (X1st) with depth (d) 5 through 10.
b. O moves first (O1st) with depth (d) 5 through 10.

Each game conditions will be tested 10 times to find the average execution time for each move. In the experiments that will be carried out, the obtained information during experiment are recorded. Those information are in the form of the chosen holes of each turn and their evaluation value, the execution time of each turn, the average execution time, the number of LN and the total of execution time. The information of these experiments then summarized and grouped by who moves first and by depth. These information are in the form of depths, the average execution time, the number of LN, the percentage of O speed compared to X, the percentage of O LN compared to X and the end result of the game. These information then analysed and compared to determine which is the fastest and the most optimal between both algorithms, to be implemented at Connect Four.

4. Results and discussion

4.1. Experimental Setup

The experiments of AB Pruning and MTD(f) algorithm at Connect Four prototype are carried out with 12 different conditions. The experiments conditions such as the computer with which algorithm which moves first (1st) and varied depth (d). The AB Pruning algorithm is implemented on X computer, while MTD(f) algorithm is implemented on O computer. In each experiment, the same depth applied to both algorithms. The tested parameters are the execution time, the number of leaf nodes, the evaluation value of chosen holes and the number of achieved wins. Each condition respectively is tested 10 times and the chosen execution time is the average. From the results of carried out experiment, the fastest and the most optimal from both tested algorithms can be determined to be implemented at Connect with deep enough depth.

4.2. Experimental Results

The summary of experimental results of each depth is shown in Table 1. At depth 8, O(MTD(f)) as well as X(AB Pruning) won once. The average execution time of each move of X is 2.3185 seconds and O is 1,5027 seconds, while the number of leaf nodes (LN) of X is 218.777 and O is 73.957. The execution time of O is 35,19% faster and evaluate 66,2% fewer leaf node (LN) than X. Parenthesis will be provided at speed percentage if O is slower than X. The number of pass/ ABMemori function calls by O is 715 times. Similarly, the other depth.

Comparison of the number of wins of each depth are summarized in the chart as shown in Figure 6, where X won 5 times, O won 5 times and draw 2 times.

Comparison of the average execution time of each depth are summarized in the graph as shown in Figure 7. Comparison of the number of LN evaluated by both algorithms at each depth are summarized in the graph as shown in Figure 8. The difference in the execution time and the number of LN of both algorithms increases with deeper depth and is seen much different at depth 9 and 10.

Table 1. Summary of experimental results of each depth

Depth	Average Execution Time (seconds)		Speed O:X Percentage	Number of *Leaf Node* (LN)		Number of LN O:X Percentage	MTD(*f*) Number of Pass	End Results (Win)
	X	O		X	O			
5	0,1317	0,1089	17,28%	12242	5451	55,47%	201	X(1), O(1)
6	0,3083	0,3089	(0,19%)	17885	8705	51,33%	255	X(1), O(1)
7	1,6981	0,6767	60,15%	175759	36906	79,00%	456	X(1), O(1)
8	2,3185	1,5027	35,19%	218777	73957	66,20%	715	X(1), O(1)
9	14,7149	3,0543	79,24%	1311365	143557	89,05%	478	X(1), O(1)
10	27,1850	5,0731	81,34%	3352837	327609	90,23%	590	Draw(2)
Sum of End Results				X win 5 times, O win 5 times, Draw 2 times				

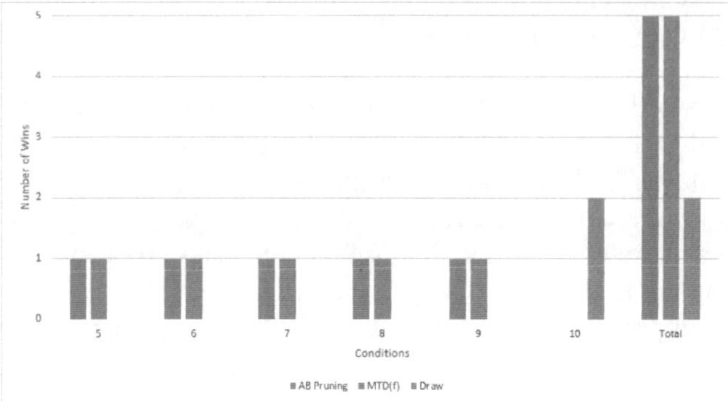

Figure 6. Number of wins comparison of each depth

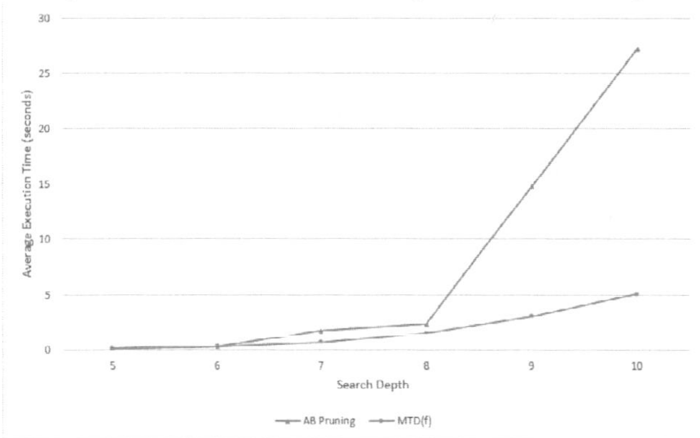

Figure 7. Execution time comparison of each depth

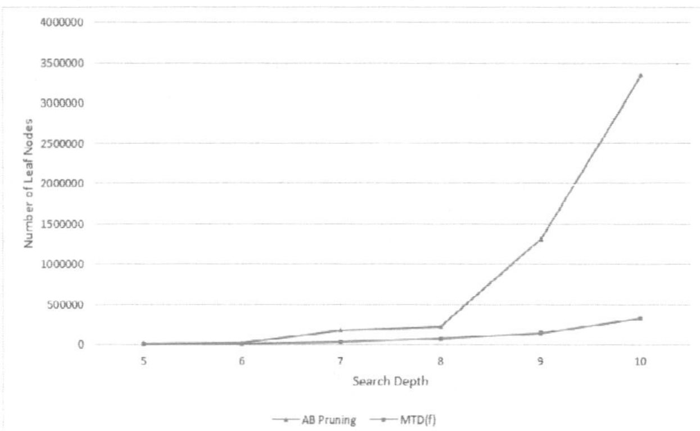

Figure 8: Number of ln comparison of each depth

4.3. Analysis and Discussion

MTD(f) chooses holes which are same as AB Pruning because both algorithms are minimax search variant and use same evaluation function.

At depth 6, the execution time of MTD(f) algorithm is a little slower (almost equal) than AB Pruning even though MTD(f) evaluates fewer LN. This is because the code complexity of algorithm MTD(f) is greater than standard AB Pruning code. As a result, the execution time of MTD(f) become slower, so that the difference in the number of LN is required even more.

The number of LN of O compared to X percentage at depth 7 is greater than depth 8. This anomaly occurs because of the number of passes MTD(f) at odd depth on average are fewer than at even depth. The number of evaluated LN will be reduced if the number of passes are few. These anomalies also occur as a result of reduction in the number of AB Pruning branching factor as square root at even depth. This leads to pruning occur more often causing the number of LN evaluated by AB Pruning are reduced.

Weighting holes as shown in Figure 5 makes computer explore holes with order from the largest to the smallest weight. The number of LN is reduced because the minimax value can be found earlier which lead to early pruning.

At Connect Four, the end result of game is not only affected by algorithm optimality but also affected by who moves first. First turn computer has greater chance to win the game. The execution time and the number of leaf nodes are also affected by the factor of who moves first.

MTD(f) on average is faster and evaluate fewer leaf nodes than AB Pruning. These are because MTD(f) using zero-window search and tranposition table. Zero-window search reduces the search space by pruning earlier than AB Pruning wide-window search. Transposition table is used to store and retrieve the evaluation value of board states so that it is not necessary to reevaluate the same board states. The addition of reflection function at transposition table make the computer does not need to evaluate board states which are as same as the reflection of previous board states. At deep enough depth (9 and 10), the execution time of MTD(f) is not slow and much faster than AB Pruning.

Based on experiments with 12 different conditions which have been carried out, the win percentage of X is 41,67%, the win percentage of O is 41,67%, while the percentage of the game ended in a draw is 16.66%.

The suitable algorithm for Connect Four from two tested algorithms are MTD(f). This is because MTD(f) is as optimal as AB Pruning, but its execution time is not slow and much faster if the depth is deep enough.

5. Closing

5.1. Conclusions

Based on the analysis results done on the problems, the problems formulation, research design and experiments in this research, the following conclusions are found:

a. MTD(f) algorithm is as optimal as AB Pruning at Connect Four. Based on experiments with 12 different conditions, the percentage achieved by MTD(f) is win 41,67%, lose 41,67% and draw 16,66%.
b. The execution time of MTD(f) algorithm on average is faster than AB Pruning at Connect Four. At depth 6, the execution time of MTD(f) is 0,19% slower than AB Pruning. At depth 5, 7, 8, 9 and 10, the execution time of MTD(f) is respectively 17,28%, 60,15%, 35,19%, 79,24% and 81,34% faster than AB Pruning.
c. MTD(f) algorithm on average evaluates fewer leaf nodes than AB Pruning at Connect Four. At depth 5, 6, 7, 8, 9 and 10, MTD(f) computer respectively evaluates 55,47%, 51,33%, 79%, 66,2%, 89,05% and 90,23% fewer leaf nodes than AB Pruning computer.
d. MTD(f) algorithm is suitable to be implemented for Connect Four. MTD(f) is as optimal as AB Pruning, but its execution time is not slow and much faster if the depth is deep enough.

5.2. Recommendations

Below are the recommendations which can be given as consideration for further research:

a. Comparing MTD(f) algorithm with newer and better algorithms than standard AB Pruning.

b. Improving the existing Connect Four evaluation function, so that AI can be more accurate and optimal in choosing holes.
c. Search algorithm optimization to improve execution time by using database. The evaluation value of each segment on board is pre-calculated and stored in form of hash table in database.

References

[1]. M. Sipper, *Evolved to Win*. Raleigh, USA: Lulu, 2011.
[2]. J. Lestari and A. Winata, "Implementasi Algoritma Minimax dengan Optimasi Alpha-Beta Pruning pada Aplikasi Permainan Connect Four," in *Prosiding Seminar Nasional Multidisiplin Ilmu (SeNMI)*, 2012, pp. C–50–C–58.
[3]. A. M. Sarhan, A. Shaout, and M. Shock, "Real - Time Connect 4 Game Using Artificial Intelligence," *J. Comput. Sci.*, vol. 5, no. 4, pp. 283–289, 2009.
[4]. M. S. Handayani, D. Arisandi, and O. S. Sitompul, "Rancangan Permainan Othello Berbasis Android Menggunakan Algoritma Depth - First Search," *J. Dunia Teknol. Inf.*, vol. 1, no. 1, pp. 28–34, 2012.
[5]. Siswanto, Laurentsia, and M. Anif, "Pengembangan Aplikasi Permainan Othello dengan Negamax Alpha Beta Pruning dan Brute Force," in *Seminar Nasional Sistem Informasi Indonesia (SESINDO)*, 2014, pp. 346–352.
[6]. I. Millington and J. Funge, *Artificial Intelligence for Games*, 2nd ed. Burlington, USA: Morgan Kaufmann, 2009.
[7]. Gunawan, Y. Kristian, and H. Andika, "Game Playing untuk Othello dengan Menggunakan Algoritma Negascout dan MTDF," in *Seminar Nasional Aplikasi Teknologi Informasi (SNATI)*, 2009, p. F–61–F–66.
[8]. A. Plaat, "Aske Plaat MTD(f), a new chess algorithm," 1997. [Online]. Available: people.csail.mit.edu/plaat/mtdf.html. [Accessed: 29-Feb-2016].
[9]. K. Shibahara, N. Inui, and Y. Kotani, "Adaptive Strategies of MTD-f for Actual Games," in *Proceedings of the 2005 {IEEE} Symposium on Computational Intelligence and Games (CIG05), Essex University, Colchester, Essex, UK, 4-6 April*, 2005.
[10]. E. Adams and J. Dormans, *Game Mechanics: Advanced Game Design*. Berkeley, USA: New Riders, 2012.
[11]. S. Russell and P. Norvig, *Artificial Intelligence: A Modern Approach*, 3rd ed. Upper Saddle River, USA: Pearson Education, 2010.
[12]. A. Plaat, J. Schaeffer, W. Pijls, and A. De Bruin, "*An Algorithm Faster than NegaScout and SSS * in Practice*," unpublished, 1995.
[13]. A. Plaat, J. Schaeffer, W. Pijls, and A. De Bruin, "*A new paradigm for minimax search*," Dept. Comput. Sci., Univ. Alberta, Edmonton, Tech. Rep. TR-94-18, Dec. 1994.
[14]. A. Plaat, J. Schaeffer, W. Pijls, and A. de Bruin, "*Best-first fixed-depth minimax algorithms*," *Artif. Intell.*, vol. 87, pp. 255–293, 1996.
[15]. M. Stenmark, "*Synthesizing Board Evaluation Functions for Connect4 using Machine Learning Techniques*," M.C.S. thesis, Dept. Comput. Sci., Østfold University College, Halden, Norway, 2005.
[16]. R. L. Rivest, "*Game Tree Searching by Min / Max Approximation*," *Artif. Intell.*, vol. 34, pp. 77–96, 1988.

Optimizing Subgroups Formation for E-MBMS Transmissions in LTE Networks

M Algharem[1], M H Omar[1], D Stiawan[2], R Budiarto[3]

[1]InterNetWorks Research Group, School of Computing, College of Arts and Sciences, Universiti Utara Malaysia, Malaysia
[2]Department of Computer Engineering, Faculty of Computer Science, University Sriwijaya, Palembang, Indonesia
[3]Dept. of Computer Information System, College of Computer Science and Information Technology, Albaha University, Albaha P.O Box 1988, Saudi Arabia

{algharem,mhomar}@internetworks.my, deris@unsri.ac.id, rahmat@bu.edu.sa

Abstract. Long Term Evolution (LTE) network provides a high throughput with low latency which make it suitable for multicast and broadcast services. In Conventional Multicast Scheme (CMS), data is transmitted according to the user with worst channel condition which results in wasting network resources. To overcome the drawback of CMS, a new subgrouping mechanism is proposed to split the multicast group into several subgroups based on users channel quality. The performance of the proposed mechanism has been evaluated using LTE simulator. The simulation results show that the proposed mechanism increase the multicast performance compared to CMS in term of goodput and spectrum efficiency, while maintain fairness index of users in an acceptable level.

Keywords: E-MBMS, Modulation and Coding Schema, multirate, multicast

1. Introduction

Long Term Evolution (LTE) network was introduced by the Third-Generation Partnership Project (3GPP) and was considered as the latest step towards the 4th generation of radio technologies. LTE offers a high throughput with low latency which make it the best choice for Multimedia Service. LTE network exploits the benefits of Orthogonal Frequency Division Multiple Access (OFDMA), in which various users data is multiplexed in frequency and time domains [1]. In OFDMA, the full frequency bandwidth is divided into orthogonal subcarriers, where each subcarrier is allocated 15 kHz. The LTE frame consists of 12 consecutive subcarriers and 10ms duration. Each frame consists of 10 subframes; each subframe is 1ms, which is equal to the Transmission Time Interval (TTI); and then each subframe is equal to two time slots, where each slot is 0.5ms in the time domain and 12 subcarriers in the frequency domain. However, each slot is composed of a resource block (RB), which is the minimal radio resource allocation unit in the LTE. Each RB consists of seven symbols when the normal Cycle Prefix (CP) is used or six symbols when the extended CP is used, as used in E-MBMS subframe [1].

Recently, mobile devices are equipped with large screen with high resolution which requires high data rate for video and has the ability to transmit and received data with higher bit rate. In addition, the using of these devices is no longer limited only on making voice calls but they are also used to browse the Internet, watch video, watching news, play online game, and watching TV. Thus, beside the bandwidth scarcity, will increase the demand on network resources and then force the network operators to efficiently utilize the network resources. A group-oriented services, such as multicast and broadcast, are an efficient way for utilizing the network resources. The increasing demand of the group-oriented services has resulted in defined and standardized a new service called Multimedia Broadcast Multicast service (MBMS), which was introduced by the 3GPP [2,3] in 2005 over the UMTS Release 6 [4]. Since 2005, many enhancements have been made to the MBMS standard which evolved into enhanced MBMS (E-MBMS) that developed over the 3GPP LTE standard network Release 9 [5].

In E-MBMS, the same data is sent to Users Equipment (UEs), whose belong to the same Multicast Group (MG), using the same channel result in efficient sharing and usage of radio network resources. On other side, it is a challenge to select the Modulation and Coding Schema (MCS) which can satisfy all UEs belong to the same group. The cell center of UEs can receive the data with high bit rate, whereas, the cell edge UEs suffer a poor channels condition. Therefore, they can receive the data only with low MCS level which result in wasting the network resources. In CMS, a single data rate is selected to transmit the data to all UEs in each MG. The data rate is selected according to the user with worst channel gain (WCG) [6,7,8]. The CMS is a reliable multicast transmission schema to deliver the data with high fairness. In contrast, the CMS reduces the system performance by forcing the users with high channel gain (HCG) to receive the data with low MCS corresponding to the users with WCG.
Several studies have been carried out to improve the multicast performance by overcome the limitations of the CMS. For example, in [9], the authors proposed a resources allocation approach called Opportunistic Multicast Scheduling (OMS), which exploits the multiuser diversity by only selects the users with HCG to be served in a time slot. In [10], the MG is split into two subgroups (cell center and cell edge subgroups), and split the data stream into two layers (base and enhanced layer). The based stream is received by both subgroups while the enhanced stream is only received by cell center subgroup. Another interesting study was proposed by [5, 11], in which the UEs in each MG are split into several subgroups depending on users channel gain. Then each subgroup are served with MCS level corresponding to the user with WCG. Tan et al. in [i12], proposed a schema in which the multicast groups are divided into a set of subgroups dynamically using a coalition game theory. However, most previous works consider a single group which not always happens in the real system. Moreover, in a multi-groups system, splitting each group to several subgroups will results in high number of subgroups which required more radio resources and reduce the spectrum efficiency (SE). Consequently, determining the number of subgroups and the amount of radio resources that should be allocated to each subgroup are still an open issue and need to be properly selected.
In this paper, an innovative Radio Resource Management (RRM) mechanism is proposed to increase the E-MBMS performance. The proposed RRM efficiently splits each group to three subgroups and allocated the resource to each subgroup. It uses two thresholds to split the MG to three subgroups (upper, lower, and medium subgroups). These thresholds are selected according to the Standard Deviation (StD) and the average of users channel quality.

2. System Modeling and Problem Formulation

In LTE network, there are 15 levels of CQI, each level associated with an MCS level. Let $L=15$ is the CQI levels, then $CQI_l=l$, where $l=\{1,2,...,L\}$. Thus, MCS_i is the MCS associated with the CQI_l. Any user sent a CQI_l feedback to its eNodeB, can successfully receive and decode the transmitted data which is transmitted with the MCS_i level where $i \leq l$.

Let consider K users belong to G groups that receiving the data using N subcarriers over a single eNodeB as illustrated in Figure 1. The set of users, groups, and subcarriers are represents by $\mathcal{K}, \mathcal{G}, \mathcal{N}$ respectively. The system bandwidth W is equally shared between all subcarriers, so the bandwidth of subcarrier n is $B_n = \frac{W}{N}$. For simplicity, we assume each subcarrier n have equal power P_n. Assume a perfect orthogonality preserved, and perfect synchronization, so there is no inter symbol interference and no inter-carrier interference. All users receive transmitted data over one or more subcarriers without any interference. It is also assumed that the eNodeB uses reliable feedback channels to receive CQI report from each user without any delay. Let $\mathcal{K}_g$ denotes the g users group set, and the cardinality $\parallel \mathcal{K}_g \parallel$ denotes the number of users in group g, where $g = \{1,2,\ldots,G\}$. Thus, for multicast $\parallel \mathcal{K}_g \parallel \geq 2$, and $\parallel \mathcal{K}_g \parallel = 1$ in unicast. Moreover, $\sum_{i=1}^{G} \parallel \mathcal{K}_g \parallel = K$, $\mathcal{K} = \bigcup_{g=1}^{G} \mathcal{K}_g = \mathcal{K}_1 \cup \mathcal{K}_2 \cup \cdots \cup \mathcal{K}_G$. Let MCS_g^v represents the MCS vector of users in group g, where $MCS_g^v = \{MCS_1, MCS_2, \cdots, MCS_M\}$, $M = \parallel \mathcal{K}_g \parallel$.

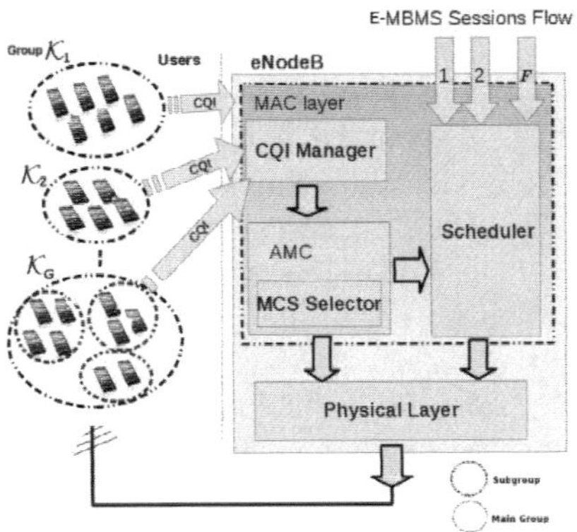

Figure 1. Subgrouping formation schema.

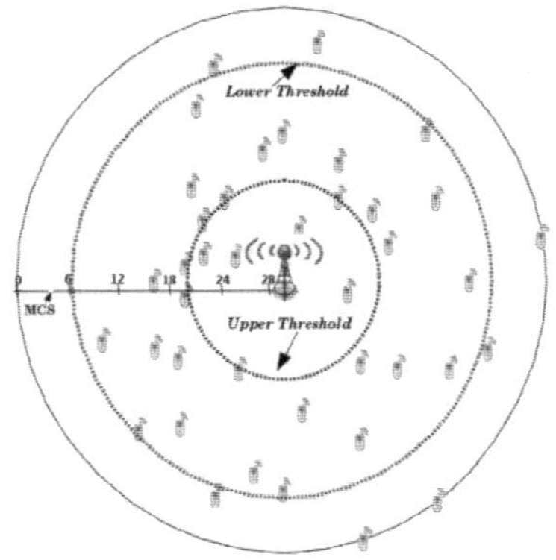

Figure 2. Upper/ lower thresholds in group g.

Let consider $r_{k,n}$ is the data rate of user k on subcarrier n. Then the data rate of user k can be expressed by equation (1).

$$R_k = \sum_{n=1}^{N} r_{k,n}\omega_{k,n} \quad (1)$$

where $\omega_{k,n}$ is a subcarrier indicator to show whether the subcarrier n is used by user k or not. is formulated as equation (2).

$$\omega_{k,n} = \begin{cases} 1, & if\ subcarrier\ allocated\ user\ k; \\ 0, & otherwise. \end{cases} \quad (2)$$

In CMS, the transmission data rate of a group is selected according to the user with WCG. Let consider $r_{g,n}$is the worst channel gain for group g on subcarrier n. The aggregate data rate (ADR) of group g can be calculated as in equation (3).

$$R_g = \parallel \mathcal{K}_g \parallel \sum_{n=1}^{N} r_{g,n}\omega_{g,n} \quad (3)$$

where $\omega_{k,n}$ indicates either the subcarrier was utilized by group g or not. The total ADR of all groups in $\mathcal{K}$can be obtained as equation (4).

$$R_T = \sum_{g=1}^{G} R_g = \sum_{g=1}^{G} \sum_{n=1}^{N} \parallel \mathcal{K}_g \parallel r_{g,n}\omega_{g,n} \quad (4)$$

The RRM allocates the system subcarriers to all groups in a way that can maximize the system throughput while keep the fairness between UEs in acceptable level. According to [13], CMS throughput is bounded by the users with WCG, and will saturate when the UEs number increases in Rayleigh and Ricean fading environments. As aforementioned, the MMS techniques were emerged to overcome the limitations of the CMS. The MSF maximize the multicast throughput by split the G groups into S subgroups with set $\mathcal{S}$, where $\mathcal{S} = \{1,2,\ldots,S\} and\ G \leq S \leq K$. In details, each group g can be splitted into $\mathcal{S}_g$ subgroups, where $\mathcal{S}_g = \mathcal{S}_{\mathrm{g}}^1 \cup \mathcal{S}_{\mathrm{g}}^2 \cup \ldots \cup \mathcal{S}_{\mathrm{g}}^l$, and $1 \leq l \leq 15$. Indeed, each group g can be splitted to 15 subgroup as maximum which equal to the number of MCS level in LTE network. Then transmit the data to each subgroup using a WCG rate. To efficiently utilize the multiuser diversity, the MG users are splitted according to their channel gains. Thus, each subgroup $\mathcal{S}_g^l$ contains a set of users with same or close channel gains.

$$max \sum_{s=1}^{S} \sum_{n=1}^{N} \sum_{k=1}^{K} r_{s,n}\omega_{s,n}\alpha_{s,k} \tag{5}$$

$$\omega_{s,n}, \alpha_{s,k} \in \{o, 1\}, \forall s \in \mathcal{S}, \forall k \in \mathcal{K}, \forall n \in \mathcal{N} \tag{6}$$

$$\sum_{s=1}^{S} \omega_{s,n} = 1, \sum_{s=1}^{S}, \alpha_{s,k} = 1, \forall s \in \mathcal{S}, \forall n \in \mathcal{N}, \forall g \in \mathcal{G} \tag{7}$$

where $\omega_{s,n}, \alpha_{g,s,k}$ are binary indicators, which indicate either subcarrier n and user k belong to subgroup s or not.

The RRC has to assign subcarriers and users to proper subgroups in order to maximize the total throughput. Thus, the problem described in equations (5)-(7) is considered as NP-Hard problem (non-deterministic polynomial-time) which does not have an optimal solution. Indeed, there is a solution for this problem by using exhaustive search algorithm, but it is usually a high complex and time-consuming computations [14]. Moreover, the complexity of NP-Hard increases exponentially with the number of subcarriers, subgroups, and groups which make it unrealistic to practical used. Furthermore, there are S^N solutions for scheduling radio resources between all subgroups S [9, 15]. Thus, it is necessary to find a suboptimal solution which can be used in real system. This paper introduces the use of StD of users' SINR to split each group to three subgroups; worst, best, and medium subgroups. The worst subgroup will contain all cell edge UEs whose MCSs are extremely low. The best subgroup will contain the cell center UEs whose MCSs are extremely high, whereas the medium subgroup will contain the remain UEs.

3. System Modeling and Problem Formulation

The proposed subgrouping mechanism uses the StD to show how the users MCS levels are distribute and deviated from the average value of all users' MCS level values. The StD with small value means that MCSs of all users are closed to each other, whereas, the big StD value means that all users are far from each other. However, in case of the users' MCSs standard deviation value is small enough, the MCS level of worst user case will be suitable for all users. The StD and average of users' MCSs will be used to divide the multicast users into several categories or subgroups by using upper and lower thresholds, as shown in Figure 2. These thresholds will be used to split each group g into three subgroups. Nevertheless, several steps should initially be performed in order to calculate the upper and lower MCS thresholds for each group g. The following steps are repeated for each group g in the $\mathcal{K}$ groups set:

- **Step 1:** Users have to measure the SINR for received signal. The SINR for each subcarrier can be calculated using equations (8) and (9) [5,16].

$$SINR(m) = \frac{\sum_{i=1}^{A} \sum_{j=1}^{B} \frac{w(\tau_i(m)+\delta_i)P_j}{q_i(m)}}{\sum_{i=1}^{A} \sum_{j=1}^{B} \frac{(1+w(\tau_i(m)+\delta_i))P_j}{q_i(m)} + N_0} \tag{8}$$

with

$$w(\tau) = \begin{cases} 1 & 0 \le \tau \le T_{CP} \\ 1 - \frac{\tau - T_{CP}}{T_u} & T_{CP} \le \tau \le T_{CP} + T_u \\ 0 & otherwise \end{cases} \tag{9}$$

where P_j is the average power associated with the j path, $\tau_i(m)$ the propagation delay from eNodeB i, δ_j the additional delay added by path j, $q_i(m)$ the path loss from eNodeB i, T_{cp} the length of the CP and T_u the length of the useful signal frame, N_0 the noise power.

- **Step 2:** mapping the SINRs of a user subcarriers into one effective SINR, (so-called Exponential Effective SINR Mapping (EESM)), using the following equation (10) as stated in [6, 17].

$$SINR_{eff} = -\beta . \ln\left(\frac{1}{RB} \sum_{m=1}^{N} e^{-\frac{SINR_m}{\beta}}\right) \tag{10}$$

IOP Conf. Series: Materials Science and Engineering **190** (2017) 012045 doi:10.1088/1757-899X/190/1/012045

where $SINR_m$ is the $SINR$ of the mth resource block assigned to the user; β is a factor which can be amended to match the $SINR_{eff}$ to a specific MCS. RB is the number of resource blocks assigned to the user.

• **Step 3:** map the obtained $SINR_{eff}$ to the corresponding CQI value which is achieved by a BLER less than 10% in Single Input Single Output transmission mode (SISO) over Additive White Gaussian Noise (AWGN) channel.

• **Step 4:** user will send the obtained CQI to its eNodeB which is responsible for selecting the corresponding MCS level, as listed in Table 1.

• **Step 5:** calculate the average $\overline{MCS_g^v}$ value of all the $g\text{-}th$ group where $\overline{MCS_g^v} = average(MCS_g^v)$. Then, calculate the standard deviation σ of MCSs values of all users using equation (11).

$$\sigma_g = \sqrt{\frac{1}{n}\sum_{i=1}^{n} (MCS_i - \overline{MCS_g^v})^2} \tag{11}$$

• **Step 6:** calculate the upper threshold T_{up} and the lower threshold T_{low} for all users using equations (12) and (13).

$$T_{up}^g = \overline{MCS_g^v} + \sigma_g \tag{12}$$

$$T_{low}^g = \overline{MCS_g^v} - \sigma_g \tag{13}$$

The T_{up}^g and T_{low}^g values will be used to select users with abnormal MCS level who are deviated away from the average of users' MCS.

• **Step 7:** each group g in $\mathcal{K}$ set will be splitted into three subgroups (S_g^b, S_g^m, S_g^w), where

$$S_w^g = \{MCS_i | \ \forall \ MCS_i < T_{low}^g\},$$

$$S_b^g = \{MCS_i | \ \forall \ MCS_i > T_{up}^g\},$$

$$S_m^g = \{MCS_i | \ \forall \ MCS_i \le T_{low}^g, MCS_i \le T_{up}^g\},$$

Table 1. CQI and their interpretations.

CQI Index	Modulation	Code rate x1024	SE [bit/s/Hz]
1	QPSK	78	0.1523
2	QPSK	120	0.2344
3	QPSK	193	0.3770
4	QPSK	308	0.6016
5	QPSK	449	0.8770
6	QPSK	602	1.1758
7	16QAM	378	1.4766
8	16QAM	490	1.9141
9	16QAM	616	2.4063
10	64QAM	466	2.7305
11	64QAM	567	3.3223
12	64QAM	666	3.9023
13	64QAM	772	4.5234
14	64QAM	873	5.1152
15	64QAM	948	5.5547

Table 2. Simulation parameters.

Parameter	Value
Carrier Frequency	2GHz
Path loss	*PL(db)=128.1+37.6 *log₁₀d*
Thermal noise	–174 dBm /Hz
Downlink Bandwidth	3 MHz
Symbols for TTI	12
Sub-Frame Length	1 ms
Frame Type	FDD
eNodeB radius	1 km
eNodeB Power transmission	43 dBm
Modulation Schemes	QPSK, 16QAM, 64QAM (dynamic)
No. of users	20-100 UEs
No. of E-MBMS group	1
User transmission power	23 dBm
User distribution	Randomly and uniformly distributed
User speed	3 km/h
User mobility model	Random direction
CQI scheme	Full Bandwidth
Application flows	Video
Video rate	440 kbps
Simulation time	20 Second

IOP Conf. Series: Materials Science and Engineering **190** (2017) 012045 doi:10.1088/1757-899X/190/1/012045

4. Simulation and Results

LTE simulator (LTE-Sim) has been used to evaluate the proposed mechanisms after an extremely modifications and extensions of its functions and classes to support E-MBMS network. LTE-Sim is an open source framework simulator developed by G. Piro and F. Capozzi [18].

4.1. Simulation scenario

The proposed RRM mechanism has been compared to the CMS in term of cumulative goodput, fairness, and spectrum efficiency (SE). For simplicity, only E-MBMS has been activated in the simulation. Thus, the whole bandwidth was assigned to the E-MBMS. For more accurate, each scenario was performed with different number of UEs who uniformly distributed. Each scenario was repeated 20 times, then the average value was calculated. A realistic video trace files with 440 Kbit/s was used, which is available in [19]. The main simulation parameters are listed in Table 2.

4.2. Simulation results

The first evaluation metric experimented is cumulative goodput and the results are shown in Figure 3. The proposed subgrouping mechanism increases the system performance in term of goodput. Simulation results demonstrate that the cumulative goodput of the proposed mechanism is better than the cumulative goodput of the CMS. This gain due to the using of subgroup technique which is no more limited by the WCG as in the CMS.

The second evaluation metric is the fairness index (FI) of all UEs [20]. The FI is defined by equation (14).

$$FI = \frac{(\sum_{k=1}^{K} r_k)^2}{K \sum_{k=1}^{K} (r_k)^2} \tag{14}$$

where r_k denotes the goodput of the kth user, and FI value is variance between $(1/K) \leq FI \leq 1$.

The maximum FI (FI=1) can be obtained when all user are served with the same rate. As shown in Figure 4, the maximum fairness is achieved by CMS. The FI of the proposed subgrouping mechanism is less than the FI of the CMS. The FI of the proposed subgrouping mechanism decreased as the number of UEs increased, because the subgroups size increases as the UEs increase. Nevertheless, the FI of proposed mechanism is close the optimal value because each group was split to only three subgroups, which means three different rates available for all UEs group.

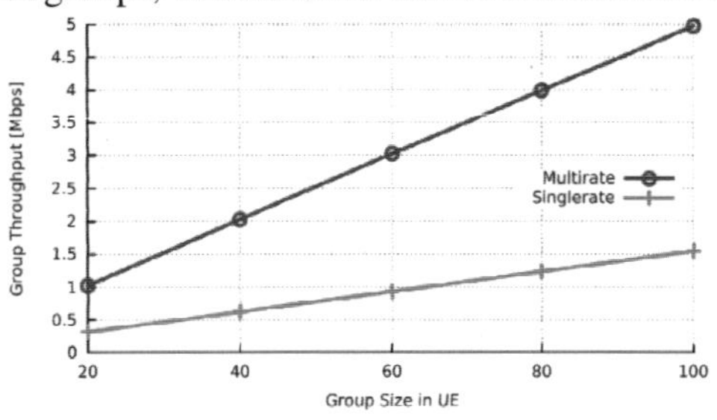

Figure 3. Cumulative Goodput difference between proposed mechanism and the CMS.

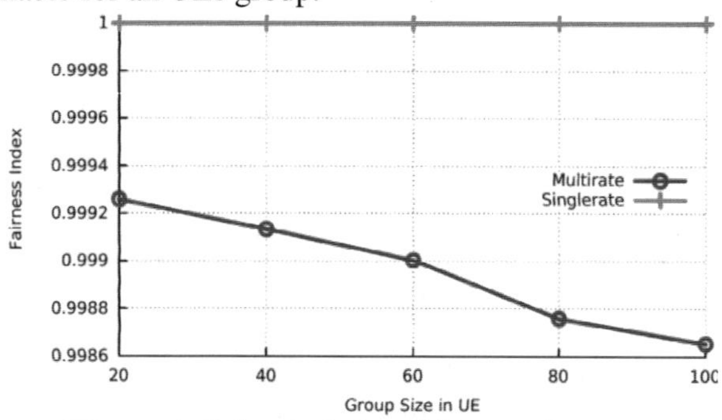

Figure 4. Fairness Index difference between proposed mechanism and CMS.

5. Conclusion

In this paper, an efficient and low complex subgrouping mechanism was proposed to improve the performance of the E-MBMS. The SINR of a group UEs have been used to split the UEs into several subgroups. The StD and average of the UEs SINR were used as a criteria to classify the UEs according to their SINR. The simulation results showed that the proposed subgrouping mechanism improves the goodput of the E-MBMS compared to the Conventional Multicast Scheme, while keeping the fairness between all UEs in acceptable level.

IOP Conf. Series: Materials Science and Engineering **190** (2017) 012045 doi:10.1088/1757-899X/190/1/012045

References

[1] Rinne M and Tirkkonen O 2010 *Computer Communications* **33** 16 pp 1894–1906.
[2] 3GPP 2009 Introduction of the multimedia broadcast/multicast service (MBMS) in the radio access network (RAN), Stage 2. TS 25.346 3rd Generation Partnership Project (3GPP).
[3] 3GPP 2012 Multimedia Broadcast/Multicast Service (MBMS); Architecture and functional description, TS 23.246, 3rd Generation Partnership Project (3GPP).
[4] 3GPP 2009 General UMTS architecture, TS 23.101 3rd Generation Partnership Project (3GPP).
[5] 3GPP 2013 Multimedia Broadcast/Multicast Service (MBMS); Stage 1, TS 22.146, 3rd Generation Partnership Project (3GPP).
[6] Kim J and Cho D-H 2005 *Proc. 62nd IEEE Vehicular Technology Conf. (Dallas, USA)* pp 725-729.
[7] Bochrini S, Bouras C, and Kokkinos V 2013 *6th Joint IFIP Wireless and Mobile Networking Conf. (Dubai, UAE)* pp 1-8.
[8] Antonios A, Christos B, Kokkinos V, Papazois A and Tsichritzis G 2010 *IEEE 6th Int'l Conf. on Wireless and Mobile Computing, Networking and Communications (Niagara Falls, Canda)* pp 361–367.
[9] Low T-P, Pun M-O, Hong Y P, and Kuo C-C J 2010 *IEEE Trans. on Wireless Communications*, 9 2 pp 791-801.
[10] Eusebio P and Correia A, 2005 *Proc. 2nd Int. Symp. on Wireless Communication Systems (Siena, Italy)* pp 777-781.
[11] Hou F, Cai I-L X, Ho P-H, Shen X and Zhang J 2009 *IEEE Trans. on Wireless Communications*, **8** 3 pp 1508-1519.
[12] Tan C K, Chuah T C, and Tan SW 2014 *Computer Networks* **64** pp180–194.
[13] Suh C and Mo J, 2006 *Proc. 25th IEEE Int. Conf. on Computer Communications (Barcelona, Spain)* pp 1-12.
[14] Sharangi S, Krishnamurti R and Hefeeda M 2011 *IEEE Trans. on Multimedia*, **13** 1 pp 102-115.
[15] Afolabi R O, Dadlani A and Kim K 2013 *IEEE Communications Surveys Tutorials*, **15** 1 pp 240-254.
[16] Koh C H and Kim Y Y, 2006 *Proc. 64th IEEE Vehicular Technology Conf. (Montreal, Canada)* pp 1-5.
[17] Araniti G, Scordamaglia V, Molinaro A, Iera A, Interdonato G, and Spano F 2011 *IEEE Int. Symp. on Broadband Multimedia Systems and Broadcasting (Erlangen, Germany)* pp 1-5.
[18] Elrabiei S M and Habaebi M H, 2010 *Proc. 5th IEEE Int. Symp. on Wireless Pervasive Computing (Modena, Italy)* pp 600-605.
[19] Video trace library. [Online]. Available: http://trace.eas.asu.edu/.
[20] 3GPP 2006 Physical layer aspect for evolved Universal Terrestrial Radio Access (UTRA), TR, 25.814 3rd Generation Partnership Project (3GPP).

IOP Conf. Series: Materials Science and Engineering **190** (2017) 012046 doi:10.1088/1757-899X/190/1/012046

The Performance Of SISO In Wireless Open-Access Research Platform (WARP) Using QAM Modulation

Jenny Putri Hapsari[1] and Ida Anisah[2]

[1]Deparment of electrical Engineering, Sultan Agung Islamic University (UNISSULA)
[2] Department of Electrical Engineering, Politeknik Elektronika Negeri Surabaya (PENS)
E-mail: [1]jenny@unissula.ac.id, [2]ida@pens.ac.id

Abstract. This paper shows the implementation of SISO communication system using QAM modulation in WARP. The performance of the proposed system is evaluated in terms of Bit-Error-Rate (BER) for different M-array level: 4, 8, and 16 in indoor and outdoor environments, both using LOS condition. In the analysis results, SISO performance with 4-QAM modulation achieves better performance compared to 8-QAM and 16-QAM for both environments. Meanwhile, SISO performance with QAM modulation in outdoor environment achieves better results compared to indoor environment for different M-array level. In indoor environment with transmitted power -26 dBm, BER achieved for 4-QAM, 8-QAM, and 16-QAM are 1.33×10^{-4}, 2.078×10^{-2}, and 8.76×10^{-2}, respectively. While in outdoor environment with the same transmitted power, BER achieved for 4-QAM, 8-QAM, and 16-QAM are $0, 1.497 \times 10^{-2}$ and 5.928×10^{-2}, respectively. SISO performance using M-QAM modulation is better than using M-PSK modulation.

Keyword: SISO, WARP, BER, QAM

1. Introduction

SDR (Software Defined Radio) is an opening architecture based on modern communication theory and digital signal processing. SDR uses standardized hardware units, connected bus, and downloading software. SDR can allow the wireless devices to be reconfigured by the software update which can implement the emerging new wireless communication standards. Thereby, it can decrease the time for developing product [1] [2].

One of the SDR platform which is being developed by Rice University in America is Wireless Open Access Research Platform (WARP). WARP is an intelligent open-access platform which capable, scalable, and extensible for wireless communication research. WARP platform is designed to meet the high level performances. WARP architecture consists of 4 main components: *Hardware Custom, Platform Support Packages, Open-Access Repository,* and *Research applications*. [3][4]

WARP implementation has been applied in SISO, MIMO, MISO, and cooperative communication. [5] [6] [7] [8] All those implementation used PSK modulation which only considering the phase shift. Thus, the previous studies, shown BER values are high enough for a high transmitted power.

In this paper, SISO communication system will be implemented on the WARP module using QAM modulation. QAM modulation considering the phase and amplitude shift, therefore BER is low [9] [10] and saving energy. [11] The research will be evaluated in terms of bit-error-rate for various transmitted power. The implementation will be conducted in real channel in indoor and outdoor environments with both using LOS condition.

The contents of this paper are organized as follows. Section II about SISO using M-QAM modulation. Section III describe the results and analysis. Section IV describe conclusions.

2. SISO Using M-QAM Modulation

SISO simulation design using M-QAM modulation is shown in Figure 1. The simulation used ideal canal with 10^3 bit transmitted. The performance analysis can be found from the comparison of transmitted and received bit as a result of BER for different Eb/No.

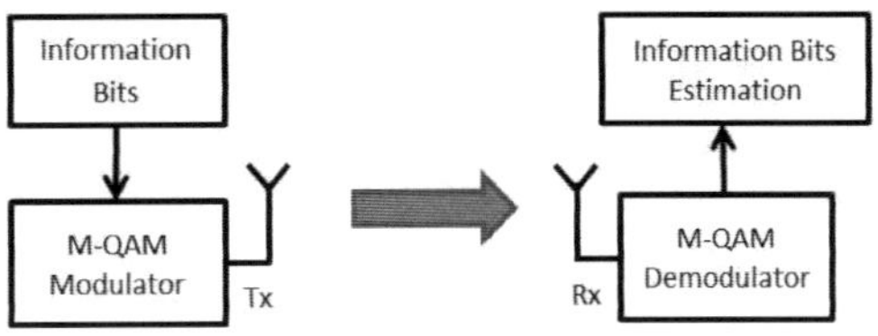

Figiure 1. Simulation block of SISO with M-QAM Modulation

The next step after simulation is the implementation SISO using M-QAM modulation in WARP module. The block diagram of SISO implementation using M-QAM modulation is shown in Figure 2. The number of bit transmitted is 30000 bit. The implementation of the system consists of M-QAM modulation, additional pilot and preamble, up-sampling, filter, up-converter, normalization in transmitter side, down-converter, match filter, down-sampling, preamble detection, canal estimation, and M-QAM demodulation in receiver side.

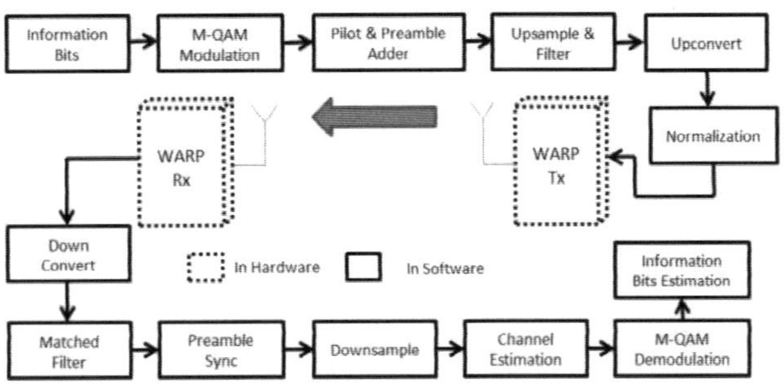

Figure 2. System implementation diagram of SISO with M-QAM Modulation

The implementation of the proposed method is following these steps. First, generating the data information in form of random bit, such as bit $u_i = [u(1)u(2) \dots u(n)]$ with n is the number of bit generated. Then form the symbols with the help of equation (1)

$$v = \sum_{i=1}^{k} u(i)\, 2^{k-i} \quad (1)$$

With k = log2 M and M is the modulation level.

The integer symbols will be processed as modulation input and translated into complex number that corresponding with the modulation in use. The results of the modulation is the complex number which have real and imaginary value. The symbols from the results of modulation can be described as

$$v_m = [v_1\ v_2\ v_3 \dots v_{(n/k)}] \quad (2)$$

The pilot symbol is added as a canal estimation response. The pilot symbol is inserted in every 12 symbol periodically. In receiver side, the process of symbol detection is done with canal estimation using pilot method. In this process, there are many assumption, such as canal is assumed has a flat

IOP Conf. Series: Materials Science and Engineering **190** (2017) 012046 doi:10.1088/1757-899X/190/1/012046

fading type and the number of tap is one. The convolution operation between transmitted symbols with response canal can be minimalized into multiplying operation because the number of tap is one. The canal estimation scheme is, in the transmitter side, the information symbols is inserted with a pilot symbol (estimator) which placed periodically. In the receiver side, the received symbols multiplied with the same scale from transmitter side. The value of symbols is known as the results of multiplying with estimator because the assumption of flat fading canal. For the canal estimation of the other symbols are done by point regression in estimator. The received symbols then used as linier point regression which is known as interpolation process.

Those symbols then added with preamble symbols to determine the location of the first information symbol in the receiver side. The number of preamble symbols are 13 symbols. The first symbol detection is done based on the higher correlation between the received signal sample and preamble sample. The next process is upsample. Upsample is used to improve the sampling rate of the transmitted data. This process is done by passing the symbols of the previous process into *Squared Root Raised Cosine* (SRRC) filter with oversampling rate 8. In receiver side, used match filter to eliminated high frequency.

The next process is upconvert in transmitter side and downconvert in receiver side. Each of signal samples from pulse-shaping filter output is then multiplied by each cosine and sine signal samples with frequency of 5 MHz. The upconvert and downconvert processes are consecutively expressed, as follow:

$$\begin{aligned} &exp\left(\sqrt{-1} \times 2\pi f_c t\right) \\ &exp\left(-\sqrt{-1} \times 2\pi f_c t\right) \end{aligned} \tag{3}$$

with $f_c$is sampling frequency, t is time.

The measurement of SISO communication system with QAM modulation is conducted in two differents environment: indoor and outdoor. Every enviroment is set to have LOS condition, The transmitter and receiver nodes are separated with 6 meters and antennas are put on 40 cm above the floor.

3. Analysis and Result

In this section, the result of SISO communication system with M-QAM simulation and implementation are elaborated. Simulation and implementation refer to diagram blok in Figure 1 and Figure 2, respectively. M-aray level modulations are 4, 8, and 16.

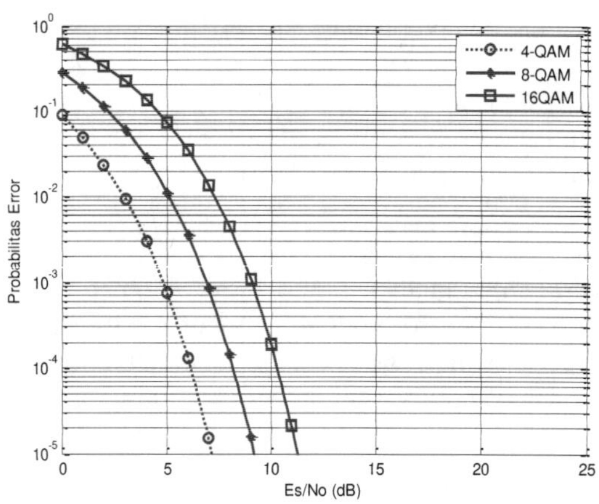

Figure 3. BER Vs Eb/No for SISO with M-QAM Modulation simulation

In the simulation, 4-QAM, 8-QAM, and 16-QAM performances are compared in terms of BER for several Eb/No, which are represented in Figure 3. From the figure 4, it can be discibed that 4-QAM is better than 8-QAM and 16-QAM, with BER values of 7.5×10^{-4}, 1.12×10^{-2}, and 7.4×10^{-2} ,respectively , at Eb/No of 5 dB. When the value Eb/No is increased, 4-QAM is better achieved. The next, the results of performance measurement of SISO communication system with 4-QAM, 8-QAM, and 16-QAM in indoor enviroment are shown in Figure 4.

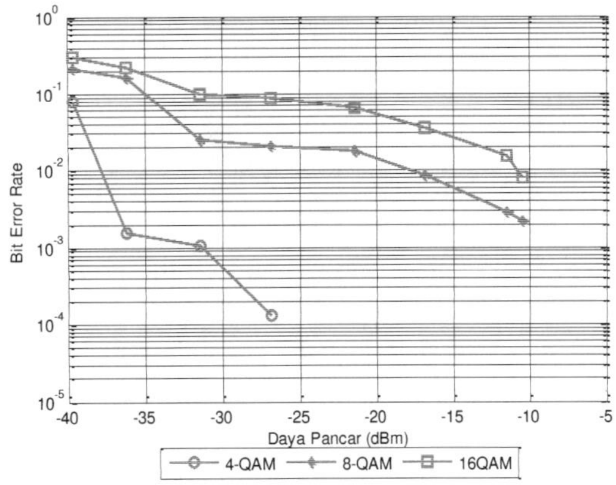

Figure 4. Performance comparison of 4-QAM, 8-QAM, and 16-QAM in indoor with LOS condition

From the measurement, it can be descibed that the performance 4-QAM is better than 8-QAM and 16-QAM for all various transmitted power. At trasmit power -26 dBm, 4-QAM, 8-QAM, and 16-QAM have BER value of each 1.33×10^{-4}, 2.078×10^{-2}, 8.76×10^{-2}, respectively. The form of slopping curve in simulation and measurement results is different because of the multipath condition that occur in measurement process. The results of performance measurement of SISO communication system with 4-QAM, 8-QAM, and 16-QAM in outdoor enviroment are shown in Figure 5.

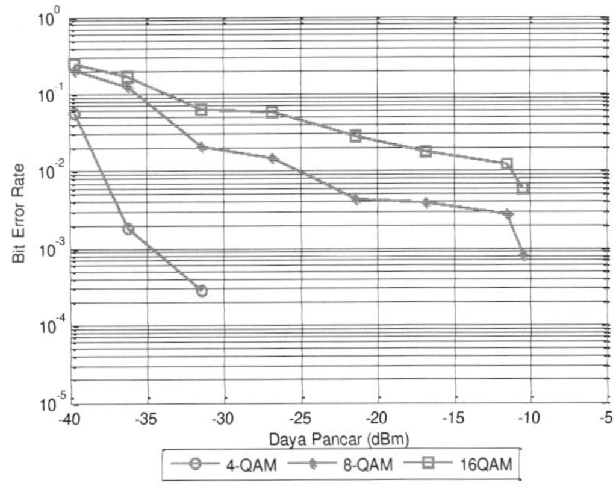

Figure 5. Performance comparison of 4-QAM, 8-QAM, and 16-QAM in outdoor with LOS condition

From the measurement, it can be described that the performace 4-QAM is better than 8-QAM and 16-QAM for all various transmitted power. At transmit power -26 dBm, 4-QAM, 8-QAM, and 16-QAM have BER value of each 0, 1.497×10^{-2}, 5.928×10^{-2}, respectively. The form of slopping curve in simulation and measurement results is different because of the multipath condition that occur in measurement process.

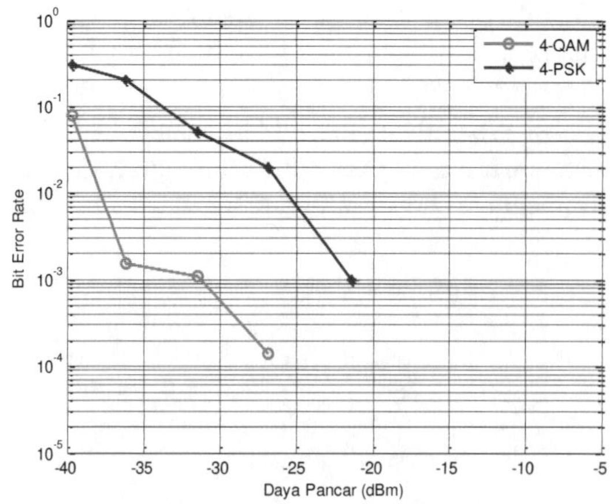

Figure 6. Performance comparison of 4-QAM and 4-PSK in indoor with LOS condition

The comparizon performance of 4-PSK and 4-QAM are shown in Figure 6. The result of 4-PSK is conducted in previous research. [5] The measurement, described that the performace 4-QAM is better than 4-PSK for all various transmitted power. At transmitter power -26 dBm, 4-QAM BER value is 1.33×10^{-4}, while 4-PSK BER value is 2×10^{-2}. At another transmitter power, BER value of 4-QAM always smaller than 4-PSK.

4. Conclusion

The implementation with SISO communication using M-QAM modulation has been shown. The performance of the system is evaluated by BER for 4-QAM, 8-QAM, and 16-QAM modulation in indoor and outdoor environment using LOS condition. From the measurements, the best performance is achieved by using SISO with 4-QAM modulation in outdoor environment. The implementation using 4-QAM produce smaller BER value than using 4-PSK modulation. Thus, SISO using M-QAM can repair the performance of M-PSK, this is indicated by low BER value and saving transmitter power.

References

[1] Li, Bo. 2011 *Analysis and Design of Software Defined Radio* IEEE International Conference on Internet Computing and Information Services p 415

[2] Grayver, Eugene 2013 *Implementing Software Defined Radio* New York Springer 978-1-4419-9331-1

[3] http://warp.rice.edu/trac/

[4] Murphy P, Sabharwal A, and Aazhang B 2006 *Design of WARP: A Wireless Open-Access Research Platform* Florence, Italy European Signal Processing Conference (EUSIPCO 2006)

[5] Hapsari JP, Suwadi, and Wirawam 2014 *Implementasi Sistem Komunikasi SISO Berbasis Wireless Open Access Research Platform (WARP)* Yogyakarta Seminar Nasional Teknologi Informasi dan Komunikasi 2089-9813

[6] Darwis RS, et all *Implementation and Performance Analysis of Alamouti Algorithm for MIMO 2x2 Using Wireless Open-Access Research Platform (WARP)* Semarang Internatinal Conference on Information Technology, Computer and Electrical Engineering (ICITACEE) 978-4799-6432-1

[7] Anisah I, Suwadi, and Wirawan 2013 *Implementasi Sistem Komunikasi Kooperatif pada Modul Wireless Open Access Research Platform* Jember Seminar Nasional Energi 3 978-602-L49L7-L-3

[8] Suwadi, Suryani T, and Anisah I 2016 *Performace Analysis Of Cooperative System Using*

Wireless Opes Access Research Platform For Indoor And Outdoor Enviroment Journal of Theoretical and Applied Information Technology Vol 83 1992-8645

[9] Chiba A, et all 2010 *16-level Optical Quadrature Amplitude Modulation using Monolitic Quad-Parallel Mach-Zehnder Optical Modulator and Signal Transmission over 75-km Sigle-Mode Fiber* IEEE

[10] Chen S, et all 2012 *Minimum Bie Error Rate Beamforming Receiver for Space-Division* IEEE

[11] Park, Sung-Joon 2007 *Triangular Quadrature Amplitude Modulation* IEEE

Quality Test of Flexible Flat Cable (FFC) With Short Open Test Using Law Ohm Approach through Embedded Fuzzy Logic Based On Open Source Arduino Data Logger

Ajar Rohmanu, Yan Everhard

Program Studi Magister Ilmu Komputer, Universitas Budi Luhur
Jl. Ciledug Raya, Petukangan Utara, Kebayoran Lama, Jakarta Selatan 12260

ajarrohmanu@gmail.com, yan.everhard@budiluhur.ac.id

Abstract. A technological development, especially in the field of electronics is very fast. One of the developments in the electronics hardware device is Flexible Flat Cable (FFC), which serves as a media liaison between the main boards with other hardware parts. The production of Flexible Flat Cable (FFC) will go through the process of testing and measuring of the quality Flexible Flat Cable (FFC). Currently, the testing and measurement is still done manually by observing the Light Emitting Diode (LED) by the operator, so there were many problems. This study will be made of test quality Flexible Flat Cable (FFC) computationally utilize Open Source Embedded System. The method used is the measurement with Short Open Test method using Ohm's Law approach to 4-wire (Kelvin) and fuzzy logic as a decision maker measurement results based on Open Source Arduino Data Logger. This system uses a sensor current INA219 as a sensor to read the voltage value thus obtained resistance value Flexible Flat Cable (FFC). To get a good system we will do the Black-box testing as well as testing the accuracy and precision with the standard deviation method. In testing the system using three models samples were obtained the test results in the form of standard deviation for the first model of 1.921 second model of 4.567 and 6.300 for the third model. While the value of the Standard Error of Mean (SEM) for the first model of the model 0.304 second at 0.736 and 0.996 of the third model. In testing this system, we will also obtain the average value of the measurement tolerance resistance values for the first model of - 3.50% 4.45% second model and the third model of 5.18% with the standard measurement of prisoners and improve productivity becomes 118.33%. From the results of the testing system is expected to improve the quality and productivity in the process of testing Flexible Flat Cable (FFC).

1. Introduction

Technological developments will affect all the changes in all fields, one of which is the development of hardware electronic device that would change very quickly and significantly, one of which is the media interface cable that connects the main board Printed Circuit Board (PCB) with the other parts of the device the electronics. Interface is now widely used in electronic devices is a Flexible Flat Cable (FFC) which has a very thin design and size so they can follow the shape of the mechanical groove design an electronic device in which case it requires a wide space for the placement of the cable in the design of electronic devices proficiency level. The manufacturing process of Flexible Flat Cable (FFC) will go through the testing process before it is sent to the company that booked Flexible Flat Cable (FFC). Testing Flexible Flat Cable (FFC) one of which is to identify whether Flexible Flat Cable (FFC) in conditions PASS (Good) or FAIL (No Good) by providing a flow of DC voltage and output indicator lamp in the form of Light Emitting Diode (LED).

The testing process is separated into two stages: the first stage is to test the condition of the Open (disconnected) for each leg of the Flexible Flat Cable (FFC) and the second stage is to test the condition Short (connected) between the legs on a Flexible Flat Cable (FFC) do manually by the operator and the determination of Flexible Flat Cable (FFC) to observe the condition of the lights light Emitting Diode[1]. This study will be made of quality test system Flexible Flat Cable (FFC) with Short Open Test using Ohm's Law through the Embedded Fuzzy Logic approach based on Open Source Arduino Data Logger. With the research and manufacture of these systems, we can get the precision measurement according to Ohm's law and gain level measurement speed and better accuracy in determining the condition of Flexible Flat Cable (FFC) PASS (Good) or Fail (No Good), in addition to the system created can be obtained reports on the results of each measurement will be stored in memory and Data Logger which can be used as a report on the production of Flexible Flat Cable (FFC).

Identification problems research is:

1. The system is used in determining the results of tests and measurements using indicators Light Emitting Diode (LED), so if there is a difference in resistance value of each foot Flexible Flat Cable (FFC) will be difficult to distinguish because there is no value measurement and display only the light intensity of Light Emitting Diode (LED).
2. The process of testing and measurement requires the operation by the operator consists of two stages: test and measurement Short (connected) and Open (disconnected).
3. There is no report in real time the results of the testing process and measurement Flexible Flat Cable (FFC).
4. Decision-making test results Flexible Flat Cable (FFC) is determined by the operator that runs the system by observing the condition of the lights Light Emitting Diode (LED).

Based on the identification of the problem, the problem is limited as follows:

1. Flexible Flat Cable (FFC) used testing has eight legs with UL Style No. 2896.
2. The method used in this research is the method of Short Open Test, Ohm's Law approach to measuring 4-wire (Kelvin) and Fuzzy Logic.
3. Embedded System that was designed and built an Arduino-based system minimum and Flow Sensor INA219.
4. In the store the measurement data using Data Logger Shield Arduino.

Based on the identification and restriction problems that have been described, it can be formulated how the test of the quality Flexible Flat Cable (FFC) can be done by an embedded system based open source arduino with the application of methods Short Open Test is using a legal approach Ohm 4-wire (Kelvin) as a process measure value detainee a Flat Flexible Cable (FFC) and fuzzy logic as decision makers the results of measurements and data logger as a data storage processing of the measurement results.

2. Related Research

In this study [2] do Arduino as an LCR meter designs using the AD5933 impedance converter for measuring reactance. The system incorporates an on board frequency generator with 12-bit, 1 MSPS and Analog to Digital Converter (ADC). Impedance response signal sample from Analog to Digital Converter (ADC) and signal Discrete Fourier Transform (DFT) is processed by on-board DSP engine. The results of processing that can be read from the I2C serial interface. Arduino controls the AD5933 via a serial interface I2C protocol. Measurements for high value capacitance and inductance values lower in impedance in the range of 1 Ω - 10 Ω by using additional circuit reference resistors are added to the reactive component of the unknown and make corrections to the measured impedance value.

In this research [4] is to create a weather station to monitor the characteristics of lightning. Data Logger will be SD Card Shield is used to store data read distance of a lightning strike, the frequency as well as the movement of the lightning from the sensor lightning and GPS data from this study has the advantage of which consumes low power, small size, greater portability and lower cost.

In this research [4] is to create a tool to be used to test the functionality of a Cable Insulation Displacement (IDC) whether there is an error or malfunction. This tool uses Microcontroller ARM7 LPC and 74HC574 octal D-latch with a test gives a value of 1 and 0 binary data and compared with the logic of Ex-OR, output using the LCD 24 * 2 as an indicator to display error messages Short or Open, the testing has been done several scenario combination of methods Short Circuit and Open Circuit.

In this research [5] is to design a measuring instrument for voltage, current and frequency of the alternating current electric single phase-based personal computer where there are two, namely microcontroller and computer hardware for implementation by generating an error average of 0.45% voltage measurement, current measurement by 2 % and a measurement frequency of 0.09%.

3. Research Methodology

3.1. Research Method

The method used in this research is the method of Short Open Test using Ohm's law approaches to measurement with 4-wire (Kelvin). Short Open Test is the method used to determine the condition of a circuit no current flow or not. In Short Open Test method, a circuit will be given voltage so the current will flow. Circuit condition is said to be "Short" if the current flowing is huge because of the resistance value close to 0 Ω (Ohm) and the circuit is said to be "open" if the current does not flow or a current value of 0 Ampere because the resistance value is very large or infinite. Ohm's Law approach method 4-wire (Kelvin) is used to measure a resistance value which is smaller than a conductor material with the aim of getting more value detainees measurement precision. The next method is fuzzy logic is implemented in an Arduino-based Embedded System Data Logger. Fuzzy logic is used to determine whether the measurement results in a condition Pass (Good) or Fail (No Good) on a Flexible Flat Cable (FFC). The resulting value is the measurement process that has been done is stored in a data logger that will be used as a place to store measurement results and used as a production report.

System development method used in this study is the waterfall method, and for testing the system using the Black-box testing other than that the system is built to be compared with the results of measurements from other instruments and compared with the results of the calculation of an equation of the resistance value of a conductor and using standard deviation and tolerance to get the value of the accuracy of the measurement.

3.2. Sampling Selection Method

The sample selection based on quota sampling method for samples taken based on the needs of researchers and predetermined criteria or condition. Samples are Flexible Flat Cable (FFC) that have been determined by the number of feet 8 specification which has a length of each is 210 mm, 420 mm and 630 mm. Samples Flexible Flat Cable (FFC) in good condition and will be prepared several scenarios for testing a system that meets the measurement function Short Open Test conductor resistivity values.

3.3. Data Collection Method

To get the data in this study used multiple methods of data collection that the method of data collection by direct observation of the test system Flexible Flat Cable (FFC) which is running at the moment, but it uses the way literature by studying various references such as document / file, datasheet, scientific papers / journals and collect data relating to Flexible Flat Cable (FFC) and study the measurement method to obtain the resistance value of a conductor.
In this study used two groups of data, data about the group's first standard specification Flexible Flat Cable (FFC) in accordance with UL2896 and eungsung-ind is used as reference for the calculation of the resistance value of a conductor. With this data we can get the value of resistance Flexible Flat Cable (FFC) on each leg by performing calculations with equations $R = \frac{\rho \cdot l}{A}$. Group second data is a group of data relating to Current Sensor INA219, Minimum System Embedded Arduino Data Logger as a test system Flexible Flat Cable (FFC) which serves to determine the resistance value of each foot on a Flexible Flat Cable (FFC) with the process of measuring the flow and voltage.

3.4. Framework Concepts

Proposed a system of quality testing Flexible Flat Cable (FFC) with Short Open Test Using Ohm's Law through the Embedded Fuzzy Logic approach is based on Open Source Arduino Data Logger. The purpose of the proposal is to change the tests and measurements manually before becoming computationally and utilize measurement data as a report of production so that it can be used as feedback to improve the quality. By way of computing all the testing, measurement and determination of the results of measurements performed by a program embedded on a board that is based on Arduino, thus avoiding errors in the determination of the quality Flexible Flat Cable (FFC).

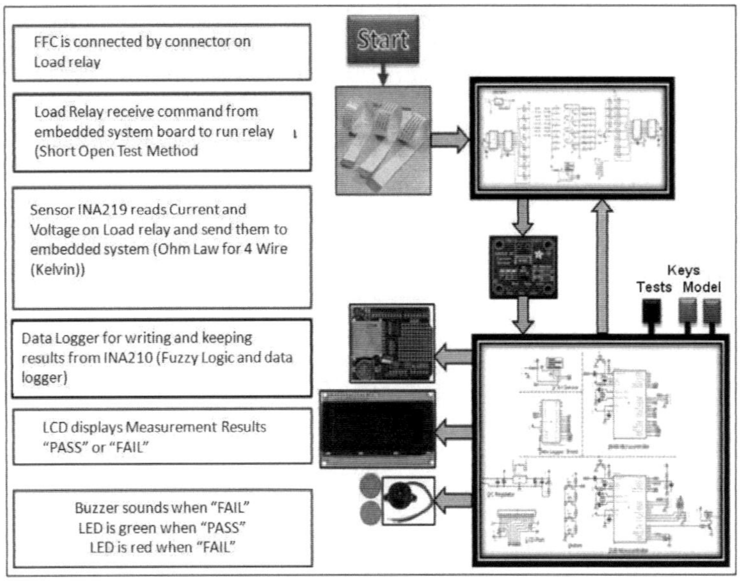

Figure 1. Conceptual framework

Figure 1 shows the overall system design. Following further clarification of the system design:

1. Flexible Flat Cable (FFC) mounted on the connectors available and connected on Load Relay Board.
2. Press the model to choose a model that will run and press the test button, the Embedded System will command Load Relay Board works according to the methods short open test, then the sensor reading INA219 current and voltage according to the instructions of embedded systems.
3. The process of testing and measurement systems performed sequentially from the first leg Flexible Flat Cable (FFC) until the last leg Flexible Flat Cable (FFC).
4. Embedded System to process data of current and voltage measurements with fuzzy logic further determines that the Flexible Flat Cable (FFC) in a test in good condition with the LCD indicator "PASS" Light Emitting Diode (LED) light green or unfavorably with the LCD indicator " FAIL "Light Emitting Diode (LED) lights up red.

4. Result of Research

4.1. Overview

Step-by-step testing process:

1. Prepare the tools used and turn the key in the ON position.
2. Press the switch up to the ON position.
3. Enter a Flat Flexible Cable (FFC) as a target to be checked on a connector that has been prepared.
4. Observe the Light Emitting Diode on the instrument, if there is one or more lights on, the Flexible Flat Cable (FFC) in conditions No Good (NG) Short.

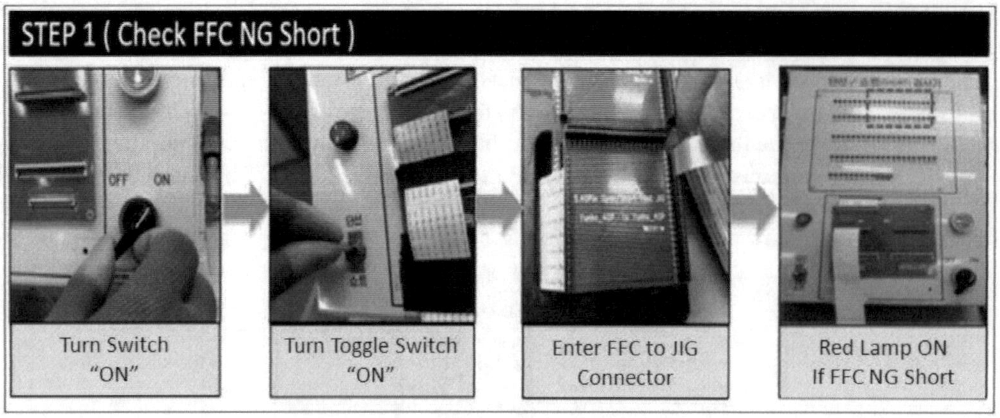

Figure 2. Short-checking process flow

Step-by-step testing process:
1. Prepare the tools used and turn the key in the ON position.
2. Press the switch up to the ON position.
3. Enter a Flat Flexible Cable (FFC) as a target to be checked on a connector that has been prepared.
4. Connect one end of Flexible Flat Cable (FFC) with an iron plate or other materials that can conduct current.
5. Observe the Light Emitting Diode on the instrument, if there is one or more lights there were dead, then Flexible Flat Cable (FFC) in conditions No Good (NG) Open.

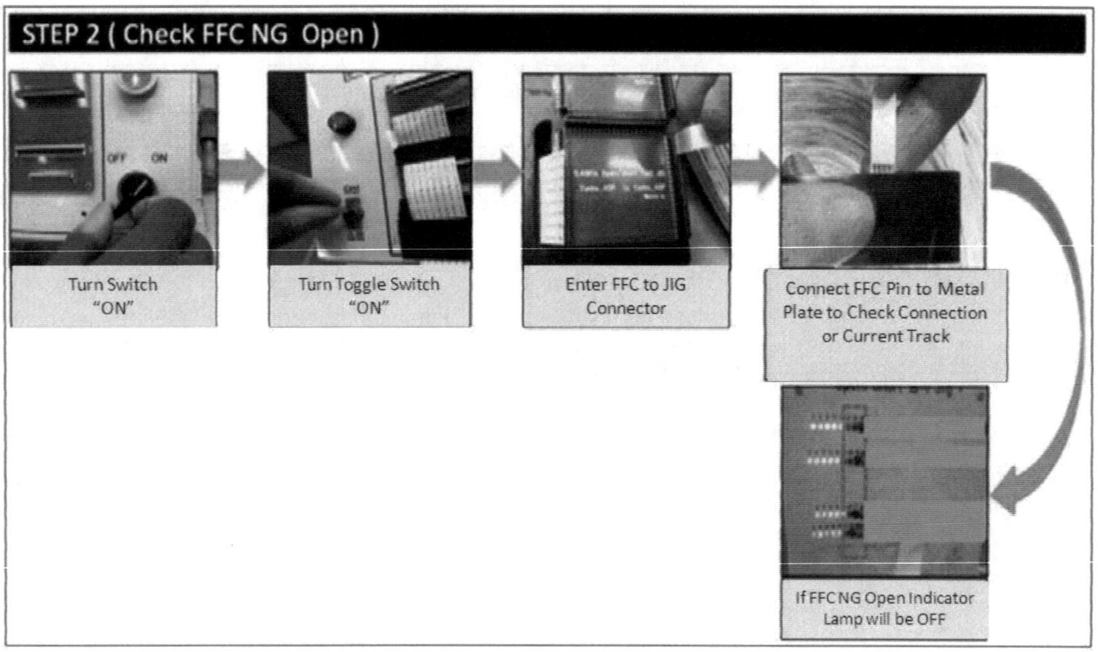

Figure 3. Open-checking process flow

4.2. Implementation System

In the implementation of the system in this study consists of two parts, namely the implementation of the hardware and software implementation on the part. Implementation of this research jointly establishes a system that works in computing. Before implementing the system must understand and process flow to load the system so that the implementation can be done in accordance design. This is a process flow system from the beginning to the end of the process to produce the output of the system.

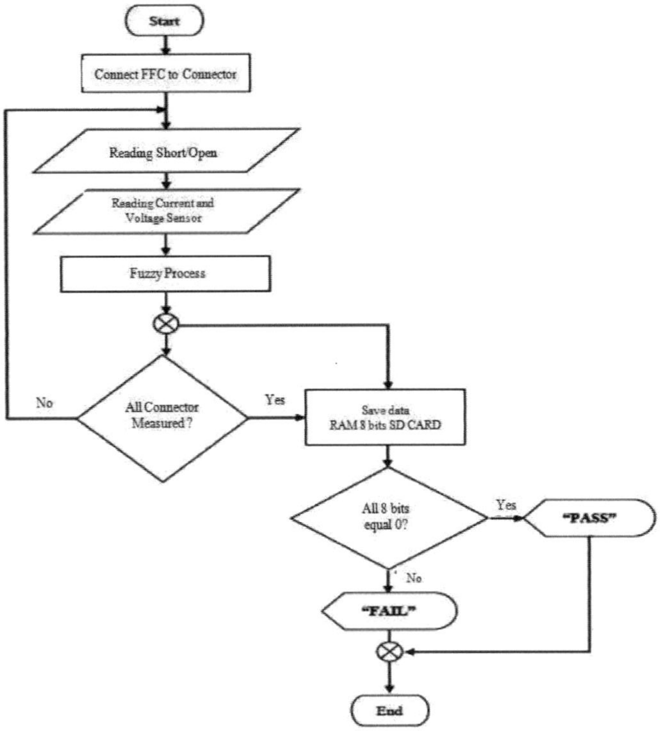

Figure 4. Process Flow System

4.2.1. Hardware

Hardware Implementation of the system in the study includes the design Schematics, PCB layout design and PCB Assembly. On the hardware is made separately into two parts which will cooperate with each other, namely the Load Relay Board and the Main Control Board Arduino. On Load Relay Board has the function of running the command from the main control that implements the method of short open reading test for voltage and current values, while the hardware implementation can be seen in Figure 5.

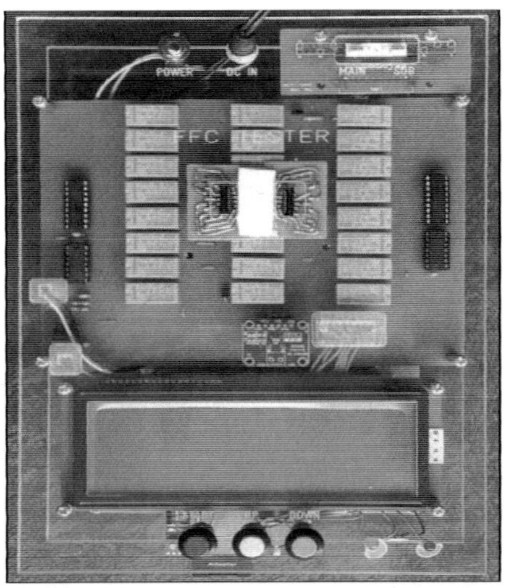

Figure 5. Hardware Implementation

4.2.2. Software
In this study determined that the current value will be streamed at 88.51 mA. In this study, designed the fuzzy logic process using Sugeno Model and Method Height defuzzification method.

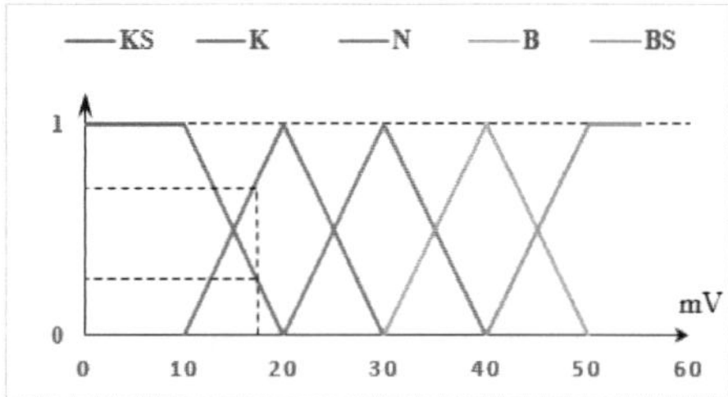

Figure 6. Domain Voltage Membership Function

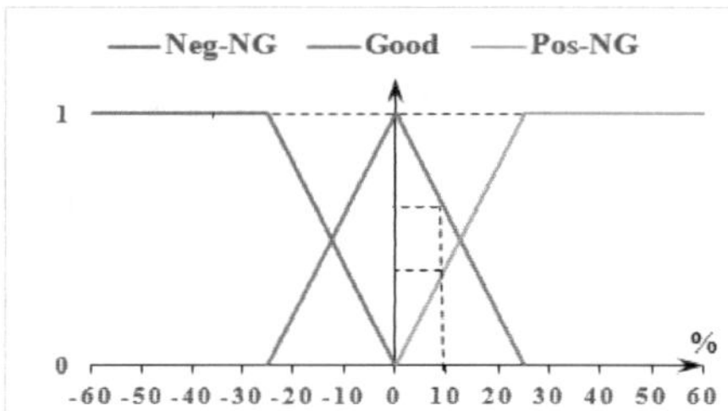

Figure 7. Domain Threshold Membership Function

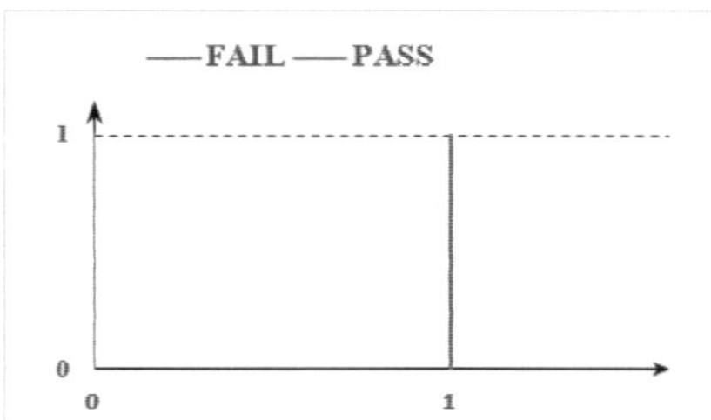

Figure 8. Output Fuzzy Singleton

Table 1. Rule Base Fuzzy Logic

		Voltage (V)				
		Very Small	Small	NORMAL	BIG	Very BIG
Thresshold [Th]	Neg-NG	Fail	Fail	Fail	Fail	Fail
	Good	Fail	Pass	Pass	Pass	Fail
	Pos-NG	Fail	Fail	Fail	Fail	Fail

In the process of fuzzy there are two variables that will be used as a variable input voltage shown in Figure 6 and a variable threshold seen in Figure 7. The variable voltage graph that there are 5 3 2 graphics of graphic triangle and trapezoid, Whereas the threshold is 1 and 2 triangular charts trapezoid graph so that it can be made fuzzy rule base as much as 15 as shown in table 1. The following is a fuzzy logic flowchart:

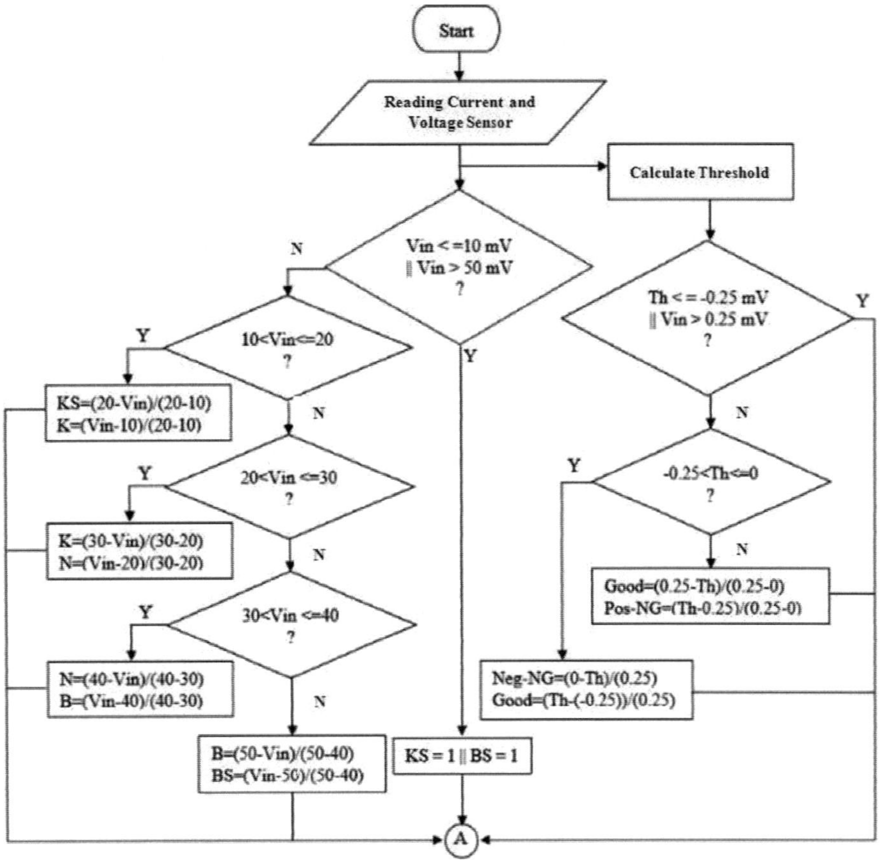

Figure 9. Flowchart Fuzzy

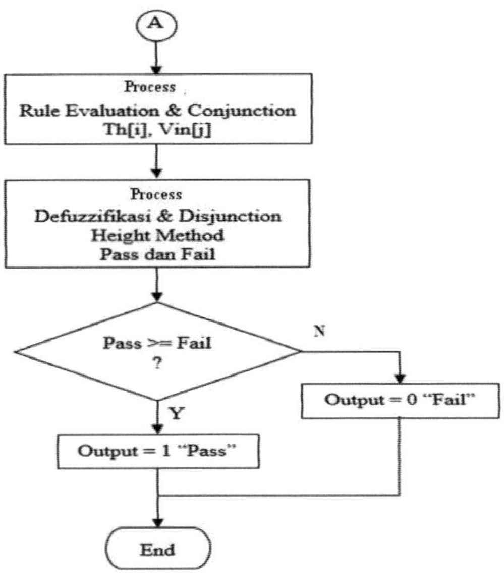

Figure 10. Flowchart Fuzzy next

From Figure 9 and Figure 10 can be seen that the variable voltage and threshold in through using fuzzy logic to output a singleton are pass or fail as shown in Figure 8.

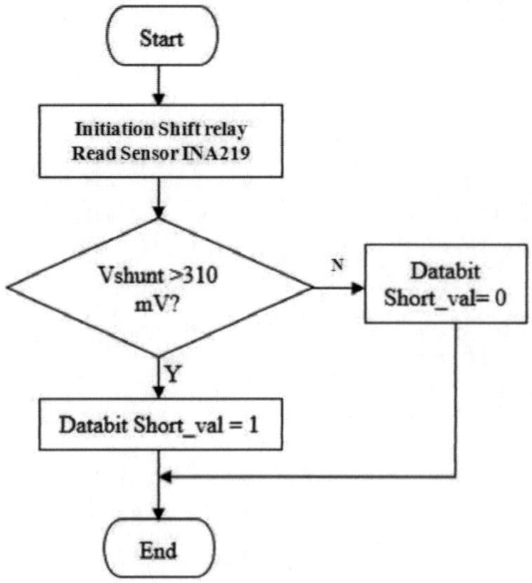

Figure 11. Flowchart Short Open

Figure 11 is a flowchart of the process Short Open test to determine whether the FFC has Short or Open conditions.

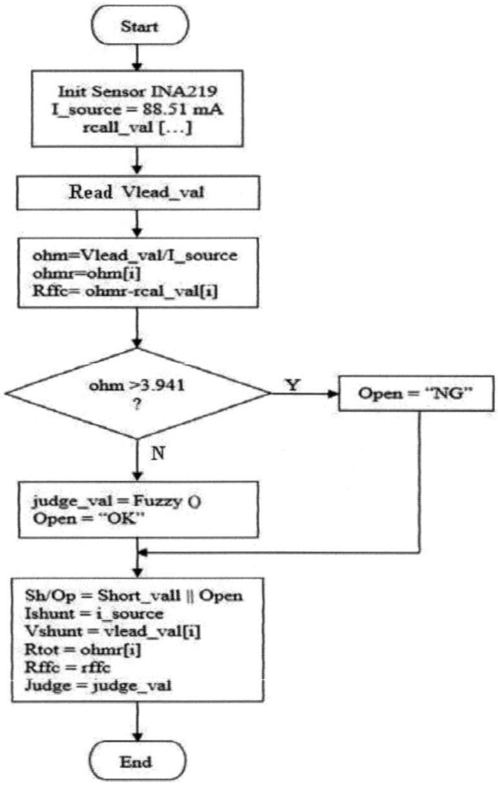

Figure 12. Flowchart Voltage Measure

Whereas Figure 12 describes a process flow measurement and data processing in order to obtain the FFC resistance value and determine whether the measurement value in accordance with the existing specifications.

4.3. *Evaluation and Testing*

4.3.1. *Accuracy, Precision and Tolerance Testing*

Testing accuracy, precision and tolerance is focused on the output data from the system functions. The method used is the standard deviation method where precision is determined by the diversity of the variables to be measured against the average, while the value of accuracy depends on the standard deviation and the number of samples and tolerance is an aberration measurements with actual values. In this test uses 40 measurement data on each model or use 5 Flexible Flat Cable (FFC) by the number of feet 8 and the value of the measurement data in units of mΩ. Here is the measurement data that has been done by the system.

Table 2. The Test Data FFC 210 mm

No	Item	K1	K2	K3	K4	K5	K6	K7	K8
1	$FFC_{(1)}$	109	109	105	112	111	106	107	108
2	$FFC_{(2)}$	111	110	105	112	112	106	108	108
3	$FFC_{(3)}$	112	111	109	107	108	106	109	108
4	$FFC_{(4)}$	111	108	109	109	108	109	110	111
5	$FFC_{(5)}$	111	109	110	110	108	109	110	111

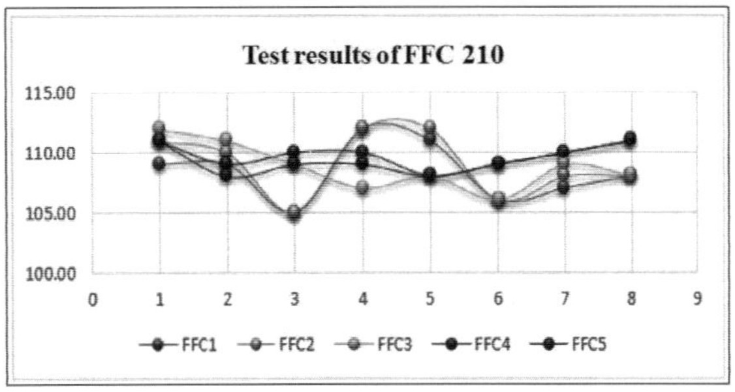

Figure 13. Graph Data Test Result FFC 210

Table 3. The Test Data FFC 420 mm

No	Item	K1	K2	K3	K4	K5	K6	K7	K8
1	$FFC_{(1)}$	236	237	233	236	233	236	239	247
2	$FFC_{(2)}$	236	236	233	236	232	235	239	246
3	$FFC_{(3)}$	231	232	236	229	232	227	232	231
4	$FFC_{(4)}$	233	234	223	233	231	238	232	236
5	$FFC_{(5)}$	232	232	223	231	229	237	231	234

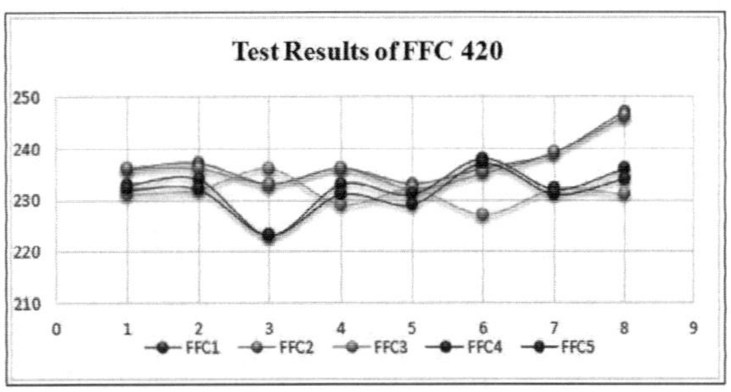

Figure 14. Graph Data Test Result FFC 420

Table 4. The Test Data FFC 630 mm

No	Item	K1	K2	K3	K4	K5	K6	K7	K8
1	$FFC_{(1)}$	369	369	348	354	356	357	363	354
2	$FFC_{(2)}$	356	356	344	355	353	363	353	360
3	$FFC_{(3)}$	358	359	353	372	369	350	352	352
4	$FFC_{(4)}$	355	355	344	354	352	363	353	359
5	$FFC_{(5)}$	355	360	355	351	356	363	360	352

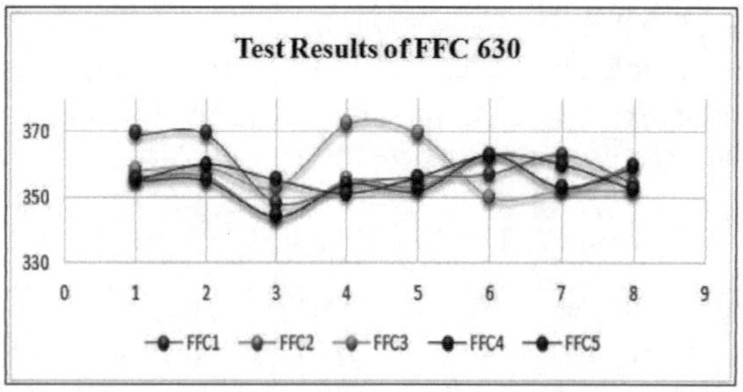

Figure 15. Graph Data Test Result FFC 630

From Table 2, Table 3 and Table 4 can be obtained values are used as the calculation of standard deviation and standard error of the mean by the following equation:

$\boldsymbol{S} = \sqrt{\frac{\Sigma(X-\bar{X})^2}{n-1}}$ dan $\boldsymbol{SE}_{\overline{X}} = \frac{s}{\sqrt{n}}$ so that the values obtained as follows:

Value mean FFC 210 mm : 4362 mΩ / 40 = 109,05 mΩ

It can be calculated standard deviation is

$$S = \sqrt{\frac{143,90}{40-1}} = \sqrt{3,690} = 1,921$$

While the value of the standard error is $SE_{\overline{X}} = \frac{S}{\sqrt{n}} = \frac{1,921}{\sqrt{40}} = \frac{1,921}{6,325} = 0,304$

Value mean FFC 420 mm : 9349 mΩ / 40 = 233,725 mΩ

It can be calculated standard deviation is

$$S = \sqrt{\frac{845,975}{40-1}} = \sqrt{21,692} = 4,567$$

While the value of the standard error is $SE_{\overline{X}} = \frac{S}{\sqrt{n}} = \frac{4,567}{\sqrt{40}} = \frac{4,657}{6,325} = 0,736$

Value mean FFC 630 mm : 14262 mΩ / 40 = 356,550 mΩ.

It can be calculated standard deviation is

$$S = \sqrt{\frac{1547,900}{40-1}} = \sqrt{39,690} = 6,300$$

While the value of the standard error is $SE_{\overline{X}} = \frac{S}{\sqrt{n}} = \frac{6,300}{\sqrt{40}} = \frac{6,300}{6,325} = 0,996$

In the method of standard deviation is stated that the smaller the standard deviation, the better and the less the value of the standard error of the average value closer to the true value. From the test results can be obtained under the standard deviation and standard error is the smallest is the measurement of Flexible Flat Cable (FFC) with a length of 210 mm with a length of 420 mm next and last is 630 mm.

Table 5. Tolerance Measurement FFC 210 mm

No	Item	FFC Test (mΩ)	$\frac{\rho.l}{A}$ (mΩ)	Difference	Tolerance
1	$FFC_{(1)}$	108,375	113	-4,625	-4,09%
2	$FFC_{(2)}$	109,000	113	-4,000	-3,54%
3	$FFC_{(3)}$	108,750	113	-4,250	-3,76%
4	$FFC_{(4)}$	109,375	113	-3,625	-3,21%
5	$FFC_{(5)}$	109,750	113	-3,250	-2,88%
Average					-3,50%

Table 6. Tolerance Measurement FFC 420 mm

No	Item	FFC Test (mΩ)	$\frac{\rho.l}{A}$ (mΩ)	Difference	Tolerance
1	$FFC_{(1)}$	237,125	226	11,125	4,92%
2	$FFC_{(2)}$	236,625	226	10,625	4,70%
3	$FFC_{(3)}$	231,250	226	5,250	2,32%
4	$FFC_{(4)}$	232,500	226	6,500	2,88%
5	$FFC_{(5)}$	231,125	226	5,125	2,27%
Average					3,42%

Table 7. Tolerance Measurement FFC 630 mm

No	Item	FFC Test (mΩ)	$\frac{\rho.l}{A}$ (mΩ)	Difference	Tolerance
1	$FFC_{(1)}$	358,750	339	19,750	5,83%
2	$FFC_{(2)}$	355,000	339	16,000	4,72%
3	$FFC_{(3)}$	358,125	339	19,125	5,64%
4	$FFC_{(4)}$	354,375	339	15,375	4,54%
5	$FFC_{(5)}$	356,500	339	17,500	5,16%
Average					5,18%

Average tolerance values obtained from tests performed on Flexible Flat Cable (FFC) in accordance with Table 5, Table 6 and Table 7 that with a length of 210 mm is - 3.50%, Flexible Flat Cable (FFC) with a length of 420 mm is 3.42%, Flexible Flat Cable (FFC) with a length of 630 mm was 5.18%. So it can be stated that the measurements made by the system met the standard measurement tolerance prisoners are allowed where the maximum tolerance is ± 25%.

4.3.2. Process Time Testing (Take Time)
At the time of the testing process (take time) this is by comparing the processing time of the system that has been running with the processing time of the system that do the research. The turnaround time on a running system is obtained by looking at the data production targets and production per hour. The following is a comparison of the processing time from a running system with a system that performed the study.

Table 8. Comparison Data Processing Time (Take Time)

Line	Item	Running System			Take Time FFC Tester		
		Target (pcs)	Result (pcs)	Take Time (ms)	Test 1 (ms)	Test 2 (ms)	Estimation Result
1	FFC_210	360	331	10.876	8.471	8.431	426
2	FFC_210	360	329	10.942			
3	FFC_210	360	328	10.976			
4	FFC_420	360	326	11.043	8.491	8.411	426
5	FFC_420	360	327	11.009			
6	FFC_420	360	325	11.077			
7	FFC_630	360	324	11.111	8.451	8.491	425
8	FFC_630	360	326	11.043			
9	FFC_630	360	323	11.146			

From table 8 for systems running data showed that the target every hour testing process Flexible Flat Cable (FFC) for 3 specifications of 360 pcs not met because the production of an average of 91% of its production target, while the processing time (take time) the mean average test every Flexible Flat Cable (FFC) on systems running for FFC_210 is 10.931 ms, for FFC_420 is 11.043 ms and FFC_630 is 11,099 ms. While the system results were obtained when the process (take time) that FFC_210 is 8.451 ms so that the estimated yield is 426 pcs per hour, while the process (take time) on FFC_420 is 8,451ms so estimates of production per hour is 426 pcs and processing time (take time) on FFC_630 is 8.457 ms so that the estimated output per hour is 425 pcs. From the data obtained that the test results using the computing system to increase the productivity of the target of 426 or 360 pcs be increased to 118.33%.

IOP Conf. Series: Materials Science and Engineering **190** (2017) 012047 doi:10.1088/1757-899X/190/1/012047

4.3.3. Implications Research
Flexible Flat Cable (FFC) is essential in electronic devices; the quality Flexible Flat Cable (FFC) is very influential on electronic products. Because of the presence of the test system quality Flexible Flat Cable (FFC) is very useful for producing quality Flexible Flat Cable (FFC) is good and improve productivity, where the resistance value on Flexible Flat Cable (FFC) measurement and testing in accordance with the standards that have been determined so that the performance of the Flat Flexible Cable (FFC) used in electronic devices can work optimally.

5. Conclusion

5.1. Conclusion
Based on the discussion of research results Quality Test Flexible Flat Cable (FFC) with Short Open Test using Ohm's Law through the Embedded Fuzzy Logic approach based on Open Source Arduino Data Logger, it can be concluded as follows:

1. Short Open Test Method, Fuzzy Logic and Data Logger can be applied to Embedded System Open Source Arduino-based test system as quality Flexible Flat Cable (FFC).
2. Black box testing results show that the system is free of syntax errors and logic errors and functionally release the results as expected.
3. Results of testing the accuracy and precision of the system with a standard deviation can be obtained results for Flat Flexible Cable (FFC) length of 210 mm is 1.921 to 4.567 lengths is 420 mm and 630 mm length was 6,300. While the value of the Standard Error of Mean (SEM) on a Flat Flexible Cable (FFC) length of 210 mm is 0.304 to 420 mm length is 0.736 and the length of 630 mm is 0.996.
4. The test results can be obtained tolerance on the system tolerance values averaging measurements prisoners on Flexible Flat Cable (FFC) for a length of 210 mm was -3.50%, length 420 mm is 3.42% and the length of 630 mm was 5.18 %.
5. The test results processing time (take time) on the system can increase production to 118.33% or 18.33% above the production target of 360 pcs per hour.

5.2. Suggestion
Based on the results of research by the author, advice to authors give is as follows:

1. The research may be continued by changing the quality of the hardware so that the level of accuracy, precision and tolerance is heading for the better.
2. Research can proceed with implementing the system using other Open Source Embedded systems and integrated with other systems based on object-oriented to support the production process.

References

[1]. I. LGein, "FFC_JIG Open Short," LGein, Bekasi, 2015.

[2]. K. Zuhri, "pengukuran akurasi dan presisi," 2015. [Online]. Available: http://www.sebatekno.com/pengukuran-akurasi-dan-presisi. [Accessed 14 May 2016].

[3]. Zonaelektro, "Besaran Listrik (Arus, Tegangan, Hambatan Dan Daya Listrik)," 2014. [Online]. Available: www.zonaelektro.net. [Accessed 12 Maret 2016].

[4]. H. Zhang, C. Z. Rengui Lu and Y. Zhao, "On-line Measurement of Internal Resistance of Lithium Ion Battery for EV and its Application Research.,"*International Journal of U-& E-Service, Science and Technology,* vol. 7, no. 4, pp. 301-310, 2014.

[5]. Wikipedia.com, "Flexible flat cable," 2016. [Online]. Available: https://en.wikipedia.org/wiki/Flexible_flat_cable. [Accessed 10 Februari 2016].

[6]. Wecorporation, Theory R, C, L Measurement & Reference Debugging for CMS-9000 Series, 02 ed., Seoul: wecorporation.inc, 2002.

[7]. S. Wasito, Vademekum Elektronika, 2nd ed., Jakarta: Gramedia Pustaka Utama, 1995.

[8]. ul.com, "Appliance Wiring Material," 2011. [Online]. Available: http://data.ul.com/link/stylepage.aspx?style=2896. [Accessed 12 Februari 2016].

[9]. S. Teli and C. Mani, "Smart Real Time Embedded Arduino Based Data Acquisition System,"*IJRET (International Journal of Research in Enggineering and technology),* vol. 4, no. 8, pp. 258-262, 2015.

[10]. a. Team, "Kelvin (4-wire) Resistance Measurement," 2015. [Online]. Available: www.allaboutcircuits.com. [Accessed 27 Desember 2015].

[11]. M. Syahwil, Panduan Mudah Simulasi dan Praktik Mikrokontroler Arduino, 1st ed., Jogjakarta: Andi Publisher, 2014.

[12]. P. Srividyadevi, D. Pusphalatha and P. Sharma, "Measurement of Power and Energy Using Arduino,"*Res. J. Eng. Sci,* vol. 2, no. 10, pp. 10-15, 2013.

[13]. sonoku, "Implementasi Fuzzy Logic Controller untuk Kontrol Kecepatan Motor DC pada Prototype Kipas Angin," 2015. [Online]. Available: http://sonoku.com/implementasi-fuzzy-logic-controller-untuk-kontrol-kecepatan-motor-dc-pada-prototype-kipas-angin/. [Accessed 4 April 2016].

[14]. F. Saftari, Proyek Robotika Keren dengan Arduino, Jakarta: Elex Media Komputindo, 2015.

[15]. RedgriftyBrick, "Electrical Engineering Short and Open Circuit," 2014. [Online]. Available: http://electronics.stackexchange.com. [Accessed 16 Maret 2016].

[16]. S. Rangkuti, Mikrokontroler ATMEL AVR Simulasi dan Praktek menggunakan ISIS Proteus dan CodeVisionAVR, Bandung: Informatika, 2011.

[17]. R. S. Pressman, Rekayasa Perangkat Lunak Pendekatan Praktisi, 7 ed., Yogyakarta: Andi, 2012.

[18]. petrockblog, "From-arduino-to-breadboard-with-a-minimum-configuration," 2012. [Online]. Available: http://blog.petrockblock.com/. [Accessed 11 Maret 2016].

[19]. P. Painjane, R. Nangare and A. Pardeshi, "Design of Solar Insolation Level Detector and Data Logger,"*INPRESSO (International Journal of Current Enggineering and Technology),* vol. 4, no. 3, pp. 2083-2087, 2014.

[20]. M. Nazir, Metode Penelitian, 6 ed., Bogor: Ghalia Indonesia, 2011.

[21]. R. Nasution, "Teknik Sampling," Universitas Sumatera Utara, Medan, 2003.

[22]. A. Mudigonda, "IDC cable Test Jig,"*IJECT ISSN 2230-9543,* vol. 6, no. 4, pp. 68-72, 2015.

[23]. S. Monk, "Measuring Voltage With An Ardunio," 2013. [Online]. Available: https://blog.the-jedi.co.uk. [Accessed 21 Maret 2016].

[24]. L. Miller, "A thermocouple datalogger based on the Arduino platform," 2010. [Online]. Available: http://lukemiller.org/index.php/2010/08/a-thermocouple-datalogger-based-on-the-arduino-platform/. [Accessed 17 Januari 2016].

[25]. A. P. Malvino, Prinsip prinsip Elektronik, 2nd ed., Jakarta: Erlangga, 1992.

[26]. F. Lukman, "SDLC Metode Waterfall Pengertian dan Alur," 2015. [Online]. Available: www.referensiinternet.blogspot.co.id. [Accessed 10 Maret 2016].

[27]. G. Loveday, Intisari Elektronika, 2nd ed., Jakarta: Elex Media Komputindo, 1992.

[28]. ladyada.net, " Logger Shield Datalogging for Arduino," 2013. [Online]. Available: http://www.ladyada.net/make/logshield/index.html. [Accessed 25 Maret 2016].

[29]. S. Kusumadewi and H. Purnomo, Aplikasi Logika Fuzzy untuk Pendukung Keputusan, 2 ed., Yogyakarta: Graha Ilmu, 2013.

[30]. I. H. Kurniawan and L. Hayat, " Design and Implementation of Voltage , Current and Frequency Measurement for Single Phase Alternating Current Based on Personal Computer,"*Techno ISSN 1410 - 8607,* vol. 15, no. 1, pp. 21-31, 2014.

[31]. D. Kho, "Teknik Elektronika Pengenalan LED," 2015. [Online]. Available: www.teknikelektronika.com. [Accessed 23 Januari 2016].

[32]. J. janesch, "Two-wire vs. four-wire resistance measurements," 2013. [Online]. Available: www.edn.com. [Accessed 20 Maret 2016].

[33]. M. Ismail, M. Haddin and A. Suprajitno, "Dengan adanya system akuisisi data yang mampu membaca tegangan dan arus pengukuran dan mengolahnya dengan metode,"*Prosiding SNATIF,* vol. 2, pp. 221-228, 2015.

[34]. I. Hariadi, Dasar-dasar Teknik Listrik Arus Searah, Jakarta: Erlangga, 1988.

[35]. M. Grusin, "Short and Open Circuits," 2013. [Online]. Available: https://learn.sparkfun.com. [Accessed 16 Maret 2016].

[36]. K. M. Gilbert, K. M. Samuel, K. I.Stanley and G. Nick, "Design of Low Cost Microcontroler-Based Lightning Monitoring Device,"*Kabarak Journal of Research & inovation,* vol. 3, pp. 32-40, 2015.

[37]. Eunsung-ind, "ULStyle No & Conductor Standart," 2015. [Online]. Available: http://eunsung-ind.com/sub4/sub4_4.php. [Accessed 04 Desember 2015].

[38]. A. Ellerbrock, A. Abdel-majeed and S. Tenbohlen, "Design and Building of a Cheap Smart Meter,"*IEEE,* pp. 1-3, 2012.

[39]. N. J. Djermanova, J. G. Kiss'ovski and V. A. Vatchkov, "Portable Arduino - Based LCR – Meter,"*Annual Journal of Electronics,* pp. 170-173, 2014.

[40]. Coys, "Penyimpanan dan Pengaksesan Data pada EEPROM AT24C128," 2011. [Online]. Available: https://electroniclib.wordpress.com/2011/01/27/penyimpanan-dan-pengaksesan-data-pada-eeprom-at24c128/. [Accessed 10 Januari 2016].

[41]. W. D. Cooper, Instrumentasi Elektronik dan Teknik Pengukuran, 2 ed., Jakarta: Erlangga, 1994.

[42]. T. M. Chung and H. Daniyal, "Arduino Based Power Meter Using Instantaneous Power,"*ARPN* Journal *of Engineering and Applied Sciences,* vol. 10, no. 21, pp. 9791-9795, 2015.

[43]. W. Blues, "macam-macam Metode Pengembangan Sistem," 2014. [Online]. Available: www.the-well-informations.blogspot.co.id. [Accessed 2 Maret 2015].

[44]. D. Artanto, Interaksi Arduino dan LabVIEW, jakarta: PT Elex Media Komputindo, 2012.

[45]. Arduino.cc, "Introduction Arduino," 2015. [Online]. Available: https://www.arduino.cc/en/Guide/Introduction. [Accessed 11 Desember 2015].

[46]. A. O. Alves, "Fuzzy for Arduino and Embedded System," 2012. [Online]. Available:

www.zerokol.com. [Accessed 20 November 2015].

[47]. Allaboutcircuit, "Kelvin (4-wire) Resistance Measurement," 2015. [Online]. Available: http://www.allaboutcircuits.com/textbook/direct-current/chpt-8/kelvin-resistance-measurement/. [Accessed 15 Februari 2016].

[48]. Adafruit, "Adafruit Assembled Data Logging shiled for Arduino," 2015. [Online]. Available: https://www.adafruit.com. [Accessed 5 Desember 2015].

Automatic Text Summarization for Indonesian Language Using TextTeaser

D Gunawan[1], A Pasaribu[1], R F Rahmat[1] and R Budiarto[2]

[1] Department of Information Technology, Faculty of Computer Science and Information Technology, University of Sumatera Utara, Medan, Indonesia
[2] Department of Information System, College of Computer Science and Information Technology, Albaha University, Saudi Arabia

Email: danigunawan@usu.ac.id, anwargoog@gmail.com, romi.fadillah@usu.ac.id, rahmat@bu.edu.sa

Abstract. Text summarization is one of the solution for information overload. Reducing text without losing the meaning not only can save time to read, but also maintain the reader's understanding. One of many algorithms to summarize text is TextTeaser. Originally, this algorithm is intended to be used for text in English. However, due to TextTeaser algorithm does not consider the meaning of the text, we implement this algorithm for text in Indonesian language. This algorithm calculates four elements, such as title feature, sentence length, sentence position and keyword frequency. We utilize TextRank, an unsupervised and language independent text summarization algorithm, to evaluate the summarized text yielded by TextTeaser. The result shows that the TextTeaser algorithm needs more improvement to obtain better accuracy.

1. Introduction

Automatic Text Summarization (ATS) is the process to reduce text in order to obtain important sentences by the machine which implements particular algorithm or method. One method to produce text summary is extraction-based summarization. Extraction-based summarization method extracts important sentences in article and then unify them into one summary, therefore the sentences yielded by this algorithm are part of the original text without modification. This method is implemented by TextTeaser algorithm to summarize text. This algorithm is available freely as open source and can be downloaded from github repository. This algorithm calculates text feature, sentence length, sentence position and keyword frequency. TextTeaser is not intended to replace original text. The result will depict the whole text. TextTeaser has been tested in articles which written in English. In this research we test the TextTeaser for articles which written in Indonesian Language. We also evaluate the TextTeaser performance compared to TextRank algorithm.

Previously, a research text summarization for Indonesian news articles by using guided summarization technique has been conducted [1]. This research utilized a supervised method called SWING [2]. They define information aspects such as WHAT, WHO_AFFECTED, WHERE, WHEN, WHY, DAMAGES according to the topic of the news to obtain the text summary. On the other hand, we do not need information aspect because TextTeaser algorithm calculates title feature and keyword frequency.

IOP Conf. Series: Materials Science and Engineering **190** (2017) 012048 doi:10.1088/1757-899X/190/1/012048

Another research utilizes Latent Dirichlet Allocation and Genetic Algorithm to summarize text for Indonesian language [3]. They obtain the best F-measure value 0.556926 with precision value 0.53448 and recall value 0.58134. The summary ratio for their method is 30%. This research requires training data before actually summarizing the text. Our research does not require training data as the TextTeaser algorithm does calculation on items mentioned previously.

Massive number of information does not only occur in online news. Twitter as one of the most popular social network overwhelms the user with the information. Previously conducted research proposed a method to explain a trending topic by topic categorization, explanation generation, distance-based clustering to build subtopics, and sentence compression [4]. This research collects users' messages related to the trending topic then summarize them. There are possibilities of similar messages among different users. Our research does not consider similarity among articles.

2. Research methodology

This research follows a few assumptions in order not to widen the discussion. First of all, the texts are written in Bahasa Indonesia. Then, the sentence pattern follows official Indonesian spelling system which is called *Ejaan Yang Disempurnakan* (EYD). The system does not handle typo that was made by the authors. The last, the summarized text will not handle cohesion and coherent among sentences.

According to figure 1, original text will be the input of several processes to obtain the summarized text. The summarized text will be evaluated with TextRank algorithm [5]. TextRank succeeds to identify most importance sentences. It does not need any training data, therefore adapting TextRank to another language is expected to be easy.

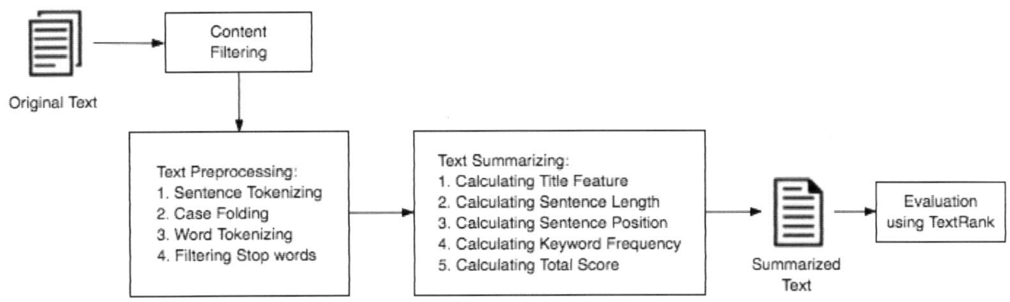

Figure 1. Text summarization process using TextTeaser and evaluated by TextRank

2.1. Preparing the Text

The required text for the research is clear from any attribution. However, the texts from the Internet are available in various data format with many different attributes for each website. Therefore, we need to prepare the text before applying TextTeaser algorithm. There are two steps of preparing the text, that are content filtering and text preprocessing.

The original texts are the main article without any extra parts from the web pages, including HTML tags, header, navigation, banners, link to other sources, and etc. However, almost all the web pages have all these extra parts to arrange the layout of the web pages. To cover this issue, there are some algorithms such as boilerplate algorithm [6] that can extract the main article from a web page. The Boilerplate algorithm does not require any inter-document knowledge, such as frequency of text block, common page layout, and etc.

Sometimes the result is still including some texts that are not directly related to the content of the article. For example, caption below the image, links to other news, author initials, or initials for news source. These extras have different structure among online newspaper publishers. Therefore, we need to adjust particular content filtering for each online newspaper publishers. This could be alleviating if the publishers follow the standard in semantic web. The publisher might embed Resource Description

Framework (RDFa) in HTML and use standardized vocabulary such as Dublin Core to make the machine easier to read. Required protocol can be developed based on available vocabulary [7].

Content filtering is a rudimentary process to obtain valid sources. The result after content filtering will be processed to the text preprocessing. The first step is case folding. This step will convert all the uppercase to lowercase in order to ease letter comparison. Next step is sentence tokenizing. This step is required as TextTeaser calculates sentence score based on its position in a paragraph. After sentence tokenizing, the next step is word tokenizing. TextTeaser will calculate word score based on its appearance in an article. Finally, we eliminates some words based on stop words list provided by Tala [8].

2.2. Text Summarization

TextTeaser uses extraction method for text summarization. It selects sentences that possess best score among others. Therefore, the calculation is done sentence by sentence. Best score sentences are obtained from the calculation regarding four elements of the article such as title feature, sentences length, sentence position and keyword frequency. Title feature is used to score the words in a sentence that have similarity to the words in the title. Sentence length is scored depends on the number of words in the sentence. Sentence position is the location of the sentence in a text. The sentence which is located in introduction and conclusion will have higher score. Keyword frequency is the frequency of the words used in the text.

The calculation of *TitleScore* is determined by calculating every word in each sentence of the article which has similarity with the words in the article title. TitleScore can be yielded by applying equation (1) to the sentence regards to the title. As shown in equation (1), K is the number of words in sentence s_i, and T is the number of words in article title t_j.

$$TitleScore(s_i) = \frac{|K \cap T|}{(|t_j| \cdot 1)} \quad (1)$$

The *SentenceLength* score is obtained by counting the number of words in a sentence. Equation (2) is used to calculate sentence length every sentence in the article. According to equation (1), $|s_i|$ is the number of words in a sentence including stop words.

$$SentenceLength(s_i) = \frac{(x - (x - |s_i|))}{x} \quad (2)$$

Table 1. Sentence position score.

Sentence Position	Distributed Probability
$0.0 < x \leq 0.1$	0.17
$0.1 < x \leq 0.2$	0.23
$0.2 < x \leq 0.3$	0.14
$0.3 < x \leq 0.4$	0.08
$0.4 < x \leq 0.5$	0.05
$0.5 < x \leq 0.6$	0.04
$0.6 < x \leq 0.7$	0.06
$0.7 < x \leq 0.8$	0.04
$0.8 < x \leq 0.9$	0.04
$0.9 < x \leq 1.0$	0.15

Another calculation regarding the sentence is the calculation of its position. The sentences which have higher scores are found at the beginning or the conclusion of the news articles rather than the other positions. Sentence position score, which is represented with $PositionScore(o, S)$, can be calculated by using equation (3). Sentence position in an article is represented with o and the number of all sentences is represented with S. $PositionScore(o, S)$ is determined based on table 1.

$$PositionScore(o, S) = \frac{o}{(S \cdot 1)} \tag{3}$$

Keyword frequency is the number of word occurrence in the whole article. Keyword frequency is calculated by using two methods, that are Density-Based Selection (DBS) and Summation-Based Selection (SBS) [9]. Before commencing the calculation of keyword frequency using DBS and SBS, article keywords and the score is determined by finding unique words. Unique words are the words that has no duplication and not included in stop words list. Next, keywords and their frequency are ordered from the highest weight to the lowest weight. We take ten top keywords to be calculated by using equation (4) to obtain keyword score (represented with $KeywordScore(k_i)$). Where k_i is the keyword that will be calculated, $|k_i|$ is the frequency of keywords occurrence in the text, and $|W|$ is the number of unique keywords in the text. DBS and SBS can be calculated after each keyword has its score.

$$KeywordScore(k_i) = (1 \cdot |k_i| \div |W|) \cdot 1.5 \tag{4}$$

Density-Based Selection is used to determine keyword rank in the text based on several parameters. Those parameters are scored keyword set, words in the sentence that will be calculated, and the words in keyword set [9]. Then the parameters will be calculated by using equation (5), where $dbs(s_i)$ is density-based selection value, K is the number of words in sentence s_i which also in top keywords list, $dbs(w_j)$ and $dbs(w_{j+1})$ are the weight of keyword w_j and w_{j+1} respectively, and $distance(w_j, w_{j+1})$ is the value of two adjacent keyword (w_j and w_{j+1}) and also not a keyword or stop word in s_i.

$$dbs(s_i) = \frac{1}{K \cdot (K+1)} \cdot \sum_{j=1}^{K-1} \frac{dbs(w_j) \cdot dbs(w_{j+1})}{distance(w_j, w_{j+1})^2} \tag{5}$$

Summation-Based Selection is used to determine the sentence weight regarding the occurrence of top keywords in a sentence. The value will be higher if there are more top keywords in a sentence. The weight is calculated by using equation (6), where $sbs(s_i)$ is summation-based selection value, s_i is a sentence in a text, $|s_i|$ is the number of words in s_i, then parameter to take the word weight (w_k) which represents the content of text with $\tau(\tau > 0)$.

$$sbs(s_i) = \frac{1}{|s_i|} \cdot (\sum_{w_k \in s_i} sbs(w_k)^{\tau})^{\frac{1}{\tau}} \tag{6}$$

Keyword frequency, represented by $KeeywordFreq(s_i)$, is obtained by summing the values of $dbs(s_i)$ and $sbs(s_i)$, then divided by 2.0 * 10.0 as shown in equation (7).

$$KeywordFreq(s_i) = \frac{(dbs(s_i) + sbs(s_i))}{2.0 \cdot 10.0} \tag{7}$$

Title feature, sentence length, sentence position and keyword frequency calculation are used to obtain sentence score. The sentence score calculation is stated in equation (8), where $TotalScore(s_i)$

represents sentence score, TS is title score, SL is sentence length, SP is sentence position and KF is keyword frequency.

$$TotalScore(s_i) = \frac{(TS \cdot 0.5 + SL \cdot 0.5 + SP \cdot 1.0 + KF \cdot 2.0)}{4.0} \quad (8)$$

2.3. Evaluation

This research uses intrinsic evaluation method with method recall (R), precision (P), and F-Score (F). Recall in the context of automatic text summarization is number of sections of text that is relevant with the original text based on the number of all sentences in the text. Precision is the number of relevant text that are obtained from all the sentences from original text. F-Score is average weight of recall and precision value. The value of F-Score is the range between 0 and 1. The summarized text from both algorithms will be considered similar if the value is close to 1.

TextTeaser algorithm will be compared with another automatic text summarization algorithm called TextRank. TextRank algorithm is a graph-based ranking model for text processing. This algorithm deciding the importance of a sentence by the vertex which represents that sentence. The higher value of vertex means the sentence is more important than other sentences. Besides, the link between one vertex to another one also has role to determine the importance of the sentence. The advantages of this algorithm is unsupervised and language independent, which means TextRank is expected to work with any language.

The calculation for Recall, Position and F-Score are shown in equation (9), where S is the result of text summarization using TextTeaser algorithm, and T is the result of text summarization using TextRank.

$$F = \frac{2PR}{(P+R)}; P = \frac{|S \cap T|}{|S|}; R = \frac{|S \cap T|}{|T|} \quad (9)$$

3. Result and discussion

This research uses 3075 samples of news articles from various online newspaper publishers in Indonesia. After content filtering and text preprocessing, the articles are summarized by using TextTeaser algorithm. We evaluate the summarized text by calculating recall (R), precision (P), and F-score (F) value between TextTeaser algorithm and TextRank algorithm.

Table 2. F-Score value after evaluation.

F-Score Range	Number of Articles	Percentage (%)
1	233	7.6
0.9 – 0.99	36	1.17
0.8 – 0.89	615	20
0.7 – 0.79	778	25.30
0.6 – 0.69	725	23.58
0.5 – 0.59	150	4.87

Table 2 shows the number of occurrences every F-Score value of 3075 summarized texts. There are 28,75% of summarized texts have F-Score $\geq$ 0.8, where 233 summarized texts have F-Score = 1. Meanwhile, 31.36% of summarized texts have F-Score between the range 0.5-0.79. Moreover, 39.89% of summarized texts have F-Score below 0.5. According to the results, TextTeaser algorithm

IOP Conf. Series: Materials Science and Engineering **190** (2017) 012048 doi:10.1088/1757-899X/190/1/012048

performance for summarizing text in Indonesian language is still average. There are a lot of rooms for improvement to obtain better text summarization result for Indonesian language.

4. Conclusion

TextTeaser is one of the text summarization algorithms that use extraction method. It selects the best sentences among text by calculating title feature, sentence length, sentence position and keyword frequency. This research evaluates TextTeaser algorithm to summarize text in Indonesian language by using TextRank algorithm. We find that utilizing TextTeaser algorithm to summarize text in Indonesian language still need more improvements. Our future work will modify the text preprocessing step and compare it with the original TextTeaser.

5. References

[1] Massandy D T and Khodra M L 2014 Guided summarization for Indonesian news articles *Advanced Informatics: Concept, Theory and Application (ICAICTA), 2014 Int. Conf. of* 140-5 (IEEE)

[2] Ng J-P, Bsyani P, Lin Z, Kan M-Y and Tan C-L 2011 SWING: Exploiting Category-Specific Information for Guided Summarization *Proc. of the Fourth Text Analysis Conf.* (Gaithersburg)

[3] Silvia, Rukmana P, Aprilia V R, Suhartono D, Wongso R and Meiliana 2014 Summarizing Text for Indonesian Language by Using Latent Dirichlet Allocation and Genetic Algorithm *Int. Conf. on Electrical Engineering, Computer Science and Informatics (EECSI 2014)* 148-153 (Yogyakarta)

[4] Winatmoko Y A and Khodra M L 2013 Automatic Summarization of Tweets in Providing Indonesian Trending Topic Explanation *Procedia Technology* **11** 1027-33 (Elsevier)

[5] Mihalcea R and Tarau P 2004 TextRank: Bringing order into texts *Proc. of EMNLP 2004* 404-11 (Association for Computational Linguistics)

[6] Kohlschütter C, Fankhauser P and Nejdl W 2010 Boilerplate Detection using Shallow Text Features *Proc. of the Third ACM Int. Conf. on Web Search and Data Mining* 441-50 (New York, ACM)

[7] Gunawan D 2015 Protocol for E-Commerce data harvesting *Technology, Informatics, Management, Engineering & Environment (TIME-E), 2015 Int. Conf. on* 1-5 (Samosir, IEEE)

[8] Tala F Z 2003 A Study of Stemming Effects on Information Retrieval in Bahasa Indonesia *Master Thesis* (Netherlands, Universiteit van Amsterdam)

[9] Hu M, Sun A and Lim E-p 2007 Comments-oriented Blog Summarization by Sentence Extraction *Proc. of the Sixteenth ACM Conf. on Conf. on Information and Knowledge Management* 901-4 (Lisbon, ACM)

Organization Readiness and ERP Implementation in Albaha University

K Alaqeel[1], M S Shakkah[2], R F Rahmat[3], A Alfageeh[4], R Budiarto[5]

[1]Commercial Observation Division, Ministry of Commerce and Industry, AlKarj Branch, Saudi Arabia.

[2] MIS Department, Faculty of Administrative and Financial Sciences, Albaha University, Saudi Arabia

[3]Department of Information Technology, Faculty of Computer Science and Information Technology, University of Sumatera Utara, Medan, Indonesia

[4]Information Technology Center, Albaha University, Saudi Arabia

[5]Smart Networked Computing Research Group, College of Comp. Sc. & I.T., Albaha University, Saudi Arabia

kaqeel@mci.gov.sa, alshakkah_11@yahoo.com, romi.fadillah@usu.ac.id, aoalfageeh@bu.edu.sa, rahmat@bu.edu.sa

Abstract. This work studies the correlation between the organizational readiness in Albaha University and the respective Critical Success Factors with regards to the Enterprise Resource Planning (ERP) implementation. The study also considers some suggestions to improve the ABU's ERP systems and roadmap towards the self –development strategy and to reduce vendor-dependency. A survey regarding ERP to the end-users, experts and developers in Albaha University was conducted. The analysis of the results in this work confirmed with the results of an existing work. The four significance success factors: Project Management, Business Process Re-engineering, System Integration, and Training and Education are recommended to be adopted to assure the smooth adoption of ERP at Albaha University.

Keywords: ERP, ORGD, Project Management, BPR System Integration

1. Introduction

Weiner [1] defines Organizational readiness (ORGD) for change as "a multi-level, multi-faceted construct. As an organization-level construct, readiness for change refers to organizational members' shared resolve to implement a change (change commitment) and shared belief in their collective capability to do so (change efficacy). The higher the organizational readiness for change, the more organizational members likely to initiate change, exert greater effort, exhibit greater persistence, and display more cooperative behaviour". Figure 1 illustrates the factors and consequences of ORGD for change.

On the other hand, Enterprise Resource Planning (ERP) is the use of software suites to pull together and systemize data of various organization's levels to provide jugdement into key performance indicators (KPIs). The ERP has wide-ranging set of achievements which assists an organization in dealing its business. In general deployment, ERP has to be integrated with other software systems with the aim of optimizing the overall system. Thus, deployment of a new in-house ERP system can involve sizeable business process reengineering and employee retraining.

Nevertheless, ERP concept in education management is not a new thing. Higher education institutions have implemented ERP systems to advancing their procedures and brandng them more transparent and well-organized.

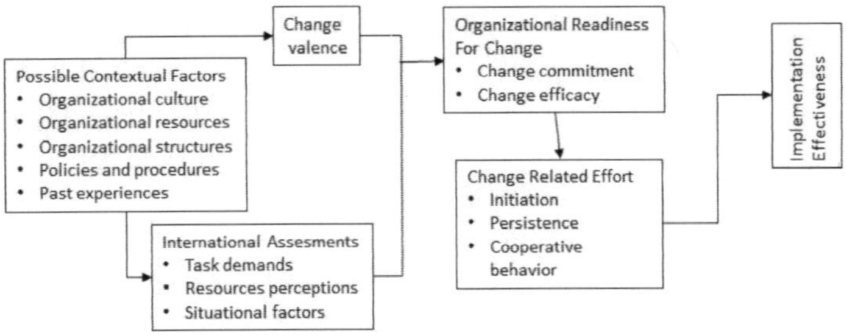

Figure 1. Factors and consequences of organizational readiness for change (adapted from [1])

The Information System and Application development in Albaha University is in general not so encouraging, especially the development of ERP systems. Among the problems faced in Albaha University (ABU) include:

- Low user satisfaction level
- High vendor dependency
- Lack of IT awareness program
- Automation of many key process is partial
- Absence of documentation of the roles, responsibilities, and accountabilities
- Absence of well-defined organizational and process framework.

However, at the same time, ABU has positive factors that are believed may support towards the improvement of the ERP systems. The positive factors include:

- Headway made by few departments by implementing ERP and student information system
- Enthusiasm to adopt e-working practices at departments/units/ centres
- Good support from ABU top management to have best technology that help university to achieve their vision.

This research work focuses on ABU organization readiness with regards to the success of ERP implementation and considers four critical success factors (CSFs): project management (PM), business process re-engineering (BPR), system integration (SI), and training and education (TED) as investigated by [2].

2. Related Works

Research works have been conducted in ERP implementation such as Ramayah and Annamalai [3] that examined the benefits of two enterprises resource planning (ERP) packages namely SAP and Oracle. King [4] concluded that the current status of ERP adoption by companies, regardless the size of the company shows a tremendous growth. Some companies have implemented successfully ERP in customer relationship management and supply-chain management systems as integrated systems.

Voordijk et al. [5] studied on the changing role of information technology (IT) in engineering consultancy firms after implementing ERP. The authors used empirical case study approach to analyze ERP implementations in the Dutsch-based engineering consultancy companies.

Grossman & Walsh [6] mentioned numerous recommendations to ERP implementation as a huge mission packed with risks. The authors gave lots of practical advices for reducing the risks. The authors also considered operational, technical and legal factors of guaranteeing an acceptable ERP implementation. The necessary softwares should be employed in a shortened timeframe with less risk, together with the potential of having the state-of-the-art of technology that embodied the common practices of companies' prior to carrying out ERP systems.

3. Research Method

3.1. Research hypotheses and the conceptual model

According to researches by Somers and Nelson [7] and Tesch, et al. [8], the organizational readiness for ERP deployment within an organization is essential to know a readiness of an organization prior to the commence of the project and to identify as early as possible probable problems. The problems may affect the implementation outcome. Therefore, the organizational readiness is expected to meaningfully contribute to the success the ERP implementation and thus, the following hypothesis is constructed.

H1: OGRD and the success of PM are positively and significantly correlated.

Works by De Soysa and Nanayakkara [9], and Capaldo and Rippa [10] note that BPR is an important dimension of the overall organizational ERP readiness framework of the organization.

The work further demonstrate that OGRD could perform a greater role in prompting the results of a BPR exercise. Thus, this paper premises the following.

H2: OGRD and the success of BPR are positively and significantly correlated.

ERP systems have modular structure and require inter-module interfacing and integration to work seamlessly. SI is one of the key strategies for successful ERP implementation [10, 11]. Thus, it is meaningful to hypothesize:

H3: OGRD and the success of SI are positively and significantly correlated.

Understanding the OGRD enables organizations to capture their users' training needs [12]. The OGRD helps organizations to gain a better understanding of the skill levels of their staff, the availability and adequacy of human resource skills and capabilities in recognizing the available training facilities, trainer capabilities and training requirements for implementing a complex system like ERP [13, 14]. Therefore, we contend that OGRD is helpful for providing an appraisal of the human and infrastructure capabilities that may be required to conduct TED successfully, and hence develop the following hypothesis.

H4: OGRD and the success of TED are positively and significantly corralated.

The proposed model was built based on the assumptions of the associated hypotheses: H1, H2, H3, and H4 as shown in Figure 2 [2].

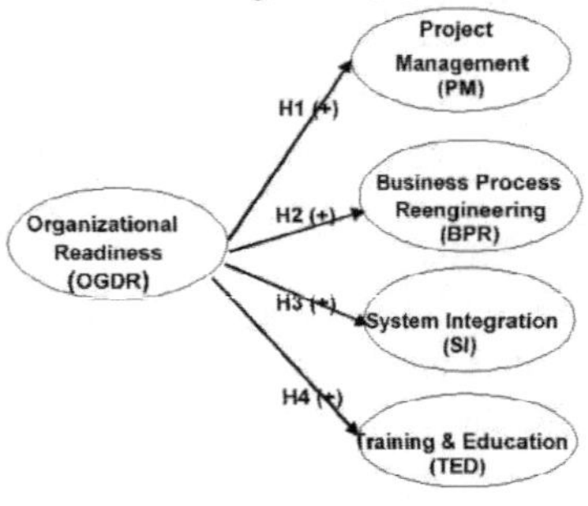

Figure 2. The conceptual ORGD - ERP model (adopted from [2])

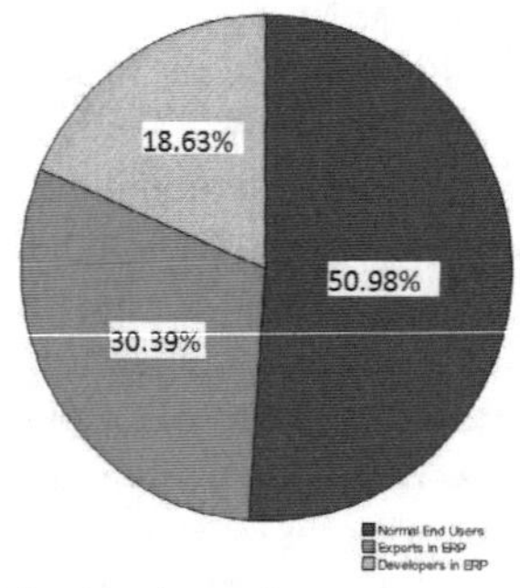

Figure 3. Distribution of respondents (End user, experts, and developers)

3.2. Instrument development and data collection

A questionnaire with five-point Likert-type scale ranging from "strongly disagree" (1) to "strongly agree" (5) was used to collect data. The questionnaire was adapted from Ram et al., [2], five constructs: OGRD, PM, BPR, SI, and TED with their respected items were surveyed. The questionnaire was validated by experts, then translated to Arabic version forth and back and checked to be in the same levels.

The survey instrument was pre-tested in two phases, initially with academics. The feedback and comments received from both sets of respondents were incorporated to improve format, presentation, content clarity and questionnaire length. SPSS software was used with data and reliability were firstly checked, they were with good values. Demographic data are: End-users, Experts, and Developers.

4. Results and Analysis

4.1. Statistical results

The distribution of respondents is shown in Figure 3. 150 questionnaires are distributed and 107 are collected and validated.

Table 1. Spearman's ρ for PM- ORGD Correlation

		PM construct	OGRD Construct
PM construct	Correlation Coef.	1.000	.585**
	Signif. (2-tailed)	.	.000
	N	102	102
OGRD construct	Correlation Coef.	.585**	1.000
	Signif. (2-tailed)	.000	.
	N	102	102
** Correlation is significant at the 0.01 level (2-tailed).			

Table 2. Spearman's ρ for BPR- ORGD Correlation

		OGRD Construct	BPR construct
OGRD construct	Correlation Coef.	1.000	.573**
	Sig. (2-tailed)	.	.000
	N	102	102
BPR construct	Correlation Coef.	.573**	1.000
	Sig. (2-tailed)	.000	.
	N	102	102
** Correlation is significant at the 0.01 level (2-tailed).			

Table 3. Spearman's ρ for SI- ORGD Correlation

		OGRD Construct	SI construct
OGRD Construct	Correlation Coef.	1.000	.651**
	Sig. (2-tailed)	.	.000
	N	102	102
SI construct	Correlation Coef.	.651**	1.000
	Sig. (2-tailed)	.000	.
	N	102	102
**. Correlation is significant at the 0.01 level (2-tailed).			

Table 4. Spearman's ρ for TED- ORGD Correlation

		OGRD Construct	TED Construct
OGRD Construct	Correlation Coef.	1.000	.493**
	Sig. (2-tailed)	.	.000
	N	102	102
TED Construct	Correlation Coef.	.493**	1.000
	Sig. (2-tailed)	.000	.
	N	102	102
**. Correlation is significant at the 0.01 level (2-tailed).			

According to Table 1 we conclude that H1: There was a positively moderate association between PM and OGDR and it is significance (sig < 0.05), so accept H1.
Referring to Table 2 we conclude that H2: There was a positively moderate association between BPR and OGDR (sig < 0.05), so accept H2.
Referring to Table 3 we conclude that H3: There was a positive moderate association between SI and OGDR, (sig < 0.05), so accept H3.
Referring to Table 4 we may conclude that H4: There was a Positive moderate association between TED and OGDR, (sig < 0.05), so accept H4.

4.2. Analysis
Overall, the results show that organizational readiness is significantly and positively associated with the four investigated CSFs as the conceptual model discussed in Section 3.1. The findings extend knowledge of the relationship among factors antecedent to different stages of ERP systems development [2].Thus, an understanding of organizational at the adoption stage of an ERP system project is important. The following are the detail discussions on the findings.

ORGD effect on PM. The results indicate that organizational readiness is crucial in accomplishing the project success as it provides better planning and control through regular meetings and reasonable deadline. Albaha unversity IT Center management should evaluate a number of factors including: organizational process assets for risk management, information collection on change management procedures, and human resources to understand the organizational readiness for project management. Moreover, Albaha unversity IT Center management is recommended to strengthen the Project Management Unit with proper staff and PM softwares and tools.

ORGD effect on BPR. The findings show that BPR exercise signifantly influences the organizational readiness. The exercise needs experts and man power that may cost more the university. It is suggested that the Albaha university management must take advantage of the expertise and skills available within the university to conduct BPR exercises.

ORGD effect on SI. The results say indirectly that considering internal skills and resources is important to ensure SI activities run easily. Consequently, the university should focus on the availability of a skilled workforce to support integrated interfaces. Albaha University is likely still relies on external support to maintain their integrated systems. Albaha University IT center management should evaluate a number of factors, including: the experience and skill set of IT staff, particularly in SI; the availability of legacy in-house systems; third-party information systems that need to be interfaced/ integrated with ERP; procedure and regulations for checking data integrity. The evaluation will help to understand well ORGD for SI.

ORGD effect on TED. The study colcudes that the university have to consider to materialize staff continuous professional developmet program. The university management should have an effective management of training programs to foster a learning and sharing culture among the staff. In-house skills development to achieve successes in ERP prjects is also recommended, to fulfill the need of special training. In addition, Albaha University management should consider a number of factors related to the TED and allocate proper budget for training programmes.

5. Conclusion and Future Works
The analysis of the results in this work confirmed with the results of work by Ram [2] that is the ORGD influences the success of ERP implementation. Alabaha University management need to study the university's organizational readiness to measure the technological, human, and infrastructure capabilities in designing and implementing ERP systems. The four significance critical success factors: PM, BPR, SI, and TED are recommended to be adopted to assure the smooth adoption of ERP at Albaha University.

Acknowledgement

The authors thank to Albaha University for the research funding under research contract no. 95-1436.

References

[1] Weiner B J 2009 *Implementation Science* **4** 67, **DOI:** 10.1186/1748-5908-4-67.
[2] Ram J, Corkindale D and Wu M 2015 *Journal of Computer Information Systems* **55** 2 pp 29-39.
[3] Annamalai C and Ramayah T 2011 *Business Process Management Journal* **17** 3 pp 495-509.
[4] King W R 2005 *Information Systems Management* **22** 3 pp 83-84.
[5] Voordijk H, Stegwee R and Helmus R 2005 *Business Process Management Journal* **11** 4 p 418.
[6] Grossman T and Walsh J 2004 *Information Systems Management* **21** 2 pp 38-42.
[7] Somers T M. and Nelson K G 2004 *Information & Management* **41** 3 pp 257-278.
[8] Tesch D, Kloppenborg T J and Frolick M N 2007 *Journal of Computer Information Systems* **47** 4 pp 61-70.
[9] de Soysa S and Nanayakkara J 2006 *Int. Conf. on Information and Automation (Colombo, Sri Lanka)* pp 27-32.
[10] Capaldo G and Rippa P 2009 *Journal of Enterprise Information Management*, **22** 6 pp 642-659.
[11] Al-Mashari M, Al-Mudimigh A and Zairi M 2003 *European Journal of Operational Research*, **146** 2 pp 352-64.
[12] Misra H, 2008 *IEEE Int. Conf. on Industrial Engineering and Engineering Management (Singapore)* pp 1719-23.
[13] Fathian M, Akhavan P and Hoorali M 2008 *Technovation* **28** 9, pp 578-90.
[14] Tan J, Tyler K, and Manica A 2007 *Information & Management*, **44** 3 pp 332-51.

Formulation of Policy for Cyber Crime in Criminal Law Revision Concept of Bill Book of Criminal Law (A New Penal Code)

Eko Soponyono[1], Brav Deva Bernadhi[2]
[1] Faculty of Law, Diponegoro University, Semarang, Indonesia
[2] Faculty of Industrial Technology, Sultan Agung Islamic University, Semarang, Indonesia

facta@elektro.undip.ac.id

Abstract. Development of national legal systems is aimed to establish the public welfare and the protection of the public. Many attempts has been carried out to renew material criminal law and those efforts results in the formulation of the concept of the draft Law Book of the Law of Criminal Law in the form of concept criminal code draft. The basic ideas in drafting rules and regulation based on the values inside the idology of Pancasila are balance among various norm and rules in society. The design concept of the New Criminal Code Act is anticipatory and proactive to formulate provisions on Crime in Cyberspace and Crime on Information and Electronic Transactions. Several issues compiled in this paper are whether the policy in formulation of cyber crime is embodied in the provisions of the current legislation and what the policies formulation of cyber crime is in the concept of the bill book of law - criminal law recently?.

1. Background

Since proclamation of independence of Republic Indonesia on August 17th, 1945, the noble national goals are to protect the people and the country of Indonesia in term of social defense, public welfare or social welfare, justice, and prosperous.

To achieve social welfare and social defense based on Pancasila and the Constitution of 1945. the implementation of values of ideology and constitution are implemented through national development in all aspects of national life [1].

Achievement of the quality of life for justice and prosperous society which are pursued by the Indonesian government is conducted through a sustainable national development including the development of the national legal system by the government of Indonesia. The national legal system is programmed in the Long Term Development Plan in the Vision and Mission of the National Legislation Program in year duration of 2005-2009 based on decree of parliament No.01 / DPR RI / III / 2004-2005 [1]

Development of the national legal system is in aim to establish public welfare and protection and it is accommodated in legislation program (Prolegnas). In any activities of Prolegnas, a legal

IOP Conf. Series: Materials Science and Engineering **190** (2017) 012050 doi:10.1088/1757-899X/190/1/012050

system must address the principle of *kenusantaraan* (diversities) which recognizes the diversity or heterogeneity of law. There are many diversities of law emerged in Indonesia such as customary law, Islamic law, the law of other religions, contemporary law and western law. Those laws are presented through unification against certain laws, partially form, and in the form of codification.

The attempts to to renew material for criminal law has resulted in the formulation of the "Draft Bill Book of the Law of Criminal Law". This attempt has been initiated because the concepts of basic thinking are no longer in line with the philosophy of Pancasila as state ideology and it does not reflects the idea of balance among the norm in society.

Preparation of concept according to Barda Nawawi Arief, is essentially a renewal effort or restructuring of the entire system of substantive criminal law contained in the previous Criminal Code of Dutch East Indies [2-3].

Based on the description above background, the issues raised in the paper consists several issues such as whether the policy formulation crime of "cyber crime" is embodied in the provisions of the current legislation and what the policy formulation crime of "cyber crime" in the concept of the bill book of law or Criminal Code recently.

2. Literature

The short meaning of policy is wise. It means that policy always uses reason, clever, and adept. Furthermore, the meaning of policy is a series of concepts and principles containing outlines and basic plans in the implementation of employment, leadership and how to act especially on the governance of the organization. Policy is also has a meaning as a statement of goals, objectives, principles or intent as guidelines for management in order to achieve the target [2,4]. The meaning of wise is means, able, smart, experienced, wise, whereas the policy means wisdom and policy [2,5].

The key to use policy which is related to the original meaning will associate with:

1. Government or polity, political wisdom or cunning,
2. Wise, prudent or expedient conduct or management, conduct or management,

A principle, plan, or couse of action, as pursued by a government, organization and individual

The concept formulation in concept draft of Criminal Code until 2015 can be regarded as the embodiment of integration between the two approaches. These two approaches are policies orientation and the value of policy and oriented integrated approach. Construction of the criminal justice system is part of the construction of the National Law System which is based on the values of Pancasila. Consequently, the renewal of the National Criminal Law should be well motivated and oriented by basic ideas of Pancasila which contains the value of balance. The value of balanced covers religious norm or moral or divinity, humanity, nationality, democracy, and social justice.

3. Method

3.1. Approach Problems

The problem is formulated in this paper include the central issue of criminal law enforcement policy or policy formulation. Therefore, the approach cannot be separated from policy-oriented approach to analyze legislative policy in defining and formulating action of crime in cyber crime. The approach in this paper is a normative legal research. In a normative legal research, there are several approaches such as statute approach, conceptual approach and the comparative approach [2,7-13].

3.2. Research Specification

Specifications research conducted in this paper is a normative legal research or legal research literature suggested by Soerjono Soekanto. Comparative law is a method for a normative legal research. In this method research is conducted on various sub-systems of law which is applicable in a given society or across sectors of the legal systems in the various different communities.

3.3. Types and Sources of Data
Data taken from literature as secondary data consists of:

a. Primary legal materials such as materials binding law. This material covers
    1) A basic rule contained in the Preamble of the Constitution of 1945
    2) Basic Regulation in the body of Constitution 1945, Provisions of the People's Consultative Assembly and Long Term Development Policy
b. Secondary law providing an explanation of the primary legal materials, such as the draft law or Draft New Criminal Code Act 2015, the results of research and scientific works of the criminal law.
c. Tertiary legal materials such as materials providing instructions and an explanation of the primary and secondary legal materials. These materials are taken from dictionaries and encyclopedias.

3.4. Method of Data Collection
Based on secondary data, the primary and secondary legal materials in formulation of the problem are set, and then they are classified according to the source and hierarchy to be studied comprehensively.

3.5. Method Data Analysis
All the data and information obtained in this study are analyzed using qualitative and normative analysis method. Qualitative analysis of the data presents the normative quantitatively and it is grounded in descriptive and predictive analysis. The policy formulation of cyber crime is formulated after analysis to find the weaknesses of juridical in policy formulation in the provisions of the current legislation. The policy formulation also considers the concept of the draft of New Law Penal Code, comparative study of criminal law, the views of scholars and theoretical studies. Analysis is also carried out for the second issue to discovery the ideal policy formulation for cyber crime.

4. Discussion

4.1. Policy formulation crime of "cyber crime" is embodied in the provisions of the current legislation
Legislation and parlement are currently formulating crime of "cyber crime" according to the Law of Republic of Indonesia No. 11 in year 2008 about transaction of electronics and information. Analysis on Cyber Crime Convention is related to the formulation of the provisions in the articles of the ITE Law No. 8 of 2011. ITE criminal provisions in the Act have not been separated by the Council of Europe Convention on Cyber Crime on November 23rd, 2001 in Budapest Hungary. The reached agreement was incorporated the convention into European Treaty Series No. 185. This Convention is activated after the ratification conducted by at least 5 (five) countries, including at least ratification by three member states of Europe Council.

Positive attitude the Government of Indonesia with the ratification is established with full awareness and it is related to the increasing intensity of the digitalization, convergence and continuing globalization of information technology as a means to do a criminal act. The positive response is shown by the formulation of Law No. 8 of 2011 on Information and Electronic Transactions or ITE. Among of articles there are several articles that related to the the convention of cyber crime i.e. Article 27 until Article 37. Formulation in Article 27 of the cconvention cyber crime is known by the term of illegal access. Illegal access is to access the Electronic Information and or Electronic Records that contain forbidden act and violated ethics charge such as child pornography

In the convention of cyber crime, forbidden acts formulated Article 31 is called illegal Interception and system interference. Analysis of the formulation of Criminal Provisions within of the Act ITE No. 8 of 2011 is more focused on the problem of juridical, because the most strategic policy in the enforcement criminal law is the formulation stage. The emergence of problems in this phase is resulted in more serious implementation of the next stages i.e. the application and the execution stage. It is asserted that the formulation stage or formulation phase of legislative policy is the most strategic phase in the efforts of crime prevention and control through the penal policy. Therefore, any the mistakes or weaknesses of legislative policy are strategic errors that may obstruct the efforts of crime prevention and control in the application and execution stage.

The analysis of juridical issues includes several items such as

1) The qqualifications of juridical
2) The guidelines for the threat of criminal penalties in criminals

The main position of Article 103 of the Criminal Code related to the first issue is that the provisions in Chapter I to Chapter VIII of the book is also applied to deeds by statutory provisions shall be punishable by others, unless by law otherwise specified .

It is related to the second issue, the policy formulation of volume of crime (Straf Maat) especially for criminal penalties is put at higher position. The high volumes of fines are not always met by the convict. Furthermore it becomes the juridis problem if the convicted person does not pay the fine imposed. Juridical problem exists if the formulation of sentencing guidelines to pay the fine imposed does not exist.

4.2. Policy of crime formulation crime of cyber crime in the concept of the new Penal Code

In the policy formulation, the concept of criminal provisions in Article 378 to Article 384 contains three Paragraph. The first paragraph is related to the use and destruction of electronics information and domain. The second is about no right to access computers and electronics systems, and the third is about the child pornography via computer

The title in the first paragraph does not seem related to actions to access a computer without the right, but the formulation of Article 378 is formulated in the elements of accessing a computer without authority, so that such formulations also covers the formulation of second paragraph about no right to access of computer and electronic systems and it is similar with the formulation of Article 381. The formulations this article also includes an element of "damage" that exist in the first paragraph. It can be concluded that the formulation between paragraphs are not consistent with the title of paragraph.

Formulation of the provisions in the second paragraph are impressed as a formal offense, but in the formulation of Article 381 in letters A, E, F and G, there is an element of a result. In letter A, there are words "may cause a disturbance". In letter E, there is a word "result, in the letter F, word "lead" is used and in the letter E, the word "result" is adopted. The same patterns are presented in the formulation of Article 383, in letter D, there are words "consequently can affect electronic systems in the Central Bank".

In the third paragraph, the formulation of the provisions of Article 384 in the explanation, it is said to be "quite obvious". However, the formulation of the letter B said, "Making available child pornography through a computer system". In the letter E, it is said, "Possessing child pornography in a computer system or a computer data storage medium", and it means there are similarities. The element of "supply contains the same meaning as having". These two elements are in possesion of criminal.

Based on entire analysis in the second issue, it is concluded that the concept has formulated a criminal offense related to "Cyber Crime". It is made as participation through ratification of the "Council of Europe Convention on Cyber Crime," held on November 23, 2001 in Budapest, Hungary.

5. Conclusion

Policy formulation crime of "cyber crime" is embodied in the provisions of the current legislation. The provisions of legislation is a form of participation of Indonesia in the ratification of Council of Europe Convention on Cyber Crime on November 23rd , 2001 in Budapest Hungary. Juridical problems found from the analysis of the criminal provisions have shown that no formulation for juridical qualification for crime or offense, and it affects juridical. The finding shows less understanding about the ideal standard in the Criminal Code.

The concept of the draft Law Book of the Criminal Law Act 2015, in Book I about the General Conditions has formulated crime of cyber crime. This general provision responded to the formulation of the criminal provisions of Book II of Crime in Chapter VIII which stated that crime that dangerous for public safety, health, products, and environment.. In Book II, it is stated that criminal offenses related to "Cyber Crime" has been formulated in the provision at Part Five of "Crime against Informatics and Electronics". Policy formulation needs to be replicated and for reasons of implementation and it must be easy to implemented.

6. References

[1] Arief, 2001, Barda Nawawi, *Masalah Penegakan Hukum dan Kebijakan Penanggulangan Kejahatan,*PT. Citra Aditya Bakti, Bandung, 2001.
[2] Arief, Barda Nawawi, 1994, *Beberapa Aspek Pengembangan Ilmu Hukum Pidana (Menyongsong Generasi Baru Hukum Pidana Indonesia)*, Pidato Pengukuhan, Diucapkan Pada Peresmian Penerimaan Jabatan Guru Besar Dalam Ilmu Hukum Pada Fakultas Hukum Universitas Diponegoro, Semarang
[3] Arief, Barda Nawawi, 1998, *Beberapa Aspek Kebijakan Penegakan dan Pengembangan Hukum Pidana,* PT Citra Aditya Bakti, Bandung
[4] Arief, Barda Nawawi, 2005, *Bunga Rampai Kebijakan Hukum Pidana*, Citra Additya Bakti, Bandung.
[5] Arief, Barda Nawawi, 2005, *Tindak Pidana Mayantara, Perkembangan Kajian Cyber Crime Di Indonesia*, Divisi Buku Perguruan Tinggi, PT Raja Grafindo, Persada, Jakarta
[6] Departemen Pendidikan dan Kebudayaan, 1989, *Kamus Besar Bahasa Indonesia,* Balai Pustaka, Jakarta.
[7] Ibrahim J, 2005, *Teori dan Metodologi Penelitian Hukum Normatif,* Bayumedia Publishing, Surabaya.
[8] Ketetapan MPR RI Nomor IV/MPR/1999, Tahun 1999-2004, Dasar Pemikiran.
[9] Soekanto S, 1985, *Penelitian Hukum Normatif, Suatu Tinjauan Singkat*, PT Raja Grafindo Persada, Jakarta.
[10] Sudarto, 1986, *Hukum Dan Hukum Pidana,* Alumni, Bandung
[11] Sudarto,1983, *Hukum Pidana dan Perkembangan Masyarakat, Kajian Terhadap Pembaharuan Hukum Pidana,* Sinar Baru, Bandung
[12] Sudarto, *Masalah Penghukuman dan Gagasan Pemasyarakatan,* Badan Penyediaan Bahan Kuliah, Semarang.
[13] Sudarto, 1979, *Suatu Dilemma Dalam Pembaharuan Sistim Pidana Indonesia,*Fakultas Hukum UNDIP, Semarang